Electromagnetic Compatibility
in Power Electronics

László Tihanyi

**IEEE
PRESS**

The Institute of Electrical and
Electronics Engineers, Inc.
New York, N.Y., U.S.A.

J. K. Eckert & Company, Inc.
Sarasota, Florida
U.S.A.

Butterworth-Heinemann Ltd.
Jordan Hill, Oxford
United Kingdom

Copyright © 1995
J. K. Eckert & Company, Inc.
3614 Webber St.
Sarasota, Florida   U.S.A.

Distribution in North America is by
  The IEEE Press
  445 Hoes Lane
  P.O. Box 1331
  Piscataway, NJ   08855-1331
  U.S.A.
  ISBN: 0-7803-0416-0   IEEE Order No.: PC0312-9

Distribution elsewhere is by
  Butterworth-Heinemann
  Linacre House
  Jordan Hill
  Oxford OX2 8DP
  United Kingdom
  ISBN: 0-7506-2379-9

Series editor: J. Eckert
Technical consultant: M. Mardiguian
Inspiration: G. Dickel
Special thanks to: Dudley Kay, Hugh Denny, Duncan Enright, Denise Gannon

Printed in the United States of America

10   9   8   7   6   5   4   3   2   1

Library of Congress Cataloging-in-Publication Data
Tihanyi, László.
    EMC in power electronics / László Tihanyi.
      p.   cm.
    Includes bibliographical references and index.
    ISBN 0-7803-0416-0 (hardcover)
    1. Electronic circuits—Noise. 2. Power electronics.
3. Electromagnetic compatibility. 4. Electromagnetic noise.
I. Title.
TK7867.5.T55   1995
621.382'24—dc20   94-43963
                                                                  CIP

This book is published with the understanding that it provides information only and does not render engineering services. The book draws information from sources believed to be reliable, but neither the publisher nor the author guarantees the accuracy or completeness of the information, and neither the publisher nor the author shall be held responsible for any damages resulting from use of the information in this book or for any errors or omissions.

This book is dedicated to my father's
memory. Through the example of his own
life, he taught me how to work.

Contents

Foreword

Electromagnetic compatibility (EMC) has come a long way from the "black magic" approach of the early 1960s to an almost exact science, with its analytical methods, measurement techniques, and simulation software. Three decades ago, all existing handbooks on EMC could be counted on the fingers of one hand, but today they could occupy several shelves in a respectable library.

However, although there is an abundance of books covering the many radio-frequency aspects of EMC (e.g., noise reduction in analog and digital circuits, shielding theory and practices, math modeling of EMI radiation and coupling, EMI testing, lightning and electrostatic discharge, and so on), only a very few books are available that thoroughly address the EMC side of power electronics.

László Tihanyi's book goes deeply into practical details of power supply components' noise generation, diode recovery, and the SCR noise spectrum. It performs a thorough, but very practical, examination of the parasitic behavior of EMI filters, capacitors, and inductances, and how they affect the filter transfer function, sometimes turning the expected attenuation into gain.

Finally, a rigorous but, again, very practical analysis is presented for the time-to-frequency conversion of single impulses, including their often-neglected energetic aspect.

—*Michel Mardiguian*
St. Rémy les Chevreuse, France
October 1994

Preface

In the course of their daily routine, experts in the field of power electronics more and more often encounter the problem of high-frequency interference. In practice, EMC issues usually are ignored until a problem is revealed by testing or in normal operation. As a result, EMC fixes tend to be applied at the test or even production stages of product development, which can lead to solutions that are unsatisfactory, unnecessarily expensive, or both.

To avoid this situation, those who are involved in design, development, production, and operation of semiconductor equipments must be able to identify and solve EMI problems as early as possible. These individuals must acquire a grasp of practical noise reduction techniques without actually becoming professional EMC engineers. Although a great deal of written material on EMC has appeared in technical journals and symposium records, these sources constitute a collection of miscellaneous subjects which do not always interrelate and are difficult to use in engineering practice. This book brings together a cohesive package of information on the somewhat specialized subject of EMC in power electronics, thereby saving the reader from the task of sorting through hundreds—perhaps thousands—of volumes of marginally related material. It is intended to be useful to newcomers to the EMC field as well as those who are already fighting stubborn EMI problems.

Laszlo Tihanyi
Budapest, Hungary
June 1994

Acknowledgments

The author extends very special thanks to Michel Mardiguian, whose remarks helped very much in the preparation of this manuscript. Thanks also to Jeff Eckert for assisting with the translation into English and guiding a long project to completion.

1

Introduction

The role of electrical energy in our everyday lives has grown by leaps and bounds during the twentieth century. In the early decades of the 1900s, experts in most fields took a direct approach to achieving their goals, without much regard for or understanding of the negative side effects of their technological innovations. Thus, the deleterious consequences of rapid development came into the limelight only later. In recent decades, some of those consequences have reached international proportions, making it necessary to study them, with the ultimate aim of reducing or eliminating the disagreeable effects.

One such problem is environmental electromagnetic pollution. Unacceptably high-level electromagnetic disturbances can prevent electrical and electronic devices, apparatus, and systems from operating properly in a common electromagnetic environment. A device is considered to be electromagnetically compatible only if its effects are tolerated by all other devices operating in the same environment. To ensure that this compatibility exists, a relatively new engineering discipline, *electromagnetic compatibility* (EMC), has evolved. EMC is the field of electrical engineering that studies, analyzes, and solves electromagnetic interaction problems.

Achieving EMC requires us to view disturbances from two distinct viewpoints: *electromagnetic emissions* and *electromagnetic susceptibility*. Because electromagnetic noise propagates by conduction and radiation, the scope of problems outlined above continues to broaden (see Fig. 1.1).

Because *electromagnetic interference* (EMI) first emerged as a serious problem in telecommunications (or, in particular, broadcasting), EMC tends to be discussed, even to the present day, within the scope of telecommunications technology. However, there are limitations to this approach.

1

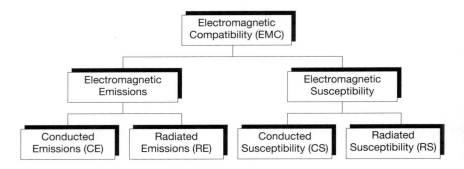

FIGURE 4.1   Areas of electromagnetic compatibility

In the first half of this century, electromagnetic disturbance sources, for the most part, were limited to motor-driven machinery and switching apparatus. But with the rapid spread of power semiconductors and power electronic systems, interference levels on power mains have increased significantly in intensity and frequency of occurrence. This trend is universally forecast to continue. At the same time, the world is becoming more densely populated with devices that are increasingly sensitive to electromagnetic disturbances. In industrial spheres, electronic control systems, data processing equipment, and other sensitive devices play an increasingly important role.

These developments have produced quite a serious situation. In telecommunications, adequate methods for computing and solving EMC problems have been developed over the years. Unfortunately, many of these methods cannot be applied directly to the field power electronics, which has its own peculiarities. The scope of this book, therefore, is to provide a wide overview EMC principles while highlighting EMC engineering practices that are specifically applicable to power electronics.

EMI produced by power semiconductor equipment decreases rapidly above the range of a few megahertz. In various standards and specifications, high-frequency disturbances that affect power mains are limited to the range of 10 kHz to 1,000 MHz. Because different EMC solutions are applicable to different frequency ranges, this book emphasizes EMI suppression in the range of 0.15 to 30 MHz. Likewise, discussion of susceptibility issues is limited to this frequency range, with particular attention given to impulse-like noise phenomena.

After a brief history of the development of EMC standards in Chapter 2, Chapter 3 provides a general description and classification of EMI. Chapter 4 deals with methods for measuring conducted high-frequency disturbances, and Chapter 5 surveys EMI specific to power electronic equipment. The characteristics of circuit elements used for noise suppression are the subject of Chapter 6. In Chapter 7 EMI suppression methods used in semiconductor and electromechanical devices are summarized. Chapter 8 explores various aspects of EMI filter circuits, and EMI filters methods are discussed in Chapter 9.

Moving into the realm of susceptibility, noise-withstand capability tests are described in Chapter 10, and Chapter 11 offers a look at the primary EMS reduction techniques for power electronic equipment. Finally, Chapter 12 addresses filter circuit design for EMS reduction, with particular emphasis on impulse-like disturbances.

2

History of EMC Standardization Efforts

Since the emergence of EMC as a field of inquiry, suppliers of electrical energy have looked for applicable solutions. Difficulties in achieving EMC have become greater with the fast proliferation of equipments that generate high-frequency EMI and electronic devices that are susceptible to these interfering signals.

For the sake of providing the proper power quality, two objectives must be achieved. On the one hand, HF emissions that can be imposed on the power mains must be limited; on the other hand, we need to elaborate on current EMS test methods with the goal of reducing the electromagnetic susceptibility (EMS) of devices on the consumer end of the power grid.

Efforts to establish acceptable HF disturbance emission levels, however, have run up against several difficulties. Many years ago, experts examined HF pollution of the mains. Initial efforts to achieve EMC in this realm focused on limiting the HF emissions of equipment connected to the mains. Preliminary EMI measurement methods were developed at that time, and several limits for HF emission of electrical equipments were determined. A more comprehensive study of EMS came into being only in the second half of this century, stimulated by more widespread use of electronic devices [33, 54, 84, 114, 151, 223, 293, 305, 306].

CISPR (Comite International Special des Perturbations Radioelectriques, or International Committee for Radio Interference) was the first international organization authorized to promulgate international recommendations on the subject of radio interference. CISPR was founded in Paris, in 1933, by representatives of countries that had become concerned with the problem of radio frequency interference. In early discussions, they agreed that the primary job would be to document standard EMI measurement methods and to determine internationally acceptable noise level limits. The founding conference proposed to establish a common commission in the IEC (International Electrotechnical Commission) and UIR (Union International de Radiodiffusion, or International Union of Broadcasting) to facilitate the preparation of recommendations. CISPR held two plenary sessions before World War II to deal with the determination of acceptable noise levels and establishment of standard EMI measurement techniques.

After World War II, the UIR was not reconvened, and CISPR became a special committee of IEC. CISPR differs from other study groups insofar as several other international organizations participate in CISPR's work with observer status. CISPR's preliminary efforts were to publish a set of documents that would describe widely applicable requirements for EMI measurement equipment and techniques. This effort was largely completed in 1961.

During a 1973 plenary session held in the USA (Monmouth College, West Long Beach, New Jersey), a decision was made to reorganize CISPR. Existing subcommittees were disbanded, and six new subcommittees were established in their place. A decision was made to hold regular subcommittee sessions in the future. The attendees also established the working methodology of the subcommittees and determined an order for publication of the subcommittees' results. The new subcommittees and their interest areas were created as follows:

- *Subcommittee A: Interference measuring devices, measurement methods.* Subcommittee A was charged with describing the requirements for detectors of EMI measuring instruments and methods for statistical evaluation of measurement results. The subcommittee made a revision of Publ. 16 which, first of all, discussed the circumstances of detectors used to measure average values. This job was required by the continuing growth of narrowband emissions. During the revision, in addition to current and voltage measurements, measurement techniques for energy and electrical and magnetic field components were also studied. Other topics for Subcommittee A included the definition of a line impedance stabilization network (LISN) applicable to the measurement of heavy currents (i.e., >25 A), a description of requirements for coaxial cable shielding and tests for cable shielding performance, as well as a review of existing methods for measuring the insertion loss of EMI filters.

- *Subcommittee B: EMI from industrial, scientific and medical apparatus (ISM).* This subcommittee is responsible for the promulgation of test methods applicable to equipment that generates intentional HF signals, and for the establishment of adequate limit values for HF disturbances. The committee has completed a revision of Publ. 11, and a warning note has been added recommending that stricter limits might be necessary for microwave ovens below 5 kW because of their widespread use in the home. The subcommittee also was charged to study whether the test methods and limits for domestic appliances were applicable in the range above 30 MHz. For terminal noise voltages in the range of 0.15 to 30 MHz, the limits were already determined to be acceptable. In addition, the subcommittee was asked to study EMI of heavy-current or high-voltage thyristors.

- *Subcommittee C: Noise caused by high-power cables, high-voltage equipment, and electrical traction.* Tasks of this subcommittee, at the time of this writing, were extended primarily to EMC problems caused by insulators applied in high-voltage energy transmission. The committee's work on Publ. 18 was finished in 1986 and, following that, its agenda was extended to performing a more thorough examination of EMC problems related to high-voltage dc energy transmission.

- *Subcommittee D: Ignition interference from motor vehicles, combustion engines, and related subjects.* Examination of noises produced by motor vehicles are included in subject field of this subcommittee. Updates of former works were issued in the report of Publ. 12. The committee conducts an ongoing effort to define EMS requirements for vehicular electronics.

- *Subcommittee E: EMS of radio and television receivers.* In Publ. 13, this subcommittee issued its standards covering EMS test methods for receivers. Its mission at the time of this writing concerned two major fields. First, they will work closely in cooperation with IEC Subcommittee 12. Second, they are in the process of setting standards related to EMS in telecommunication systems, focusing on HF impulses in the frequency range up to 30 MHz. Results will be issued as Publ. 20.

- *Subcommittee F: EMI in domestic appliances, fluorescent tubes and similar devices.* This subcommittee issued an update of its former efforts in Publs. 14 and 15. In subsequent periods, they intend to take a special look at solutions for EMC problems caused by portable hand tools and fluorescent tubes. In the framework of this mission, they will review safety questions and the effects of HF disturbances peculiar to impulses. This subcommittee's field of interest also covers the determination of requirements for narrowband measurements. Studies will be made to determine new limits for narrowband emissions.

Through diligent effort, these subcommittees have issued some very useful recommendations. Limits for EMI emission of electrical equipments first were established only for the frequency range 0.15 to 30 MHz, thus meeting broadcast requirements. These limits were later extended downward to the frequency range of 10 to 150 kHz. CISPR requires the measurement and attenuation of HF emissions in the frequency range of 30 to 300 (1,000) MHz.

Increasing attention is being devoted to the study of electromagnetic disturbances in the frequency range of 1 to 18 GHz. The examination of emissions in this frequency range is driven primarily by the growing need for space telecommunications.

With semiconductor equipment (ranging from power electronics to household appliances) coming into general use, the requirements for clean mains power have become increasingly strict. The degenerating situation also has hampered international trade. In the framework of CENELCOM (Comite Coordination Européen des Normes Electriques pour le Marche Commun, or European Coordination Committee of Electrical Standards in the European Common Market), a decision was made to establish a Common Standardisation Committee to create a standard for electrical equipment emission limits.

The Common Standardisation Committee, formed in 1970, immediately linked itself with representatives of electrical energy suppliers and electrical household appliance manufacturers. The Common Market was enlarged in 1973 and, following that, CENELCOM was reorganized under the name of CENELEC (Comite European de Normalisation Electrotechnique, or European Electrical Standardisation Committee).

CENELEC established three Subcommittees. Subcommittee CC3 or, according to the German abbreviation (Normen Komission) NK3, dealt with EMI and semiconductor household appliances. The standard issued in 1975 under EN 50006 limits the harmonics of numbers 3 to 15 and the disturbances responsible for flickering. The draft standard deals separately with phase-controlled thyristor equipment. Additionally, the standard draft touches upon applied technologies, the validity sphere, and calculation methods. The standard draft EN 50006 was published in CENELEC countries immediately after its acceptance, without any alteration. After the preparation of the standard draft, Subcommittee CC3 was terminated.

Thereafter, several countries turned to the IEC with a request to take up the problem of EMI with regard to semiconductor apparatus connected to mains. For surveying this special field, the IEC established a new subcommittee under the mark TC77. The Subcommittee TC77 works in cooperation with several nations and CISPR. To deal with the heterogeneity of the subject matter, duties were distributed between five special groups as follows:

- Working Committee #1: Terminology
- Working Committee #2: Mains impedances and LISNs
- Working Committee #3: Harmonic and nonharmonic electrical noise produced by electrical household and other similar equipment, included dc components
- Working Committee #4: Voltage fluctuations caused by electrical household and other similar equipment
- Working Committee #5: Harmonic and nonharmonic electrical noise produced by television sets

During statutory meetings, the groups created a detailed work outline covering the entire scope of EMC subject matter.

Many EMI problems exist that cannot be solved merely through examination and measurement of electrical equipment emissions and determination of acceptable emission levels. Therefore, there is also a need to examine the susceptibility of electrical apparatus to ensure that a transient or distortion on the mains cannot cause a malfunction or, at the extreme, a breakdown.

After World War II, the CEE (Commission International de Reglementation en vue de l'Approbation de l'Equipment Electrique, or International Committee of Electrical Equipment Approbation Standard) began to deal with the subject of electrical equipment EMS. Twenty-two European countries are included in CEE. In addition, two observers take part in its work. Through the early 1970s, CEE made an effort to summarize safety requirements for household appliances and other similar equipment. In the middle years of the 1980s, its work expanded to studying the EMS of electrical equipment containing electronic units.

CEE published a recommendation related to the protection of electrical equipment against transient disturbances, titled CEE 229-SEC UK 101 F 72 (1972-Sept). The recommendation covers specific data for test voltages and methods, as well as the test signal generator circuit. The recommendation prescribes measurement methods for several kinds of disturbing signals. Voltage pulses imitate transients formed by switching processes. The tests offered included the addition of an ac voltage with a frequency differing from the mains frequency. This test answers whether the electrical apparatus can be operated safely on mains with a remote control system. Another test, employing a short-duration voltage cutoff, serves to simulate disturbances generated during a momentary short-circuit on the mains (during which time all users suffer a voltage cutoff until a fuse or other device clears the fault).

IEC joined the study of EMS requirements in the early 1960s. This subject was first addressed by IEC Subcommittee TC65; later, Subcommittee TC77 joined in the effort. In the course of their work, these subcommittees lent support to CEE in many respects.

The study of the EMS of electrical equipment and articulation of measurement methods, as well as the compilation of recommendations and standards for this field of EMC, has been the specialty of IEC Subcommittee TC65. According to plans, the results of this group's work will be published in a standard draft IEC 1000 in the early 1990s. Subcommittee TC77 also will take part in compiling this draft standard. The secretariat document TC77B(4) reveals existing and planned EMS test methods that will be involved in the final text. No doubt, also, CISPR will include the examination of EMS problems in electrical equipments in its program. Signs of this opening are evident in Publications 13 and 20, which deal with EMS in radio and television sets.

Subcommittee TC65, at the time of this writing, is working on EMC problems in industrial automation and process control systems. The result of this work has been issued in IEC 801. Fields of study are indicated in the list of chapters, which is as follows:

801-1 General Introduction
801-2 Electrostatic Discharge (ESD)
801-3 Radiated Electromagnetic Fields
801-4 Fast Transients
801-5 Surges

In the early 1990s, several subcommittees of the IEC continue to study EMI phenomena. The Subcommittee TC110 (set up by CENELEC) and the organization ACEC (Advisor Committee on Electromagnetic Compatibility) strive jointly for harmony in that comprehensive work.

3

Description of
Electromagnetic Disturbances

Any electrical equipment (but especially semiconductor circuits) qualifies as a potential source of EMI. In general, electrical apparatus functioning as man-made EMI sources can be classified into two principal groups: equipments whose primary function is to generate and utilize intentional high-frequency (HF) signals (radio transmission excluded), and those that generate HF electromagnetic energy as an unintended by-product of their primary functions. In the former case, HF signals radiated into the environment will be qualified as "undesired disturbing signals." However, these ISM equipments may generate HF signals not only in the fundamental or intended frequency range, but also over a wide frequency range—usually on both sides of the fundamental carrier. Such equipments include industrial induction heat furnaces, high-frequency medical equipment, and so forth.

The other electrical equipment group generates HF signals that are not required for normal operation. These signals appear as electromagnetic noise in the environment. The EMI of these equipments can be traced back to quick voltage and current transitions. Although the levels of incidental signals are usually relatively low, they often are major causes of EMI. Some of the more important incidental EMI sources include switches, electrical cleaning equipment, fluorescent tubes, and power electronic equipment.

The electromagnetic emissions of electrical equipment are not easy to precisely specify and classify, but we can attempt to do so if we know some of the characteristics of the offending signal. To a degree, classifications are arbitrary, but they can help us understand the electromagnetic emission of power electronic equipment. Generally, the character, frequency content, and transmission mode provide the basis for classifying man-made electromagnetic disturbances, but it is not unusual to categorize them in terms of energy content, waveform, and other factors.

3.1   CLASSIFYING DISTURBANCES BY FREQUENCY CONTENT

Figure 3.1 illustrates the assorted HF disturbances by frequency content. Electromagnetic disturbances with an upper limit of 0 to 1,250 Hz or 2,000 Hz increase losses on the mains and distort the waveform of the voltage. Therefore, examination and elimination of this type of electromagnetic noise is a sphere independent of the issue of HF disturbances.

At the time of this writing, the frequency range of 1.25 to 150 kHz is not often examined by EMC engineers, although electromagnetic disturbances in this frequency range are coming to light more and more. The acceptable level of harmonics in this frequency range is specified in many national standards.

The range of radio frequency disturbances starts at 150 kHz. This range is generally divided into the bands of 0.15 to 30 MHz and 30 to 300 (1,000) MHz. The reason for this division is that different transmission modes and measurement methods apply to the HF disturbances.

However, one cannot sufficiently classify HF disturbances in terms of the frequency content only; the character must also be examined. A major distinction is between narrowband and broadband signals, and the difference can be the determining factor in the appropriate application of emission level standards and in solving EMI problems.

A noise signal is narrowband if its spectrum components are found at discrete frequencies, with very narrow (ideally, zero) bandwidth. In general, devices that produce HF signals as an operational necessity will create some narrowband disturbances.

The frequency spectrum of a broadband electromagnetic disturbance is continuous and covers a relatively wide range. Broadband disturbances are divided into

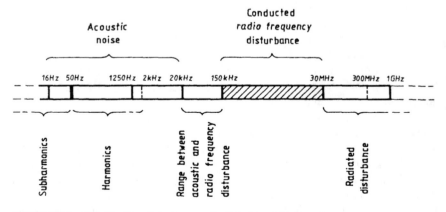

FIGURE 3.1   Classification of electromagnetic disturbances by frequency

two additional groups, namely *coherent* and *noncoherent* signals. A signal or emission is said to be coherent when neighboring frequency increments are related or well defined in both amplitude and phase. For broadband situations, neighboring amplitudes are approximately equal. The amplitude and phase of the components of a noncoherent noise signal are random with regard to neighboring frequency increments.

As the designation also shows, classification according to bandwidth means the ratio of the EMI to a reference bandwidth. In EMC practice, this reference bandwidth is the measuring bandwidth, but it can be associated with a potentially susceptible victim receptor, too. This ratio can be given in a manner derived from both the measuring bandwidth and the characteristics of the disturbances [32].

An electromagnetic emission is qualified as broadband under the following conditions:

- While tuning the measuring bandwidth of the EMI instrument over a range of ±2 impulse bandwidths around its center frequency, a change in peak response is detected of 3 dB or less.
- In the case of an impulse series, the repetition rate of the emission is less than or equal to the measuring bandwidth of the EMI instrument.

An electromagnetic emission is qualified as narrowband if either of the following is true:

- Tuning the measuring bandwidth of the EMI instrument over a range of ±2 impulse bandwidths around its center frequency, a change in peak response is detected of more than 3 dB.
- In the case of an impulse series, the repetition rate of the emission is higher than the measuring bandwidth of the EMI instrument.

If an emission is determined to be broadband in one test and narrowband in another, specification requirements apply for both emission types.

The above definition of narrowband and broadband categories can also be expressed mathematically. An electromagnetic emission is broadband if:

$$\left| 20\log \frac{a\,(f_0)}{a\,(f_0 \pm 2B_i)} \right| < 3 \qquad (3.1)$$

where

$$a\,(f_0) \;=\; \frac{a\,(\omega)}{2\pi} = \text{the amplitude density function}$$

$$a(\omega) = \left| \int_{-\infty}^{\infty} f(t) \times e^{-j\omega t} dt \right| \tag{3.2}$$

f_0 = center of the measuring bandwidth

B_i = bandwidth of examination

If the inequality stated in Eq. (3.1) is not valid, the electromagnetic emission is narrowband.

A mathematical expression can be given for an impulse series to determine the measuring bandwidth, B_m, at which an electromagnetic disturbance will be seen as a broadband emission. In case of a pulse train consisting of impulses of width T_p and a repetition rate f_p, the emission is of broadband character if:

$$B_m > \frac{0.15}{T_p} \times \frac{a(f_0)}{\int_{-\infty}^{\infty} |f(t)| dt} \tag{3.3}$$

where

$\omega_0 = 2\pi f_0$ and

f_0 = the measurement center frequency

The HF emission is of narrowband character if:

$$B_m < \frac{0.15}{T_p} \times \frac{a(f_0)}{\int_{-\infty}^{\infty} |f(t)| dt} \tag{3.4}$$

Let us consider the HF emission of a switched-mode power supply (SMPS) for CISPR EME measurement. The SMPS runs at frequency f_0, but the generated pulse train consists of impulses with 5 μs width (small load). The measurement bandwidth is $B_m = 9$ kHz. As derived from Eq. (3.3), the HF emission will be seen as broadband if:

$$9 \times 10^3 > \frac{0.15}{5 \times 10^{-6}} \times \frac{a(f_0)}{\int_{-\infty}^{\infty} |f(t)| \, dt} \qquad (3.5)$$

This yields:

$$\frac{a(f_0)}{\int_{-\infty}^{\infty} |f(t)| \, dt} < \frac{(9 \times 10^3)(4 \times 10^{-6})}{0.15} \sim 0.3 \qquad (3.6)$$

Given that the first spectral sections of the supposed pulse train consist of a horizontal line descending at –20 dB/decade (see Chapter 5, Section 5.3.1), the HF emission will be seen as broadband at frequencies where the ratio of $a(f_0)$ and the horizontal spectra is less than 0.3. As the break frequency is about 70 kHz, this is greater than 200 to 250 kHz. As the pulse width becomes narrower (less load), this frequency limit increases.

The emission of semiconductor equipment is of the character of a pulse train, which can be qualified as broadband and coherent. The frequency spectra of the electromagnetic emission of most power electronic equipment covers the range from the mains or operational frequency up to a few megahertz.

3.2   CLASSIFYING DISTURBANCES BY CHARACTER

Electromagnetic disturbances will often affect the mains voltage. These disturbances can be of long or short duration. Changes of long duration usually are not included in the domain of EMC, as they mainly cause alterations in the rms value of the mains voltage. The duration of short changes runs from a few seconds down to less than a microsecond. Short-duration electromagnetic disturbances appear as distortions on the mains voltage.

Electromagnetic disturbances of short duration can be divided into three groups:

1. *Noise* (N), which is a more or less permanent alteration of the voltage curve. The noise is of periodic character, and its repetition frequency is higher than the mains frequency. Electric motors, welding machines, and so forth are regarded as characteristic noise sources. The amplitude of noise is less than the peak amplitude of the mains voltage itself. With switch-mode power supplies, noise often takes the form of a few volts of ripple at the switching frequency, riding over the mains voltage.

2. *Impulses* (S), which are positive and negative peaks superimposed on the mains voltage. Impulses are characterized by having short duration, high amplitude, and fast rise and/or fall times. Impulses can run synchronously or asynchronously with the mains frequency. Noises, created during various switching procedures, can exist between impulses. Typical devices that produce impulses are switches, relay controls, rectifiers, and SCR circuits.

3. *Transients* (T), and other transitional processes. Here, the time period can range from a few periods of the industrial frequency to a few seconds. Most commonly, transients are generated by high-power switches.

The question arises as to what electromagnetic disturbances can be regarded as isolated transients or impulses, and what causes it to become a broadband noise of pulse train character with continuous spectra? Impulses are distinguished, in addition to their high amplitude and slope, by a low duty cycle. The duty cycle, δ, is defined as [1]:

$$\delta = \tau \times f_r \qquad (3.7)$$

where

τ = the impulse width measured at the 50% height

f_r = the pulse repetition rate, or average number of pulses
or impulses per second, for random occurrences

An electrical equipment having a lower duty cycle (δ) than 10^{-5} can be regarded as a source of transients. The factor δ of some characteristic equipment producing transients is included in Table 3.1. The duty cycle of the most incidental emit-

TABLE 3.1   Characteristics of transient noises produced by electrical equipment [1]

Electrical Equipment	Repetition Rate (Hz)	Impulse Width (s)	Duty Cycle (δ)
Fluorescent tubes	100	10^{-7}	10^{-5}
Brush-commutator motors	10^3	10^{-8}	10^{-5}
Relays and solenoids:			
Industrial control units	10	10^{-7}	10^{-6}
Slot machines	1	10^{-7}	10^{-7}
Casual use	10^{-3}	10^{-7}	10^{-8}
Switches:			
Wall switches	10^{-3}	10^{-6}	10^{-9}
Business machines, household appliances	10^{-4}	10^{-6}	10^{-10}

ting sources is much lower than 10^{-5}. When the duty cycle becomes significantly higher than 10^{-5} (as with switched mode power supplies), the emitting source is no longer regarded as transient or impulse but as continuous.

Another concept, noise potential, was introduced to quantify the effect of electromagnetic disturbances [103]. To grade these values of N, S, and T, they were characterized by numbers of 1 through 5. The noise potential, as a definition, is:

$$\gamma = (2N + 5S + T)^2 \qquad (3.8)$$

In this relationship, all noise phenomena (and particularly impulses) were weighted. Weighting symbolizes, first of all, the degree of difficulty of filtering out the disturbances. The weighting numbers were developed through measurements and practical experiences. The noise potentials of some noise-generating equipment are included in Table 3.2. It can be seen from that table that thyristor equipments are considered to be the strongest noise sources.

3.3 CLASSIFYING DISTURBANCES BY TRANSMISSION MODE

Electromagnetic disturbances travel by conduction on wiring and by radiation in space. Electromagnetic disturbances below approximately 10 MHz spread primarily by conduction; at higher frequencies, radiation becomes dominant. Also,

TABLE 3.2 Sources of noise, impulses, and transients and their weighting factors [103]

Electrical Equipment	N	I	T	γ
SCR motor controllers	3	5	3	1156
SCR regulators	2	5	2	961
Arc welders	3	4	4	900
Converters	4	4	3	900
SCR power supplies	3	5	1	900
Solenoid valves	1	5	2	841
Line printers	3	4	2	784
Office machines	3	4	2	784
Portable drill	4	3	2	625
Contactors, power relays	1	4	1	529
Fluorescent lamps	1	4	1	529
Switching power supplies	3	3	1	484
Inverters	2	3	2	441
Diathermy X-ray equipment	3	2	1	289
Automatic sanders	2	2	2	256
Air compressors, air conditioners	2	1	5	196
Heavy-duty motors	2	1	5	196
Electroplating equipment	3	1	2	169

disturbances may get into circuits that are closely spaced via inductive and capacitive coupling. The different characteristics pertaining to the transmission mode are explained by the fact that in the frequency range of 0.15 to 30 MHz, only the conducted EMI usually must be measured and suppressed. In this frequency range, measurement of radiated EMI is only required by certain standards and recommendations. Until the frequency content of the semiconductor-generated EMI reaches some specified number of megahertz, only the conducted EMI of these equipments generally should be measured.

Electromagnetic disturbances that get into the power mains can bring one other danger: the noise signals may be modulated by some particular frequency. Such a modulating frequency can be generated by rectifier and SCR circuits or saturated iron transformer cores. In the former case, the modulation frequency depends on the circuit of the rectifier. In the latter case, it typically is three times the mains frequency. The modulated HF signal can be demodulated by other circuits, but this demodulated signal can also disturb the operation of receivers and other similar devices.

4

Conducted EMI Measurement

Electromagnetic emissions produced by power electronic equipments are usually broadband and coherent, in the range from the operating frequency up to a number of megahertz. Conducted EMI usually should be measured within this frequency range. As most national and international standards address conducted emissions only in the frequency range of 0.15 to 30 MHz, this chapter is limited to measurement methods for signals within these boundaries. EMI measurement and suppression in the range of 10 to 150 kHz are increasingly important, and the measurement methods are similar to those for 0.15 to 30 MHz. However, that subject is beyond the scope of this book.

Electromagnetic disturbances can appear in the form of *common-mode* (often called *asymmetrical*) and *differential-mode* (*symmetrical*) voltage and current components. The definition of common-mode and differential-mode components is illustrated in Fig. 4.1. The common-mode and differential-mode components are defined by the voltages and currents, measured on the mains terminals, as follows:

$$U_d = U_1 - U_2 \quad \text{and} \quad I_d = \frac{I_1 - I_2}{2} \tag{4.1}$$

$$U_c = \frac{U_1 + U_2}{2} \quad \text{and} \quad I_c = I_1 + I_2 \tag{4.2}$$

where

U_d = the differential-mode voltage component

I_d = the differential-mode current component

U_c = the common-mode voltage component

I_c = the differential-mode current component

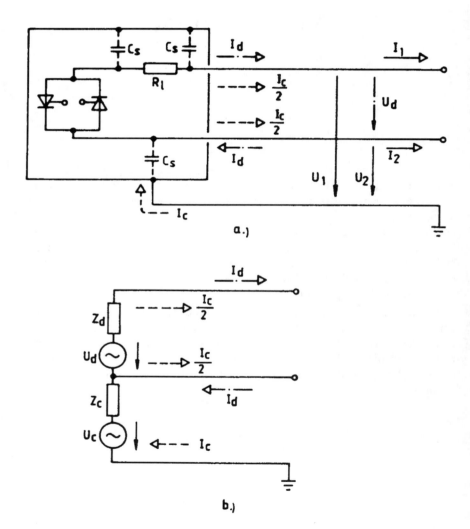

FIGURE 4.1 Differential-mode and common-mode EMI voltage and current components; (a) typical EMI source and (b) the HF substitution circuit of the EMI source (R_1 = load)

The HF equivalent circuit of an EMI source is shown in Fig. 4.1b. The differential-mode current component flows in the supply wires (including the neutral wire). The differential-mode voltage component can be measured between phase conductors as well. The common-mode current component flows from the phase and neutral conductor toward the earth. The circuit for the common-mode component is closed by the impedance Z_c, representing the stray capacitances between the earthed parts and the circuit. As seen from the figure, there is no simple relationship between the common-mode EMI components and the voltage of the EMI source, because the measured EMI depends on the mains impedances and different parasitic effects (Z_c).

4.1 EMI MEASURING INSTRUMENTS

EMI components in the range of 0.15 to 30 MHz (and, eventually, down to 10 kHz) are measured by EMI receivers. The EMI receiver usually measures the output voltage of a suitable sensor. These measuring instruments are tunable, frequency-selective voltmeters of accurate amplitude response. The instrument shown in Fig. 4.2 is typical. While requirements for mixers, oscillators, and bandpass filters used in EMI receivers can be described relatively simply, the case of detectors is not so. For EMI receivers, specifications for the detectors, including the display instruments, have been established to allow valid comparison of EMI measurement results.

Various kinds of detectors are known in EMI measurement techniques, including *peak, slide-back, average, effective (RMS),* and *quasi-peak* detectors [1]. The least commonly used detection method is RMS. Although the displays of most EMI receivers are calibrated to RMS, one should not infer that these instruments measure true effective value. Generally, only the display is calibrated to the effective value of an equivalent sinusoidal signal.

Some EMI instruments are designed to measure the average value of an emission. The reason is that many receptor victims are predominantly affected by the average of the culprit EMI signal rather than by other characteristics. The average detector, sometimes called a *field-intensity* (FI) detector, has a very long integration time constant—on the order of one second.

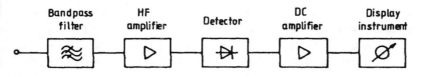

FIGURE 4.2 Functional diagram of classic CISPR EMI instruments

The detection of the peak value of the emission is usually very important in EMI measurements because many EMI sources are impulsive in character, and because an increasing number of potential EMI victims are particularly sensitive to such disturbances. Partly for this reason, several EMI specifications have their limits based on peak value. Detectors measuring true peak value are characterized by short charging and very long discharging time constants. The charging time constant is about 100 ns, and a discharge time constant can even approach 100 s.

Although average and peak value measurements give useful information about EMI, quasi-peak detection was developed and recommended for common use by CISPR. It was observed that by weighting the time constants of the peak detector, a better correlation could be made between the EMI receiver readings and the broadcast disturbances heard by human ear. Therefore, the specifications for quasi-peak detectors were established with reference to the human ear. The charging time constant was chosen to be on the order of 1 ms, and the discharging time constant set at 160 ms. Although quasi-peak detectors measure neither the average nor the peak value correctly, they are nevertheless quite suitable for characterizing EMI noise sources.

The relation between the quasi-peak detector and true peak response, as a function of receiver parameters, is given by the following relationship:

$$\alpha = \frac{\pi R_c B_6}{R_d f_r} \tag{4.3}$$

where

R_c = the charge resistance in Ω

R_d = the discharge resistance in Ω

B_6 = the 6 dB bandwidth that is approx. $0.95 \times B_i$, where B_i
   = the impulse bandwidth in hertz

f_r = the repetition rate of the measured impulse series

The above relationship is shown in Fig. 4.3.

Since the value indicated by an EMI instrument depends, among other things, on the specification of the detector and the conducted disturbances to be measured, it was necessary to coordinate technical specifications for EMI instruments. CISPR elaborated requirements for EMI instruments and published them in its Publ. 1, 1972 [342]. In Europe, at the time of this writing, only values measured by an EMI instrument matching the CISPR specifications are acceptable for civilian EMC testing.

Several arguments tend to favor the quasi-peak approach required by CISPR. Certainly, electromagnetic noise can be well characterized by quasi-peak measurements. But CISPR's decision was influenced by the fact that this method was al-

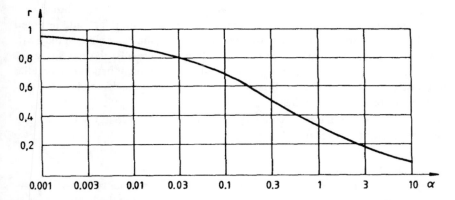

FIGURE 4.3 Ratio of quasi-peak to true peak detector outputs vs. receiver parameters and impulse repetition frequency; r = ratio of indication from true peak to quasi-peak detector, α = per Eq. (4.3)

ready known and in use during the early development of EMI measurement standards. It should be noted that CISPR has not affixed itself permanently to quasi-peak detectors and is looking at other possible solutions.

The other important characteristic of EMI instruments, based on international agreement, is the measurement bandwidth. The indicated value depends on the measurement bandwidth of the EMI instrument. Many times, EMI generated by semiconductor equipment takes the form of broadband coherent signals. Because of the coherent character (see Chapter 3), when the measurement bandwidth is somewhat less than the bandwidth of the EMI to be measured, the noise components within the measurement bandwidth can be summed up as follows:

$$U_m = \int_{f_c - B/2}^{f_c + B/2} A(f)\, df = A_c \times B \tag{4.4}$$

where

$\qquad$ B = measurement bandwidth

$\qquad$ f_c = the center frequency of the measurement bandwidth

$\qquad$ A(f) = the amplitude density function as a function of frequency

$\qquad$ A_c = the value of the amplitude density function in the center of the measurement bandwidth, considered as constant

The above relationship shows that, when measuring a coherent broadband emission, the reading is proportional to the receiver bandwidth. It should be noted that this characteristic can be also used for determining the bandwidth of the EMI. To allow a comparison of the results of EMI measurements, the measurement bandwidth must be specified. As it happens, the measurement bandwidth and the response curve of the bandpass filter were chosen to meet the demands of the medium-wave broadcasting industry.

CISPR has set up the following requirements as applicable to EMI measuring instruments for the 0.15 to 30 MHz range:

1. Bandpass filter:
   • bandwidth at 6 dB = 9 kHz (The required bandpass filter response curve is shown in Fig. 4.4. The actual response must be within the hatched area.)
2. Time constant:
   • electrical charge time constant = 1 ms

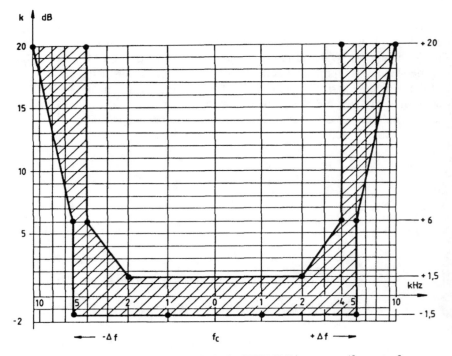

FIGURE 4.4   Overall passband selectivity limits for CISPR EMI instruments (f_c = center frequency of the measurement bandwidth, k = relative input in decibels for constant output)

- electrical discharge time constant = 100 ms
    - mechanical time constant of critically damped display instrument = 160 ms
3. Overload capability (The overload capability must always referred to the maximum deflection of the indicating instrument.):
    - circuits preceding the detector = 30 dB
    - dc amplifier inserted between detector and indicating instrument = 12 dB
4. Measurement accuracy in case of sinusoidal input signal = ±2 dB

In practice, HF disturbances are seldom purely sinusoidal; they have the character of a pulse train. The behavior of EMI instruments in the case of impulse-like signals can be best described by the amplitude relationship. CISPR recommendations give the required value and measuring method for this amplitude relationship. The measurement method is shown in Fig. 4.5. The output resistance of the impulse generator (IG) is equal to the input resistance of EMI receivers. The generated pulse train must consist of impulses with uniform spectra and impulse strength of $U_o \times RC = 0.316$ µVs. The repetition rate of the impulse generator must be set with an accuracy of at least 1 percent for the values of 1 Hz, 2 Hz, 20 Hz, 100 Hz, and 1 kHz. The spectra must be flat at least up to 30 MHz, within ±2 dB. But, because of cross-modulation distortion, the spectra above 30 MHz must be reduced at least 10 dB at 60 MHz.

The output resistance of the sinusoidal generator (SG) is also equal to the input resistance of the EMI receiver. Setting the internal voltage of the sinusoidal generator (U_g) to 2 mV (66 dBµV), the voltage on the input of the EMI instrument will be 1 mV (60 dBµV). The amplitude relationship gives the ratio between

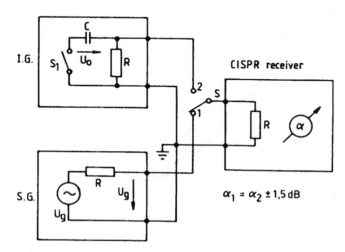

FIGURE 4.5 Method of measuring amplitude relationship (I.G. = impulse generator, S.G. = sine generator)

the deflection of the EMI instrument if the sinusoidal generator (switch S in position 1) and the impulse generator set to a 100 Hz repetition rate (switch S in position 2) are connected to the EMI receiver. The EMI instrument reading also changes as a function of the repetition rate of the pulse train. The amplitude relationship and the difference due to the change in repetition rate are given by the ratio of input values belonging to identical output instrument deflections. The amplitude relationship corresponding to CISPR recommendations must be within ±5 dB. The differences due to the change in repetition rate of the impulse series are shown in Fig. 4.6 and, in another form, in Table 4.1.

It is very difficult to meet the accuracy recommendations required by CISPR [342], particularly for low repetition rates. At a low repetition rate, and for high amplitudes, saturation causes problems; for low amplitudes, the threshold and noise make life difficult. In fact, CISPR requirements cannot be met using only one detector stage. The dynamic range of EMI instrument can be increased by applying more measurement channels [304].

Such a solution for an EMI detector is shown in Fig. 4.7. Every single measuring channel contains an independent quasi-peak detector. The saturation and threshold of every single channel differs from every other. The output signals of the independent detectors are added up by a circuit that operates the display readout.

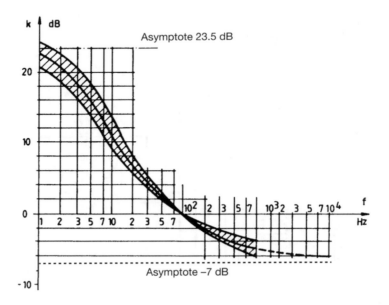

FIGURE 4.6   EMI instrument pulse response curve (k = relative input in decibels for constant output)

TABLE 4.1   Amplitude Relationship of Quasi-peak Detectors in EMI Measuring Instruments Corresponding to CISPR Recommendations

Repetition Rate (Hz)	Relative Input Level (dB)
1,000	−4.5 ± 1.0
100 (base)	0
20	+ 6.5 ± 1.0
10	+10.0 ± 1.5
2	+20.5 ± 2.0
1	+22.5 ± 2.0
single pulse	+23.5 ± 2.0

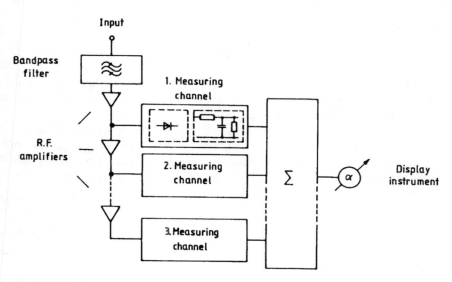

FIGURE 4.7   Functional diagram of quasi-peak detector with wide dynamic range [304]

4.2   BASIC TERMS AND CONDUCTED EMI REFERENCES

In electrotechnical engineering, the fundamental unit of signal amplitude measurements is usually *power*. This practice has been adopted in the RF community as well. To facilitate discussion of large power ranges, the decibel (dB) system is used, in which the reference is the watt (W). When converting a power level into the dB system, the result is stated in units of dBW. Thus, the conversion relationship is as follows:

$$P_{(dBW)} = 10 \log P_{(W)} \tag{4.5}$$

where

$P_{(W)}$ = measured power in watts

The milliwatt (here abbreviated as m rather than mW; hence, dBm) is more convenient to use as the power reference in many applications. The reference in this system is:

$$0 \text{ dBmW} = 0 \text{ dBm} = 1 \text{ mW} = -30 \text{ dBW} \tag{4.6}$$

The power in dBW can be converted to the dBm system using Eqs. (4.5) and (4.6) as follows:

$$P_{(dBm)} = P_{(dBW)} + 30 \tag{4.7}$$

Some examples where the dBm system is commonly used include signal generator output calibrations, receiver sensitivity, and transmission losses. In some specifications, EMI limits are also given in dBm.

In EMC applications, power is rarely used as a reference; signal or noise amplitude measurement is stressed. There are many reasons for this, not the least of which is to simplify measurement technology. In EMC, voltage is commonly used as the reference unit for conducted measurements. The voltage is derived from power as shown below:

$$P = \frac{U^2}{R} \tag{4.8}$$

where

U = voltage in V

R = resistance in Ω across which the voltage was measured

Combining Eqs. (4.9), (4.5), and (4.6), we can write:

$$P_{(dBW)} = 10 \log \left(\frac{U^2}{R} \right) \tag{4.9}$$

$$P_{(dBm)} = 10 \log \left(\frac{U^2}{R} \right) + 30 \tag{4.10}$$

The voltage ratio of either two networks or one network under different conditions is obtained from their power ratios:

$$K_P = \frac{P_1}{P_2} = \frac{\dfrac{U_1^2}{R_1}}{\dfrac{U_2^2}{R_2}} = \frac{U_1^2 R_2}{U_2^2 R_1} \tag{4.11}$$

Accordingly, the power ratio expressed in dB is:

$$K_{P\,(dB)} = 10\ \log\left(\frac{P_1}{P_2}\right) = 10\ \log\left(\frac{U_1}{U_2}\right)^2 + 10\ \log\left(\frac{R_2}{R_1}\right) \tag{4.12}$$

If $R_1 = R_2 = R$, which does not constitute a major limitation, the voltage ratio can be given as follows:

$$K_{U\,(dB)} = 10\ \log\left(\frac{U_1}{U_2}\right)^2 = 20\ \log\left(\frac{U_1}{U_2}\right) \tag{4.13}$$

In Eq. (4.13), U_2 is the reference voltage. Taking 1 V as the reference, the above relationship becomes:

$$U_{m\,(dBV)} = 20\ \log U_m \tag{4.14}$$

For EMC measurements, 1 μV is more commonly used as a reference voltage. Thus,

$$1\ \mu V = 10^{-6}\ V = 0\ dB\mu V = -120\ dBV \tag{4.15}$$

Equations (4.14) and (4.15) yield a relationship to convert the voltage from the dBV system to dBμV:

$$U_{m\,(dB\mu V)} = U_{m\,(dBV)} + 120 \tag{4.16}$$

Known values in dBm and dBμV must be transformed into each other in engineering practice. Using Eqs. (4.10), (4.14), and (4.16), the transformation formula is:

$$P_{(dBm)} = U_{(dB\mu V)} - 120 - 10 \, \log (R) + 30$$

$$= U_{(dB\mu V)} - 90 - 10 \, \log (R) \tag{4.17}$$

Equation (4.17) is plotted in Fig. 4.8 for frequently used resistance values. The chart simplifies the conversion between values known in dBm and dBμV. The above formula takes the form shown below for R = 50 Ω system, which is common in EMC measurement technology:

$$P_{(dBm)} = U_{(dB\mu V)} - 107 \tag{4.18}$$

We often use the amplitude density function (or other spectral qualities) to characterize broadband coherent signals. A spectrum is usually given in units of volts per hertz (V/Hz), often called a *broadband unit*. In EMC techniques, the measurement unit of μV/MHz is better suited and, therefore, is generally used instead of V/Hz. Values in dBV/Hz can be converted to dBμV/MHz as follows:

$$A_{(dB\mu V/MHz)} = A_{(dBV/Hz)} + 240 \tag{4.19}$$

The measured voltage of a broadband coherent signal is a function of the measurement bandwidth. The exact value of the measurable voltage can be calculated

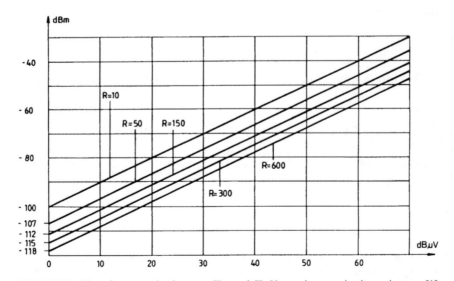

FIGURE 4.8   Chart for conversion between dBm and dBμV at various termination resistances [1]

by integration, but Eq. (4.4) provides a good approximation. Converting this relationship into the dB system, the relationship between the measurable voltage, U_m (now as narrowband unit), and the broadband unit is as follows:

$$U_{m\,(dB\mu V)} = A_{(dB\mu V/MHz)} + 20 \log B_{(MHz)} \qquad (4.20)$$

The CISPR measurement bandwidth is 9 kHz. Using the approximation of B = 10 kHz, the measurable noise voltage by means of the amplitude density function is:

$$U_{m\,(dB\mu V)} = A_{(dB\mu V/MHz)} - 40 \qquad (4.21)$$

Several EMI conducted specification limits are given in units of current. The reference is usually 1 μA. It is as if a small resistor, R_s, was added in series with the test lead, and the voltage drop across the resistor then measured with an EMI instrument to determine the unknown current:

$$I_{m\,(\mu A)} = \frac{U_{(\mu V)}}{R_{s\,(\Omega)}} \qquad (4.22)$$

The above relationship rewritten to 1 μA reference is:

$$I_{m\,(dB\mu A)} = U_{m\,(dB\mu V)} - 20 \log (R_s) \qquad (4.23)$$

For $R_s = 1\ \Omega$, the value of the noise current in dBμA is equal to the value of noise voltage measured in dBμV.

However, noise currents are rarely measured by means of a shunt resistance because (1) it can disturb the test item, (2) there is no galvanic separation, and (3) it is not practical where many cable wires are involved. Consequently, current probes are used as sensing devices, with the EMI receiver acting as a tunable voltmeter. Current probes can be characterized by their terminal-pair transfer impedance:

$$Z_T = \frac{U_m}{I_m} \qquad (4.24)$$

where

U_m = output voltage across the current probe when terminated by 50 Ω (e.g., the EMI instrument)

I_m = the unknown current flowing in the wire around which the current probe is placed

In units of dB, Eq. (4.24) becomes:

$$Z_{T(dB\Omega)} = U_{m(dB\mu V)} - I_{m(dB\mu A)} \tag{4.25}$$

Or, the unknown current is:

$$I_{m(dB\mu A)} = U_{m(dB\mu V)} - Z_{T(dB\Omega)} \tag{4.26}$$

4.3   MEASURING THE INTERFERENCE VOLTAGE

EME generated by electrical equipment can be well characterized by the disturbances' power but, for practical reasons, EMI voltage (or sometimes current) is the standard measurement unit. As a result, EMC standards and recommendations usually limit the maximum acceptable EMI voltage on the mains. Viewed as EMI sources, electrical equipments can be considered as voltage or current generators.

A problem encountered in EMI voltage measurement can be analyzed using the simplified equivalent circuit shown in Fig. 4.9. Measured on mains terminals, the EMI voltage, U_n, depends not only on EMI source parameters, but also on the mains impedance, Z_m. The HF mains impedance is not nearly constant; rather, it is highly variable in both time, location, and frequency. The variations in time are mainly the result of other equipments being switched on and off.

A wide range of measurements have been made in many countries to determine the HF impedance of the public distribution mains [184]. In each case, the HF impedance was measured between each live conductor and earth. In most cases, the absolute value of the impedance was measured, but some measurements also included magnitude *and* phase. Figure 4.10 shows the measured absolute values of mains impedances in the U.S.A. and Europe.

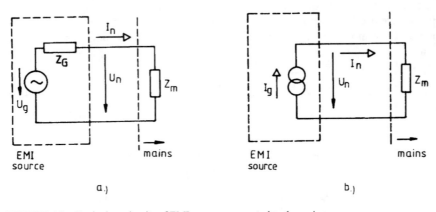

FIGURE 4.9   Equivalent circuits of EMI sources connected to the mains

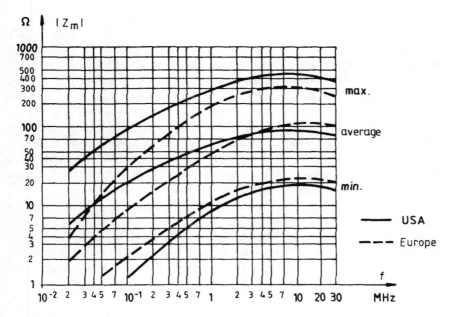

FIGURE 4.10   Power line impedance vs. frequency [184]

To ensure repeatability and comparability of EMI measurements, a standard load impedance must be created at the measurement point. This condition is met by inserting an interface circuit between the EMI source and the mains, as seen in Fig. 4.11. The interface circuit is called a *line impedance stabilization network* (LISN), also known as a *line stabilization network* (LSN), *power line impedance stabilization network* (PLISN), or *artificial mains*. The latter is consistent with

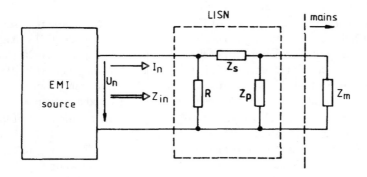

FIGURE 4.11   Interface circuit for conducted EMI voltage measurement

CISPR terminology. By specifying the serial impedance, Z_s, of the LISN to an appropriately high value, the impedance seen from the EMI source will be largely independent of variations in the mains impedance, Z_m. The LISN establishes a standard profile of load impedance (resistance) toward the EMI source and also filters out HF disturbances on the mains that could create measurement problems.

Some power electronic devices (e.g., on aircraft, motor vehicles, in telecommunication systems, etc.) are fed from dc power supplies. Measuring emissions from these units presents the same problem as illustrated in Fig. 4.11, so to ensure repeatability and comparability, measurements must be made under known impedance conditions for these devices as well. Although the LISN requirement was established for measuring emissions from equipment supplied by 50/60 Hz mains, CISPR LISNS are also used for measurements on dc-fed devices.

The most important technical characteristics for the LISN were originally specified in CISPR Publ. 1, 1972 [342] and now by CISPR Publ. 16 (1983) [349]. The superseded CISPR recommendation specified a 150 Ω ± 20 Ω load resistor. As investigation proved that the mains impedance is less than 150 Ω (see Fig. 4.10), CISPR Publ. 16 now specifies a 50 Ω load resistor for use in EMI voltage measurements to obtain a more realistic result.

4.3.1 Measuring Differential-Mode Interference Voltage

CISPR specifies the measurement of purely differential-mode noise voltages only in the frequency range of 150 to 1,605 kHz (long and medium wave). The differential-mode noise voltage is to be measured by an EMI instrument with a symmetrical (balanced) input. The input of the EMI instrument should be isolated from the terminals to be measured by a shielded, balanced transformer. The input impedance of the balance transformer should be higher than 1,000 Ω in the measuring frequency range. To make the measurements with a symmetrical-input EMI instrument, CISPR recommends the use of a Δ-LISN.

The scheme of a one-phase Δ-LISN for measurement apparatus is shown in Fig. 4.12. Setting the switch, S, to position C, the common-mode EMI voltage also can be measured. The phase shift of the LISN cannot be higher than 20 electrical degrees. The serial impedance of the LISN (which is generally a choke coil) should be higher than 1,000 Ω in the entire measurement frequency range, but at main frequency and stated current, the voltage drop shall not exceed 5 percent.

The common-mode EMI voltage must not affect the measurement of the differential-mode noise voltage. The suppression of common-mode EMI voltage should be higher than 26 dB (20:1) in the entire measuring frequency range. Common-mode EMI voltage suppression can be measured as shown in Fig. 4.13. The signal generator, G, should have an internal resistance of 75 Ω. The generator voltage, U_g, is injected between earth and the common point of two resistors (each of

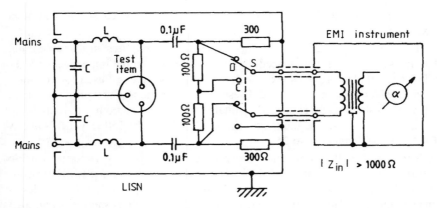

FIGURE 4.12 CISPR-standard LISN for conducted EMI voltage measurements using an EMI meter with a balanced (isolated) input

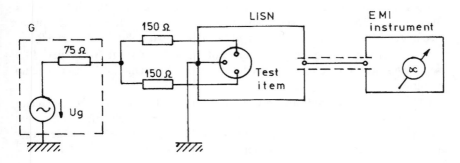

FIGURE 4.13 Common-mode voltage suppression measurement setup

150 Ω) and the terminals of the LISN to be connected to the test item. The resistors are of 1 percent accuracy. The ratio of U_g to the measured voltage (U_m) in the differential-mode measurement position should be greater than 20:1 (26 dB).

The circuit of a Δ-LISN for unbalanced EMI instruments is given in CISPR recommendations [349]. CISPR also gives the value of circuit elements for different 50, 60, and 75 Ω EMI instrument input impedances.

4.3.2 Measuring Common-Mode Interference Voltage

CISPR recommends the measurement of the common-mode EMI voltage only above 1,605 kHz but will accept it at lower frequencies. When measuring the EME of power electronic equipment, the common-mode EMI voltage generally dominates, although differential EMI also exists, which is consistent with most EMI standards.

The measurement of common-mode EMI voltage requires an instrument that has an asymmetrical (unbalanced) input. The preferred input impedance is 50 Ω. CISPR [349] also provides specifications for LISNs that are used for common-mode EMI Voltage measurement. These devices are the 50 Ω/50 μH V-LISN, the 150Ω V-LISN, the 150 Ω Δ-LISN (suitable for measuring both differential- and common-mode voltage), and the 50 Ω/5 μH V-LISN. The 150 Ω V-LISN and Δ-LISN are for currents up to 25 A.

The new CISPR recommendation specifies the use of a 50 Ω/50 μH V-LISN instead of the former 150 Ω LISN for conducted common-mode EME measurements. Figure 4.14 shows a one-phase CISPR 50 Ω/50 μH V-LISN. This device ensures the impedance between each conductor and earth as shown in Fig. 4.15, curve A. Special care is required in grounding the measuring set because of the 1 μF capacitor connected between the phase conductor and earth. The 50 Ω/50 μH V-LISN is used for currents up to 100 A.

For mains currents above 100 A but less than 400 A, the 50 Ω/5 μH V-LISN is used. Figure 4.16 shows the schematic layout of this device. It ensures an impedance match between each conductor and earth as shown in Fig. 4.15, curve B.

Figure 4.17 shows the schematic layout of a modified CISPR 50 Ω/5 μH LISN used for test items with low-impedance power supplies. This LISN is suitable up to 100 A.

LISNs that do not conform to CISPR recommendations are used in several countries and communities. Because the measured EME of electrical equipment depends on the LISN, the type or circuit of the LISN used for EMI voltage measurement should be indicated in the applicable test specification.

In many cases, LISNs may not be connected between the EMI source and the power supplies, or the current rating of the appliance to be measured may be higher than 400 A. To measure the EMI voltage without a LISN, CISPR recommends

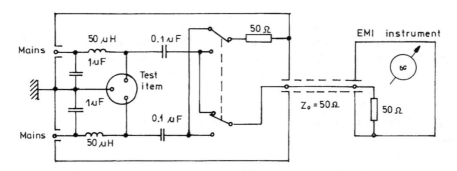

FIGURE 4.14   One-phase CISPR 50 Ω/50 μH V-LISN

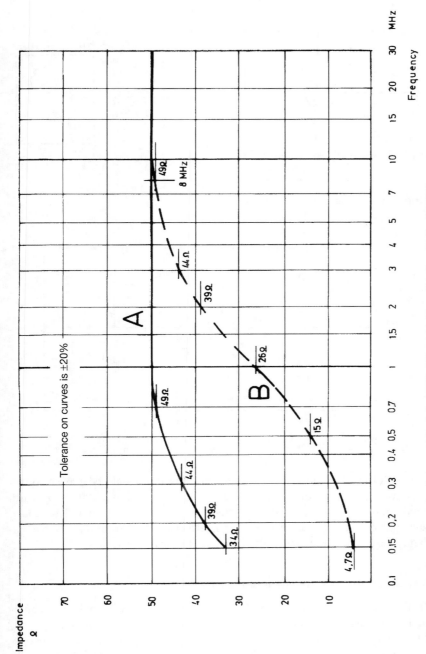

FIGURE 4.15 Impedance between Conductors and Earth for LISN of Fig. 4.14 (Curve B is for a 50 Ω, μH LISN)

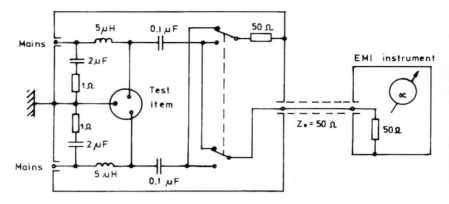

FIGURE 4.16 One-phase CISPR 50Ω/5 µH LISN for 100–400 A

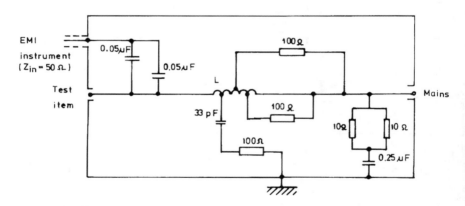

FIGURE 4.17 Modified CISPR 50 Ω/5 µH LISN for low-impedance supply

the method shown in Fig. 4.18. The isolating capacitor should be chosen so that the resulting impedance of the measuring circuit approximates 1,500 Ω. The earth ground should be of good quality. Often, special protection should be added against induced currents or other disturbing effects. CISPR recommends that disturbances cause no more than 1 dB change in the measured values.

4.3.3 Connecting the Test Item to the LISN

The measured EME when using a LISN depends on how the electrical equipment is connected to the LISN and how the EMI measuring set is grounded. Earthing the appliance through the supply conductor screening (shielding) is the only correct solution. For standard AM broadcast frequencies (<1.6 MHz), practically

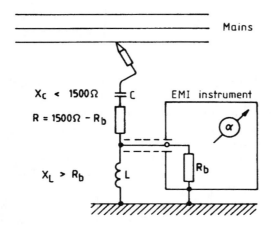

FIGURE 4.18 Circuit for conducted EMI voltage measurement above 100 A load current rating

the same measurement result can be obtained by earthing through a short, straight lead that is laid parallel to the mains. The earthing conductor length should not exceed 10 cm. For frequencies above a few megahertz, this simplified solution should be applied with great care. At higher frequencies, it is strongly recommended that shielded conductors be used, as when measuring the EME of switch-mode power supplies.

Direct earthing should always be used when testing nonradiating appliances and inadequately filtered devices. Conversely, direct earthing is absolutely prohibited when testing well filtered but radiating equipments such as medical apparatus, arc welders, and so forth. With direct earthing, the disturbance voltages across the LISN become very small; otherwise, they may be quite large.

4.3.4 Problems Involved in Measurements Using a LISN

Electrical equipment generates EMI as common-mode and differential-mode voltages and currents. In the case of high-power equipment, including power electronic equipment, noise suppression can be costly to implement. Since different solutions are required to reduce the common-mode and differential-mode noise components, it is often necessary to characterize the EMI source. Noise sources can be simulated by six kinds of idealized generators, but real EMI sources are combinations of any or all of them, with the effects combined. The six idealized noise generators are arranged in three combinations, as follows:

1. Differential-mode voltage generator (UD) and current generator (ID) connected symmetrically between the phases of the mains
2. Common-mode voltage generator (UCS) and current generator (ICS) connected symmetrically between the phases of the mains and the earth

3. Common-mode voltage generator (UCA) and current generator (ICA) connected asymmetrically between the every single phase of the mains and the earth

According to CISPR recommendations, the common-mode EMI voltage can be measured by a V-LISN, and the differential-mode voltage by a Δ-LISN. It follows, then, that making measurements with V- and Δ-LISNs, one can determine the character of the EMI source (i.e., using the above-mentioned idealized noise generators, one can determine the dominant noise mode). However, it can be demonstrated theoretically that these LISNs are not suitable for this purpose [25]. Suppose that the EMI voltage measured with V-LISN employs the idealized noise generators as shown in Fig. 4.19. Suppose also that the EMI voltage measured with a Δ-LISN employs the ones shown in Fig. 4.20. It can be seen from these figures that, in the case of ID and ICS EMI generator schemes, double deviation can occur in the results of measurements made by the two types of LISN. In the case of UCA and ICA EMI schemes, the differences between measurement results are also large, and false conclusions can be drawn about the EMI source. Supposing idealized UCA and ICA schemes, the Δ-LISN will also show a nonexistent differential-mode component.

More accurate results can be obtained by a simple modification of the V-LISN. J. Holownia [143] recommends the application of a switch in the V-LISN as shown in Fig. 4.21. With this LISN, which he calls a V/I-LISN, adequate information concerning the structure and the character of the EMI source can be obtained faster and at considerably lower cost than by other measuring methods. With the

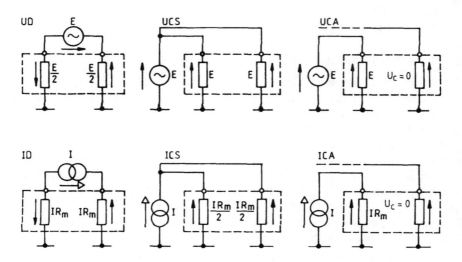

FIGURE 4.19   Basic types of idealized equivalent voltage and current sources [143]

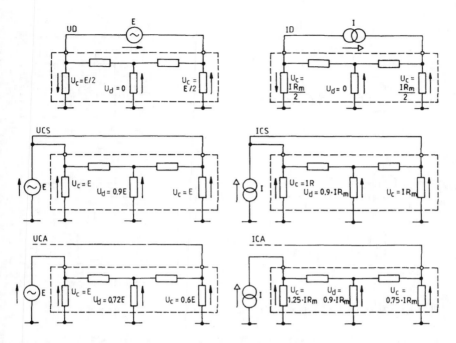

FIGURE 4.20 EMI meter indications obtained with CISPR Δ-LISN

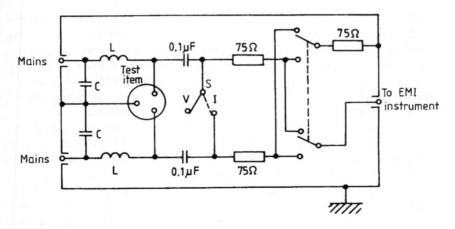

FIGURE 4.21 Modified CISPR V-LISN (V/I-LISN)

switch in the V position, EMI voltage measurement results will be the same as seen in Fig. 4.19. With the switch in position I, measurements will show the values seen in Fig. 4.22. From Figs. 4.19 and 4.22, it can be seen that the true character of the EMI source can be more precisely identified with the proposed V/I-LISN.

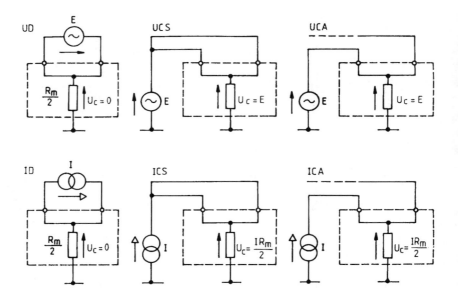

FIGURE 4.22   EMI meter indications obtained with V/I-LISN [143], switch in position I

4.4   MEASURING THE INTERFERENCE CURRENT

CISPR recommendations, and international standards developed in accordance with CISPR, do not require the measurement of EMI currents. There are several exceptions (e.g., MIL-Std. 461/462 and most other military, aeronautic, and vehicular industry standards) where the acceptable level of the interference currents is specified. Even so, to fully determine the character of an EMI source may require the measurement of EMI currents, and the HF impedance measurement of the mains cannot be made without measuring HF current components.

EMI currents can be measured via the use of special current probes. The current probe is a two-terminal-pair network that delivers a prescribed voltage across a 50 Ω output resistance for a specified input current. Manufacturers usually calibrate the current probe in transfer impedance units.

Two types are commonly used. One uses a Hall generator, and the other measures the magnetic field of the current. The current probe containing a Hall generator operates on the basis of compensation. Its measuring range extends in amplitude to 60–80 dB, and in frequency range from dc to 50–100 MHz. The measuring range can be changed by setting the transfer impedance with the aid of the compensation amplifier.

The second type of HF current probe senses the magnetic field induced by current flowing in the conductor. The induced voltage in the secondary coil of the current probe is the function (among other factors) of the rate of current change (i.e., the frequency of the measured current). The transfer impedance changes at a rate of 20 dB/decade as a function of the frequency. To make the measurements easier, with this kind of current probe the transfer impedance is kept at a constant level by passive or active matching circuitry. The passive matching circuit is used for high transfer impedances. The active matching circuits are commonly special active filters. In general, these current probes are suitable for measuring current in the frequency range of 30 Hz to 1 GHz, but a single model covers only one-third of this frequency range.

To facilitate the use of EMI current probes, they are made in toroidal form, consisting of two halves. By opening the iron core, one can insert the conductor under examination into the current probe. If two or more conductors are inserted through the current probe, it will measure the resultant current. Therefore, when all phase conductors are inserted, the common-mode component of the EMI emission will be measured.

The transfer impedance can be controlled (i.e., the current probe can be calibrated) as seen in Fig. 4.23. Setting the HF signal generator output voltage to U_g and loading it with a resistor R, the current flowing through the current probe is:

$$I_g = \frac{U_g}{R} \tag{4.27}$$

If the calibration load is 50 Ω, the transfer impedance can be expressed with the voltage measured by the EMI instrument:

$$Z_T = 50\left(\frac{U_m}{U_g}\right) \Omega \tag{4.28}$$

During the current probe calibration process, the transfer impedance should be calculated at many frequencies within the measuring frequency range.

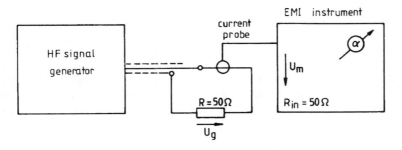

FIGURE 4.23 Test setup for measuring the transfer impedance of a current probe

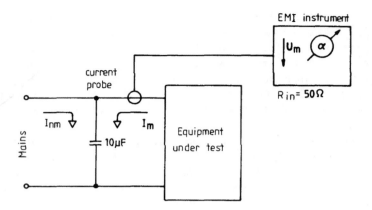

FIGURE 4.24   10 μF decoupling capacitor for measuring EMI currents of EUT

An EMI current measurement method generally used in military EMC testing employs the so-called decoupling capacitor. This procedure, widely accepted elsewhere, uses the well-known 10 μF capacitor [1]. As can be seen in Fig. 4.24, the decoupling capacitor should have an appropriately small parallel impedance both for the noise currents flowing from the mains, (I_{nm}), and the EMI current, (I_m), produced by the equipment under test (EUT). In this case, only I_m will flow across the current probe.

The HF decoupling effect of the 10 μF capacitor cannot be stepped down to arbitrary low frequencies, as the reactance of the 10 μF capacitor at the mains frequency is about 265 Ω in the case of a 60 Hz mains frequency, and 320 Ω in the case of a 50 Hz mains frequency. The lower frequency limit of the decoupling effect also depends on the power consumed by the equipment to be measured. When the electrical equipment takes a current, I_n, from the mains having the voltage, U_n, this equipment can be replaced by an impedance given by $Z_n = U_n/I_n$. The lower frequency limit, f_1, can be considered to be the value when the impedance of the 10 μF capacitor will be just Z_n. The value f_1 is as follows:

$$f_1 = \frac{I_n}{2\pi \times C \times U_n} \tag{4.29}$$

The approximate lower frequency limit of the EMI current component that is measurable with the 10 μF capacitor can be calculated with the following simple formula from the mains current of the load (i.e., of the EMI source) in amps. In the case of $U_n = 220$ V:

$$f_1 = 75 \times I_n \text{ Hz} \tag{4.30a}$$

In the case of $U_n = 127$ V:

$$f_1 = 125 \times I_n \text{ Hz} \qquad (4.30b)$$

If particularly accurate measurements are required for calculating the lower frequency limit, the above formulas cannot be used. The American standards prescribe that parallel impedance should be lower than 1 Ω. This means that the EMI current measurements can be made with the 10 µF capacitor only in the frequency range above 16 kHz. In this case, the lower limit of the measuring frequency range can be decreased only by inserting a serial choke coil between the mains and the 10 µF capacitor. The reactance of the serial choke coil should be selected so that, at the lower frequency limit, the impedance of the serial choke coil and the 10 µF capacitor are equal. After the necessary conversions, the inductance of the serial choke coil can be calculated by replacing f_1 in Hz in the following approximate formula [1]:

$$L = \frac{2500}{f_e^2} \text{ H} \qquad (4.31)$$

At low measurement frequencies, a relatively high serial impedance is needed so the voltage drop at mains frequency can be increased to an unacceptable value. However, the mains current of the EUT cannot be high by any amount, as Table 4.2 shows.

The 10 µF capacitor for measuring EMI currents can be obtained as an accessory. These decoupling capacitors are a special type of HF device, usually produced in feedthrough configuration (see Chapter 6, section 6.2.5). The 10 µF capacitor must be capable of handling mains voltage for general use up to 230 Vac RMS and at least 50 A current peak in its feedthrough [1]. The capacitor must behave no parasitic effects from dc to 50 MHz (i.e., any parasitic effects must appear well above 50 MHz—see Chapter 6, section 6.2).

TABLE 4.2   Measurement of Low-Frequency EMI Current with 10 µF Capacitor and Serial Choke Coil

Lower Measuring Frequency (f_1)	Serial Choke Coil Inductance	Serial Impedance of Choke Coil (50–60 Hz)	Maximum Load Current (I_n)
100 Hz	250 mH	78–93 Ω	150 mA
400 Hz	16 mH	5–6 Ω	2 A
1 kHz	2.5 mH	0.8–1 Ω	15 A
3 kHz	250 µH	negligible	150 A
10 kHz	25 µH	negligible	no limit

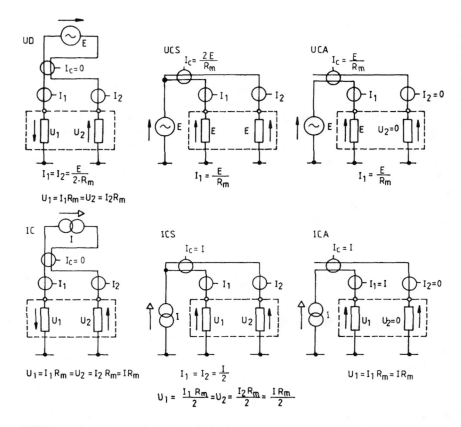

FIGURE 4.25   EMI meter indications obtained with V/I-LISN [143], switch in position V

Making measurements with the current probe, the character of the EMI source can be brought to light. The basic problem was discussed in section 4.3.3. Good results can be obtained when measuring the differential-mode and common-mode components of the EMI current generated by the EUT. The measurement results, in the case of a V/I-LISN, are shown in Fig. 4.25. The common-mode component of EMI current can be measured with the switch in the "V" position, and the differential-mode component can be measured with the switch set to "I".

Another way to characterize the EMI source is by measuring with a decoupling capacitor and current probe. This scheme is shown in Fig. 4.26. In cases of UD, UCS, and UCA EMI generators, the results will differ slightly from measurements using the V-LISN, but the difference will be no more than a few decibels.

Beginning with the early part of the 1980s, the so-called *absorptive clamp measurement technique* has gained wide acceptance for EMI measurements. Normally, the radiated interference produced by an electrical system should be meas-

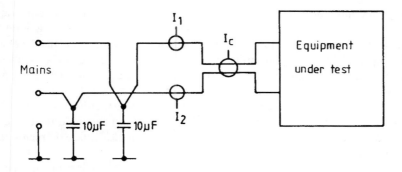

FIGURE 4.26 Differential- and common-mode measurements with a 10 μF decoupling capacitor

ured in the frequency range of 30 to 1,000 MHz to meet CISPR recommendations. These measurements are made using antennas set up inside anechoic shielded enclosures or open sites at distances of 10 m (about 30 ft) and 30 m (about 100 ft) from the EUT. During the measurement process, environmental electromagnetic disturbances must be minimized. Interference field strength measurements of this kind are costly and difficult. Since field strength measurement can be traced back to HF EMI current measurements where radiated emissions come primarily from EUT power leads, an alternate measurement technique was developed by the Swiss post office and has been accepted by CISPR for radiated emission measurements. The basic approach stems from the observation that most radiation from domestic and similar appliances radiates from the first meter or two of the power cable.

The absorbing clamp consists of a large number of ferrite rings for absorbing the power produced by the EMI source. The other task of the ferrite rings is to prevent unwanted currents from flowing on the coaxial cable that connects the absorbing clamp to the EMI instrument. The ferrite rings are split into half-rings to enable the power mains cable of the test item to be passed through the tunnel of the ferrite rings.

EMI measurement with an absorbing clamp is relatively simple. As the test technician slides the clamp along the length of the cable, the EMI instrument will show one or more high readings. The highest reading is adopted as characterizing the EMI source. The manufacturer of the clamp provides a calibration curve that relates the dBμV of an instrument reading to the dBpW (decibel picowatt) of equivalent radiated power seen by the probe.

4.5 SPECTRUM ANALYZERS

Spectrum analyzers are used for many applications, with EMI measurements being only a small percentage of the total. Their importance for EMI measurement is becoming more important because a spectrum analyzer provides a direct view of the interference levels over the entire frequency range, thereby saving time.

Spectrum analyzers are often used for expedient EMI reduction in the development stage for EMI filters and similar circuits, and also for EMI qualification testing. Because the instruments allow us to examine the entire frequency spectrum, occasional resonances can be recognized immediately.

Spectrum analyzers can be used quite effectively for qualitative EMI measurements. Using digital signal storage, the effect of any hardware modification can be traced visually and easily, as the stored and real-time EMI measurements can be compared directly.

The scale of the frequency axis is usually linear, but in some instruments it can be switched to logarithmic. In the former case, a direct comparison of the measured curves to EMI specifications may be difficult, as the logarithmically scaled frequency axis is commonly used in the EMC world. Spectrum analyzers usually display in reference to 0 dBm (1 mW). Therefore, a dBm-to-dBV conversion is required for comparing measured values to EMI specifications.

Spectrum analyzers have true peak or average detectors and are calibrated on rms. This characteristic results in a reading that is correct only for sinusoidal interference signals. When the harmonics of an EME signal can be measured separately (i.e., for narrowband signals), the results obtained by a CISPR receiver match those of a spectrum analyzer.

For impulse disturbances, spectrum analyzers give the true peak value, whereas CISPR quasi-peak detectors are recommended for EME measurements. For broadband measurements, spectrum analyzer measurements can be corrected [226]. The correction factor, F_t, consists of two parts: F_d, because the true peak value must be converted to quasi-peak value, and F_b, because of the difference in measurement bandwidth.

Several bandwidths are available in spectrum analyzers but, quite often, the CISPR bandwidth of 9 kHz (B_6) is not included. Therefore, for broadband EME, we adjust the measurement bandwidth to 10 kHz, which normally is available and is the closest to 9 kHz. The portion of the correction factor that is required because of differing characteristics of the bandpass filters is $F_b = 20 \log (10/9) = 1$ dB.

The other part of the correction factor takes into account the different detector parameters. From Fig. 4.6, we can obtain the correction factor F_d in dB to convert the true peak value into quasi-peak. This curve gives the relative input for constant output, but it is the same as the relative output for constant input. To obtain the correction factor value, we must consider the -7 dB asymptote, as the curve in Fig. 4.6 is referred to a 100 Hz repetition rate.

For example, consider an EME measurement with a spectrum analyzer. First, we measure the EME of a switch-mode power supply (SMPS). In the lower frequency range, when the spectrum lines of the generated HF disturbance can be seen separately, the spectrum analyzer reading must be corrected only because of the differing bandwidth (i.e., $F = F_b = 1$ dB must be subtracted from the readings).

At higher frequencies, as the HF disturbance becomes broadband (see Chapter 3, section 3.1), the reading also must be corrected by F_d. At about 10 kHz, SMPS F_d does not exceed 1 dB (see Fig. 4.6). Therefore, from the spectrum analyzer readings we must subtract $F = F_b + F_d = 2$ dB.

Measuring the EME of a single-phase rectifier with a spectrum analyzer, given that the repetition rate of the HF disturbance is 100 Hz, F_d is 7 dB. Therefore, the total correction factor is $F = 7$ dB $+ 1$ dB $= 8$ dB. For three-phase rectifiers, with the disturbance repetition rate being 300 Hz, the total correction factor is $F = 4$ dB $+ 1$ dB $= 5$ dB.

Spectrum analyzers offer automatic operation, but measurement times also can be adjusted manually. For manual operation, special care should be taken that the measurement time is not to short. The approximate minimum time for a scan can be calculated as follows:

$$(T_s = \frac{\Delta F}{B^2}) \text{ s/scan} \tag{4.32}$$

where

ΔF = the frequency range to be measured in hertz

B = the adjusted measuring bandwidth of the spectrum analyzer in hertz

Measurement results from spectrum analyzers can be regarded as authentic and can be compared to those made by CISPR-specified EMI instruments when (1) the response curve of the spectrum analyzer bandpass filter is within the hatched area in Fig. 4.4 and (2) the parameters of the detector meet CISPR recommendations.

Spectrum analyzers usually measure average or true peak value, but CISPR and most other specifications refer to quasi-peak detection. For emission measurements of broadband nature, the values indicated by spectrum analyzers must be corrected. The correction factor consists of two parts [226]—one because the true peak value must be converted to a quasi-peak value, and the other because of the difference in measurement bandwidth.

For example, examining the EMI of an electric motor with a spectrum analyzer, the measured values have to be corrected with 2 dB $+ 1$ dB $= 3$ dB, as the repetition rate of the pulse train is about 1 to 3 kHz.

4.6 EMI MEASUREMENTS FOR CONSUMER APPLIANCES

Potential sources of environmental electromagnetic pollution (and, for present purposes, EMI on power mains) can be divided into two major groups: industrial equipment and household appliances. EMI caused by very large power electronic equipment in industrial applications can be measured at the installation and, even in the worst case, custom-designed EMI "fixes" can be applied to control emissions.

In contrast, EMI problems caused by household appliances can be more difficult to overcome. Such appliances are mass produced and can be purchased over a wide geographical area. Many have been in use for years and, as a result, were manufactured before EMC emerged as a serious design consideration. Even though only a small percentage of existing consumer appliances have inadequate EMI suppression features, several hundred faulty devices could conceivably exist in close proximity.

For mass-produced equipment, acceptable EMI limits should be given careful consideration, and applicable EMI evaluation techniques differ somewhat from those devised for industrial equipment. One approach is to test each unit as it rolls off the assembly line to ensure that EMI limits are not exceeded. This guarantees 100 percent compliance, but it is usually considered to be prohibitively expensive.

The alternative is to perform tests on randomly selected samples. Test results will vary, but it is possible to estimate overall EMI levels within a reasonable probability range using statistical methodology. The typical (average) EMI level of the sampled equipments should be remain under a calculated EMI level to qualify the appliances as suitable for common trade. The number of units to be subjected to EMI measurements, the sampling method, and the mathematical relationships for evaluation of the measured EMI results are usually prescribed by national standards.

4.7   MEASURING IMPULSE-LIKE EMI

International recommendations and national standards tend to focus on acceptable levels of continuous electromagnetic emissions. But impulse-like disturbances should be considered, measured, and controlled as well. Most specifications require impulse-like emissions to be measured as if they were continuous, after which some type of correction is applied. Unfortunately, this really is inadequate. For solving EMI problems, it is very important to characterize impulse-like disturbances directly rather than relying on some kind of weighted amplitude figure.

4.7.1   Characterization of Impulse-Like Disturbances

The observation and examination of surge voltages on the mains began in the mid-1930s, and diagrams were created to show the rate of surge occurrences in reference to voltage levels [362]. Measurements were made by instruments registering the peak value of impulse-like disturbances.

An intensive examination of electronic device malfunctions and failures has revealed that, for the characterization of impulse-like disturbances, one should examine not only the maximum amplitude but also the energy content and maximum change rate. As of this writing, there are no recommendations or standards that offer limitations on these characteristics. Even so, they are of increasing importance.

In addition to maximum amplitude and impulse strength, considerations including energy content, RMS time duration, and RMS frequency duration have been introduced [30]. These impulse characteristics can be given by means of both the time function f(t) and the amplitude density function as follows:

$$a(\omega) = |F(j\omega)| = \left| \int_{-\infty}^{\infty} f(t) e^{-j\omega t} dt \right| \tag{4.33}$$

The five impulse characteristics of interest can be calculated as follows:

1. Peak value

$$PV = f(t)_{max} \tag{4.34}$$

2. Impulse strength

$$IS = \int_{-\infty}^{\infty} f(t)\, dt \tag{4.35}$$

3. Energy content

$$E = \int_{-\infty}^{\infty} f^2(t)\, dt = \frac{1}{2\pi} \times \int_{-\infty}^{\infty} a^2(\omega)\, d\omega \tag{4.36}$$

4. RMS frequency duration (band)

$$\Omega^2 = \frac{\displaystyle\int_{-\infty}^{\infty} \left[\frac{d\, f(t)}{dt}\right]^2 dt}{\displaystyle\int_{-\infty}^{\infty} f^2(t)\, dt} = \frac{\dfrac{1}{2\pi} \times \displaystyle\int_{-\infty}^{\infty} [\omega \times a(\omega)]^2 d\omega}{\dfrac{1}{2\pi} \times \displaystyle\int_{-\infty}^{\infty} a^2(\omega)\, d\omega} \tag{4.37}$$

5. RMS time duration

$$T^2 = \frac{\displaystyle\int_{-\infty}^{\infty} [t \times f(t)]^2 dt}{\displaystyle\int_{-\infty}^{\infty} f^2(t)\, dt} = \frac{\dfrac{1}{2\pi} \times \displaystyle\int_{-\infty}^{\infty} \left[\dfrac{d\, F(s)}{ds}\right]^2 d\omega}{\dfrac{1}{2\pi} \times \displaystyle\int_{-\infty}^{\infty} a^2(\omega)\, d\omega} \qquad (4.38)$$

As of this writing, instruments exist for measuring peak value and impulse strength, but none are currently marketed to measure the other three impulse characteristics.[*] The energy content, the RMS frequency duration, and the RMS time duration can be derived from the time function of the transient.

4.7.2 Measuring the Impulse Strength of Transients

The first instruments for measuring not only the peak value but also the voltage-time (impulse strength) of transients appeared on the market in the early 1980s [312]. Although impulse-like disturbances cannot be adequately characterized by impulse strength, more information can be obtained about impulse-like transients if we know their amplitude. Knowledge of impulse strength can be important in designing LC lowpass EMI filters.

As seen from Eq. (4.35), impulse strength is the area under the signal curve, so the impulse strength can be measured by a high-precision integrator circuit such as shown in Fig. 4.27 [312]. The operation of the integrator circuit is controlled by a logic unit. The mains frequency voltage is separated by a highpass filter, so only the impulse-like transient reaches the input of the integrator circuit. When a pulse is detected by the logic unit, the amplitude will be monitored at the same time. If the pulse amplitude rises above a given adjustable level, pulse integration is initiated. The impulse strength measurement is continued only if the pulse reaches another threshold level. After this second level is reached, the threshold level will be increased. This prevents secondary pulses from affecting the measurement. When the pulse amplitude decays below a given (and also adjustable) threshold level, the integration is terminated, and the logic unit fixes the integrator circuit output. After an analog-to-digital conversion, the logic unit prepares the integration circuit to measure the next incoming pulse by setting its output to zero.

Impulse strength measuring instruments are usually able to measure pulses of 100 to 5,000 V amplitudes and of 1 to 1,000 µs durations.

---

[*]First attempts, however, have been made to develop instruments for measuring the energy content of impulse-like disturbances.

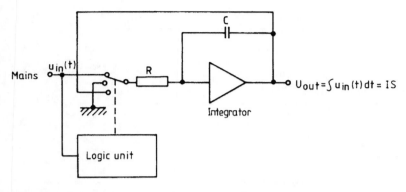

FIGURE 4.27 Functional scheme of impulse strength measuring unit

By means of impulse strength, the spreading of impulse-like signals through filter circuits can be accurately described. An example is shown in Fig. 4.28. Figure 4.28a shows a two-stage lowpass filter circuit. Resistances connected in series to capacitors represent the resistance of the cable and the losses of the high-capacitance capacitors. The input signals, $u_{in}(t)$, were 100 V peak/1,000 µs duration, 200 V peak/500 µs duration, and 1,000 V peak/1 µs duration. This means that all the examined input signals had the same impulse strength of 100 mV-s. Assuming that the circuit elements are linear and the source impedance is negligible, the output signal, $u_{out}(t)$, will be as shown in Fig. 4.28b. Referring to Fig. 4.28b, note that, despite the ten-to-one range over which input signal amplitude changes, the output signal is almost constant in peak magnitude. Furthermore, it is to be noted, the impulse strength of the output signal is almost constant.

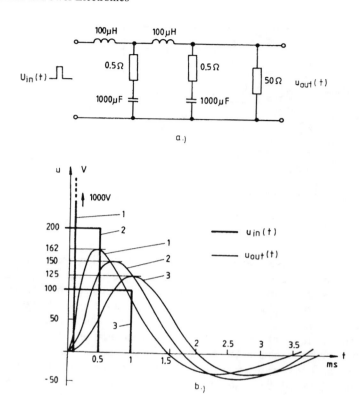

FIGURE 4.28   Impulse propagation through a filter [312]; (a) filter circuit and (b) time function of output pulses

5

EMI in
Power Electronic Equipment

Most electrical equipment that draws power from the mains generates HF signals that could negatively affect the operation of other connected equipment. To minimize the effects of high-level disturbances, there has been considerable interest in examining and understanding how industrial, commercial, and consumer products generate noise. As a result, much information is available about the character of emissions from common EMI sources such as brush-type electric motors, electrical switching equipment, fluorescent tubes, and various household appliances [1].

Beginning in the 1970s, there has been a rapid proliferation of semiconductor devices in industrial and commercial equipment. To find the most effective and economical solutions for EMI control in this rugged environment, we must understand how semiconductor devices and circuits generate and are affected by electromagnetic noise.

5.1 EMI FROM POWER SEMICONDUCTORS

Power semiconductors produce HF disturbances as an operational by-product. Although power semiconductors can be regarded as switches, their generated EMI is different from that of electromechanical switches. The characteristic difference is that the repetition rate of the HF noise produced by electromechanical switching apparatus is low, while power semiconductors generate HF disturbances with a high repetition rate. These repetition rates range from the mains frequency to tens of kilohertz. The bulk of EMI spectra of power semiconductor devices generally does not exceed a few megahertz.

5.1.1 EMI from Rectifiers

A rectifier (diode) can be regarded as a switching element. A rectifier acts as a short circuit for forward bias and as an open circuit for reverse bias. In practice, rectifier switching from one state to the other does not occur instantaneously.

The switch-on operation of a power rectifier is shown in Fig. 5.1. The voltage on the rectifier changes from the off-state to the on-state in the time t_0. The on-state current increases quickly but, in the first moment, a relatively high on-state voltage appears on the rectifier. This voltage falls back to its nominal value in the time interval t_f. This time is needed for charge carriers to get into the depletion (p-n) region. The voltage spike is actually a broadband emission.

The rectifier produces a much higher emission during its switch-off operation. The switch-off voltage and current curves are shown in Fig. 5.2. The on-state current decreases to zero during the time t_0. In contrast to an ideal rectifier, the current turns on to negative, due to the stored charge in the form of minority current carriers in the depletion region. The reverse current flows until charge carriers are present (t_s). The amount of the stored charge carrier is characterized by the diffusion capacity. Diffusion capacity values are usually given in catalogs and data sheets. This reverse current snaps very quickly ($t_r \ll t_s$) to zero (or, more precisely, to the nominal value of the reverse current). The amplitude, duration, and shape of the reverse pulse current is a function of the rectifier characteristics and circuit

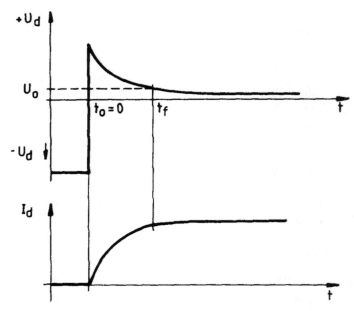

FIGURE 5.1   Rectifier switch-on

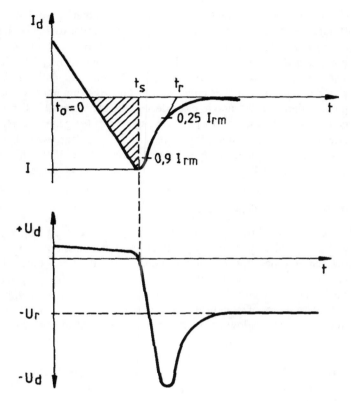

FIGURE 5.2 Rectifier switch-off

parameters. Since the amplitude of the reverse current, I_{rm}, can be quite large, and the snap-off time is very short (usually lower than 1 μs), high-voltage transients with wide frequency spectra can appear in the inductances of conductors and connecting circuits.

The interference effect of rectifiers can be reduced partly by limiting the magnitude of the surge current and partly by decreasing the slew rate of the negative surge current. The magnitude of the surge current depends on the falling rate of the load current. Therefore, it is advisable to limit this value. The emission of rectifiers at switch-off can be reduced effectively by RC snubbers connected in parallel with the rectifiers (see Section 7.3).

5.1.2 EMI from SCRs

Silicon controlled rectifiers (SCRs), like diode rectifiers, generate HF disturbances during both switch-on and switch-off operations. But in contrast to rectifiers, the HF noise levels are much higher at switch-on than at switch-off. SCR switch-off noise generation and hence EMI levels are similar to that of diodes.

The voltage and current curves of SCRs during switch-on are shown in Fig. 5.3. The firing signal arrives on the control electrode at the moment t_0. The depletion region looses its insulation capability only after a given time, called *delay time*. Afterward, the voltage on the SCR collapse very quickly—but not instantaneously. The quick collapse of the anode-cathode voltage generates HF disturbances. The switching time is about 0.5 to 2 μs, so the spectra of the generated HF noises can be quite wide. The switching time increases with the decrease in the anode-cathode voltage and also depends on the rated current of the SCR. This function is characterized in Fig. 5.4.

5.1.3 EMI from Power Transistors

Interference generated by power transistors is similar to that of SCRs. Figure 5.5a shows the typical voltage and current curves at switch-on of power transistors. The switching time of power transistors, t_{on}, particularly for power field-effect transistors (FETs), is significantly shorter than that of SCRs.

Figure 5.5b shows the curve of the collector-emitter voltage and collector current of a power transistor at switch-off. The switch-off begins at time t_0. The collector current continues to flow until a given time, t_s, called *storage time*. During

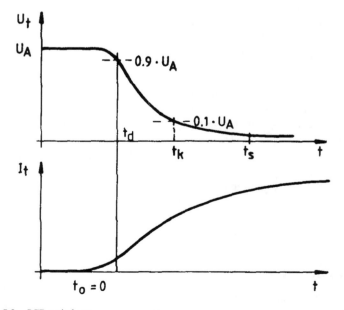

FIGURE 5.3   SCR switch-on

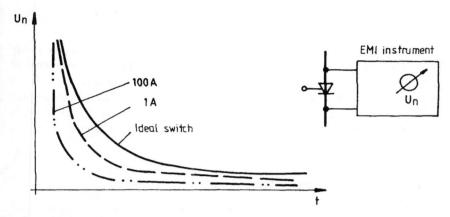

FIGURE 5.4 EMI from SCRs with different current ratings

this interval, the charge carriers will be removed from the depletion region. After the storage time, the collector current falls to zero. The fall time of the collector current is rather short, usually in the range of approximately 10 ns to 100 μs, depending primarily on the rated power of the transistor. This explains why the spectra of the EMI produced by power transistors can be much wider than that of rectifiers or SCRs.

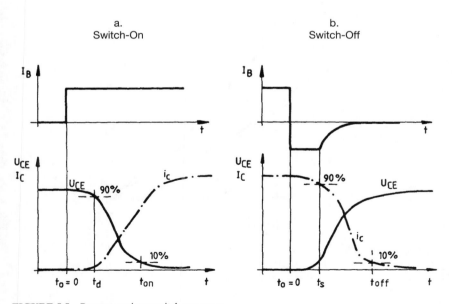

FIGURE 5.5 Power transistor switch processes

5.2   EMI FROM CONTROLLED RECTIFIER CIRCUITS

In general, the reduction of EMI from controlled rectifier circuits requires only EMI filters. To gather preliminary data for solving EMI problems in controlled rectifiers, interference measurements were made on a controlled rectifier circuit model arrangement. The generated EMI was measured as a function of the load rating and character of the load and firing angle. These measurements showed that the EMI of the power control circuit, even with a small power rating, can be unacceptably high.

5.2.1   EMI as the Function of Output Power

The EMI was measured on a three-phase, full-wave controlled rectifier on a three-phase mains (quasi-peak measurement, 9 kHz CISPR bandwidth). After setting the firing angle (α) to 0 and 70 electrical degrees, the resistive load was changed. The measured EMI is shown in Fig. 5.6. Curve 1 was measured without any delay angle and with nominal load; curve 2 was also measured without delay angle but with one-fifth of the nominal load. Curve 3 was measured with a 70° fir-

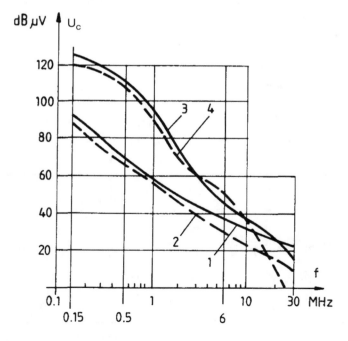

FIGURE 5.6   EMI of SCR control at different output powers and firing angles measured with CISPR receiver

ing angle and nominal load. Finally, curve 4 was also measured with a 70° firing angle but at one-third load. For 3 and 4, the EMI seems to be nearly independent of the load current. The explanation is that the switch-on process of the SCR, being a voltage ramp, is principally independent of the load current. The cause of the difference between curves 1 and 2, measured without delay angle (i.e., with the rectifier circuit) is that the interference is generated only by SCR switch-off, which depends on the load current.

5.2.2 EMI as a Function of the Character of the Load

During the EMI measurements, a single-phase ac chopper and a three-phase half-controlled rectifier were loaded by two resistors and a choke coil in series. The firing angle was constant (90°) for all measurements. Curves 1 and 2 in Fig. 5.7 show the EMI of the chopper circuit. Curve 1 shows the case of a resistive load, and curve 2 the case of a resistive-inductive load. The curves illustrate that the EMI of the chopper scarcely depends on the character of the load. Curve 3 shows the EMI of the rectifier with resistive load, and curve 4 shows the same with a resistive-inductive load. The emissions of the SCR control did not depend on the

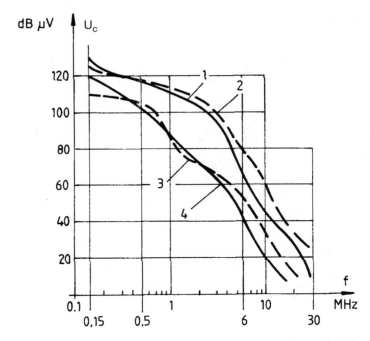

FIGURE 5.7   EMI of SCR control at different load type (CM, measured with CISPR receiver)

character of the load, although, in case of resistive-inductive load, the emissions were slightly higher.

5.2.3 EMI as a Function of the Firing Angle

One would expect the EMI of an SCR dc power controller to depend on the firing angle. This is because, the higher the firing angle, the higher will be the cathode-anode voltage that must collapse in a very short time. EMI measurements confirm that expectation, as shown in Fig. 5.8. Curves 1 and 2 were measured for a three-phase, full-wave controlled rectifier. Curve 1 was the result of a 30° firing angle, and curve 2 from a 90° angle. It can be seen clearly from the curves that, by increasing the firing angle to 90°, the EMI generated by controlled rectifier increases significantly. With a further increase in firing angle, the EMI levels begin to drop off. The EMI of the SCR dc power control is the same at 30° and 120° firing angles. As a function of firing angle, maximum EMI emission occurs at exactly the 90° point.

Curves 3 and 4 are measurements for an ac chopper. Curve 3 corresponds to a 30° firing angle, and curve 4 to 90°. As seen from the curves, the EMI of the ac

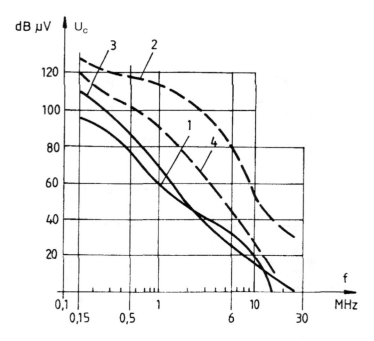

FIGURE 5.8   EMI of SCR power control as a function of firing angle

chopper also increases with the firing angle and begins to decrease when the firing angle exceeds 90°.

5.3   EMI CALCULATION FOR SEMICONDUCTOR EQUIPMENTS

The EMI generated by power electronic equipment is that of a pulse train, since rectification involves repetitive switching from conduction to cutoff. The repetition rate is some multiple of the operating (mains) frequency, depending on circuit configuration. The electromagnetic emissions of semiconductor devices are broadband and coherent. Thus, for studying EMI from semiconductor circuits, the amplitude density function is a useful tool [see Eq. (4.33)].

The amplitude density function of these emissions can be easily calculated with suitable approximations. One must remember that semiconductor devices produce differential-mode noise, but EMI standards limit the acceptable level of common-mode noise as well. As there is a close relationship between differential- and common-mode noise components (see section 7.5.2), establishing the characteristics of the differential-mode noise produced by power semiconductors allows for further calculations.

5.3.1 Spectra of Impulse-Like Signals

For estimating the EMI of power electronic equipments, the amplitude density function is sufficient to find the worst-case maximum amplitudes of the noise components. The EMI evaluation generally can be performed with knowledge of the broken-line envelope of the amplitude density function—in short, the spectrum. An approximate evaluation of the spectrum is made easier by further simplifying our assumptions. The EMI of power electronic equipment is qualified, in reference to the measuring bandwidth, as broadband. Therefore, instead of performing a discrete harmonic analysis of the pulse train, it is sufficient to define the spectrum of a single impulse in the pulse train. This is a significant simplification because, in most cases, the calculations become much easier.

The above simplifying approximation flows from the following train of thought. The graphic derivation is presented in Fig. 5.9, which shows a typical pulse train and its related harmonics. The spacing of the spectral lines is determined by the repetition rate of the pulse train (see Fig. 5.9a). Therefore, in Fig. 5.9b, as the repetition rate decreases, the spectral lines come closer together within the $\sin x/x$ envelope. The shape of the envelope is determined by the width of the impulse. A single impulse would have all the spectral lines infinitesimally close together. As seen from Fig. 5.9c, $\sin x/x$ function and its $1/x$ envelope are coming closer together as the width of the impulse increases. Ultimately, a step function would have the envelope of $1/x$ completely filled with spectral lines.

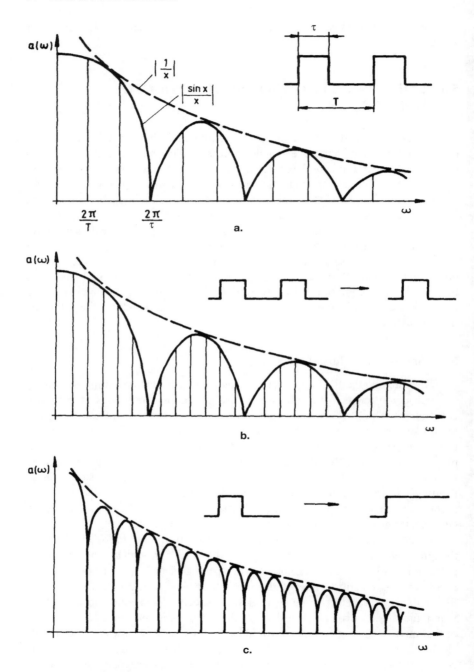

FIGURE 5.9   Spectrum of impulses [87]

In several publications, the spectra of various impulse-like signal can be found in chart form [77, 87, 102]. Figure 5.10 shows the spectra of trapezoidal pulses, Fig. 5.11 of a full-wave rectified sine wave, and Fig. 5.12 of a fractional sine wave. Although Fig. 5.11 refers to a full-wave rectified sinewave signal, it is applicable to case of half-wave rectification—in that case, the half-value should be taken (i.e., 6 dB should be subtracted from values read on the vertical axis).

As the measuring bandwidth of EMI instruments that comply with CISPR recommendations is 9 kHz, the value indicated by them can be calculated using Eq. (4.21) with a knowledge of the spectra of HF disturbances.

Figures 5.10 through 5.12 refer to voltage pulse trains. While the charts are referenced to voltage amplitudes, they apply just as well to current amplitudes. The amplitude of the EMI current would then be expressed in terms of dBμA/MHz.

It is here to be noted that, knowing the spectrum of EMI voltage or current, the radiated electromagnetic field can be also calculated with good approximation

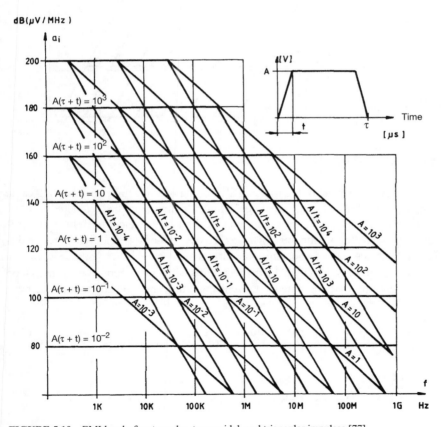

FIGURE 5.10 EMI level of rectangular, trapezoidal, and triangular impulses [77]

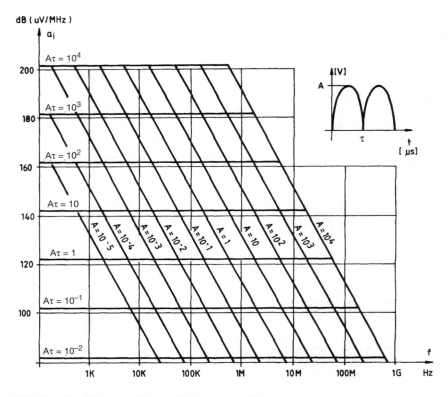

FIGURE 5.11   EMI level of fully rectified sine wave [77]

[77]. A rough estimate of the radiated electric field at one meter from a long cable carrying the CM EMI voltage can be obtained by reducing the conducted level by a factor of 33 dB. In this case, the unit of the radiated electric field strength is dBμV/m/MHz, for broadband emissions. To obtain the electrical field levels at different distances, subtract an additional $20\log(d)$, where d is the distance in meters.

More accurately, to obtain the magnetic field strength generated by a small loop of EMI current, first compute the current spectrum, and then add to it the factor

$$20\log\left(\frac{N \times A}{4\pi d^3}\right) \tag{5.1}$$

where

$$N = \text{the number of loop turns}$$

$$A = \text{the loop area in } m^2$$

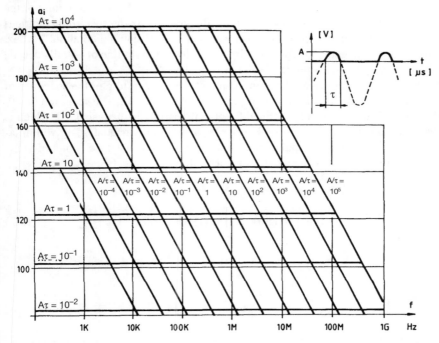

FIGURE 5.12   EMI level of fractional sine wave [77]

d = the distance from the loop in m

This formula will give exact values of H-fields in dBμA/m/MHz (for a current spectrum in dBμA/MHz) as long as we remain in near-field conditions; that is, if d < $\lambda/2\pi$, or $d_m < 48/F_{MHz}$.

The measuring unit of the magnetic field strength, calculated by the above procedure, is dBμA-turn/m/MHz.

The technique for using the chart can be demonstrated best by an example. Plot the spectrum of an trapezoidal impulse series with the following characteristics:

$$\text{amplitude, } A = 50 \text{ V}$$
$$\text{time duration of a single pulse, } \tau = 5 \text{ μs}$$
$$\text{rise time and fall time, } t = 0.05 \text{ μs}$$

Study the emission in the range of 10 kHz to 30 MHz. The spectrum can be derived from Fig. 5.10. The value for selecting the horizontal section is $A(\tau + t) =$

250 μ(V-sec), corresponding to a 168 dBμV/MHz line (8 dB above the 160 dB mark), and goes to the line of the A = 50 parameter. The intersection point of the second section is at about 70 kHz. Above this frequency, the spectrum continues with –20 dB/decade slope down to cross an A/t = 10^3 line, i.e. to about 7 MHz. Above this frequency, the spectrum is formed by a straight line of –40 dB/decade slope. This spectrum is plotted in Fig. 5.13.

For highly technical applications, it may be necessary to plot the spectrum of signals not included in the charts of Figs. 5.10 through 5.12. The spectrum of any high-speed function can be defined by graphic harmonic analysis [29], which is based on inequality related to a Fourier transform. The inequalities give limitations for the spectra of impulse-like signals from the "worst-case" point of view. For general cases, these inequalities are as follows:

$$|F(j\omega)| \leq \int_{-\infty}^{\infty} |f(t)|\,dt \qquad (5.2)$$

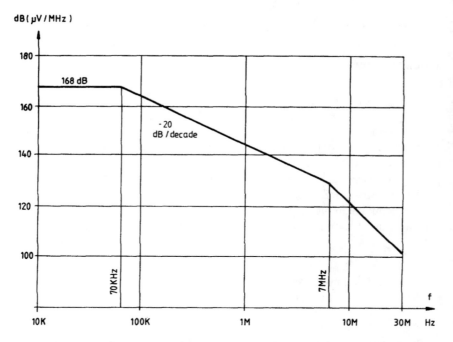

FIGURE 5.13   Spectrum of a trapezoidal impulse with 50 V Amplitude, 5 μs duration, and 0.05 μs rise time

$$|F(j\omega)| \leq \frac{\displaystyle\int_{-\infty}^{\infty} |f^n(t)|\, dt}{\omega^n} \qquad (5.3)$$

Applying the above bounds for determining the spectrum of an emission, the following rules should be taken into consideration:

1. For drawing up the derivatives, the generalized one should be considered— i.e., the impulse-function (Dirac-impulse) should be taken into consideration also.
2. The derivation should be continued only while the first impulse-function appears in the derivatives.
3. The area of the impulse-function should be calculated only when the derivatives consist of impulse-functions only.
4. For calculating the integrals, the multiplication factor 2 introduced in the bibliography, because of omitting the negative angular frequency range, should not apply; it can be proven by function analysis.

Let us demonstrate the use of the above inequalities for determining the spectra of impulse-like disturbances. First, determine the spectrum of a triangular signal as shown in Fig. 5.14. The knowledge of this spectrum is important in technical practice because the 1.2/50 μs test pulse used for EMS tests (see section 10.2) is generally substituted by that time function.

The low-frequency part of the spectra can be calculated using Eq. (5.2). Omitting the absolute value marks and knowing that during graphic harmonic analysis the equality mark covers a $\leq$ mark, it can be written:

$$F_1(f) = \frac{U_1(t_r + t_f)}{2} \qquad (5.4)$$

The other parts of the spectrum can be determined using Eq. (5.3). With n = 1, the mid section of the spectrum can be calculated. Given that the derivatives are rectangles:

$$F_2(f) = \frac{\dfrac{U_i}{t_r} t_r + \dfrac{U_i}{t_f} t_f}{2} = \frac{U_i}{\pi} \times \frac{1}{f} \qquad (5.5)$$

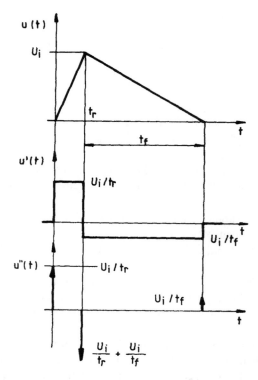

FIGURE 5.14 Time function and its derivatives of triangular time function for graphic harmonic analysis

The crossover point of the sections $F_1(f)$ and $F_2(f)$ (i.e., the first cutoff frequency) can be calculated by solving the Eqs. (5.4) and (5.5) to find f:

$$f_1 = \frac{2}{\pi (t_r + t_f)} \tag{5.6}$$

The high-frequency part of the spectrum can be calculated by Eq. (5.3) also, with n = 2. Since the integral of the impulse-function is unity, the integral of the impulse-function of amplitude A is also A. After the necessary calculations:

$$F_3(f) = \frac{U_i (\frac{1}{t_r} + \frac{1}{t_f})}{2\pi^2} \times \frac{1}{f^2} \tag{5.7}$$

The second cutoff frequency can be calculated by solving Eqs. (5.5) and (5.7):

$$f_2 = \frac{\dfrac{1}{t_r} + \dfrac{1}{t_f}}{2\pi} \tag{5.8}$$

To make the further calculations more convenient, the above functions are worth converting to logarithmic form. Measuring the amplitude in volts, the time in microseconds, and applying the approximation of $20 \log \pi = 10$, the functions describing the spectral sections are:

$$A_1 = 114 + 20 \ \log \ [\, U_i \, (t_r + t_f)\,] \quad \text{dB}\mu\text{V/MHz for } f \leq f_1$$

$$A_2 = 110 + 20 \ \log \ (U_i) - 20 \ \log \ (f) \quad \text{dB}\mu\text{V/MHz for } f_1 \leq f \leq f_2 \tag{5.9}$$

$$A_3 = 94 + 20 \ \log \left[U_i \, (\frac{1}{t_r} + \frac{1}{t_f}) \right] - 40 \ \log \ (f) \quad \text{dB}\mu\text{V/MHz for } f_2 \leq f$$

The spectrum described by the above functions is shown in Fig. 5.15.

It is worth examining the case of $t_r = t_f$ (i.e., the isosceles triangle). As seen from Eqs. (5.6) and (5.8), which provide the cutoff frequencies, the values of f_1 and f_2 are equal—i.e. the section with a -20 dB/decade slope shrinks to zero. This

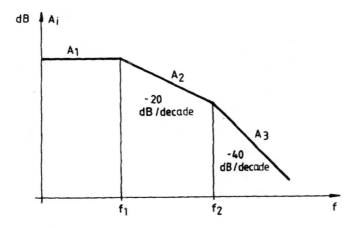

FIGURE 5.15   Spectrum of the time function shown in Fig. 5.14

also means that the horizontal section is followed by a section of –40 dB/decade slope.

Let us now determine the spectrum of a more complex signal form. The time function and its derivatives are shown in Fig. 5.16. The selection of this signal form is based on the reasoning that a similar signal is often encountered from SCRs during product development. Commutation processes usually produce sinusoidal (or truncated sinusoidal) forms, but for describing the spectrum of the emission, a straight-line approximation is permissible. The error produced by this approximation is very small, as can be seen by comparing the spectra of sinusoidal and isosceles triangle signal forms.

To establish relatively simple and clear relationships, it is reasonable to use a smaller approximation by calculating the integrals. The equations derived from graphic spectrum analysis are as follows:

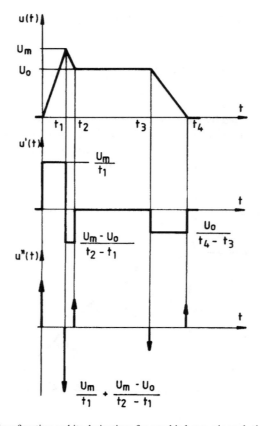

FIGURE 5.16 Time function and its derivatives for graphic harmonic analysis

$$F_1(f) = U_m t_1 + U_o (t_4 - t_1)$$

$$F_2(f) = \frac{U_m}{\pi} \times \frac{1}{f} \qquad\qquad (5.10)$$

$$F_3(f) = \frac{\dfrac{U_m}{t_1} + \dfrac{U_o}{t_4 - t_1}}{2\pi^2} \times \frac{1}{f^2}$$

The spectrum is formed by three sections again: a horizontal one, a straight line of –20 dB/decade slope, and another line of –40 dB/decade slope. It is evident that the spectrum is similar to that shown in Fig. 5.15. The cutoff frequencies can be calculated easily by solving the proper equations.

Finally, an example will cast light on how to apply some basic rules for graphic spectrum analysis. Determine the spectrum of the signal shown in Fig. 5.17. As seen from the figure, signal variations during the rise time and fall time are non-linear. Let the time function be described by a square. This means that third derivatives appear.

Considering the four rules of graphic spectrum analysis given previously, although the third derivatives exist, the spectrum again consists of three sections: horizontal, –20 dB/decade slope and –40 dB/decade slope. This is because an impulse-function appears in the second derivative; thus the third derivative need not be considered.

Example

Let us draw up the exact and calculated envelope of the waveform shown in Fig. 5.17. The parameters of the examined waveform are: $U_o = 100$ V, $T = t_2 - t_1 = 1$ ms, $t_r = t_1 = 12.5$ μs, and $t_f = t_3 - t_2 = 50$ μs. The exact spectrum is designated by $a(\omega)$ in Fig. 5.18. The envelope, calculated by the modified harmonic analysis, is drawn with a continuous line. By the former harmonic analysis, the envelope line would be as drawn with the dashed line. This envelope has a – 60 dB/decade section, as the time function has a third derivation. As can be seen – from the figure, the modified harmonic analysis gives a better approximation.

An interesting comparison is shown in Fig. 5.19, which shows the spectra of three different types of disturbances—but with identical amplitude, width, period and rise time—in one coordinate system [87]. It can be seen from the figure that, at lower frequencies, the spectrum of an exponential impulse series with high-pass characteristics is the lowest. Conversely, at high frequencies, this spectrum is the highest. This figure can offers some direction, if such is possible, for selecting the most applicable signal form for HF disturbances.

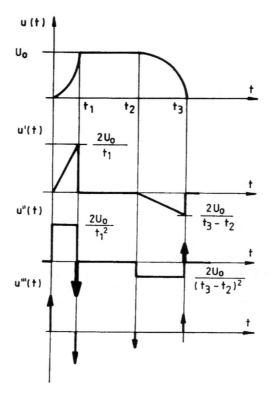

FIGURE 5.17   Time function and its derivatives of a time function with soft rise for graphic harmonic analysis

5.3.2 Predicting EMI from a Power Supply with Rectifiers

In general, separate power supplies produce the supply voltage for control units in power electronic equipment. Switching power supplies are universally recognized as inherently noisy, but EMI can also reach unacceptably high levels from traditional ac-to-dc power supplies, consisting of rectifying circuit, a smoothing capacitor, and a subsequent analog voltage regulator. In both types, a significant EMI-generating area is where the rectifying circuit connects to the smoothing capacitor.

If the rectifier circuit is connected directly to a smoothing capacitor, the energy delivered to the load is replenished periodically. This replenishment, occurring during a relatively short period, is in the form of current pulses of a magnitude that generally is many times higher than the dc load current. EMI levels caused by the rectifying circuit can be estimated from the amplitude, pulse width and rise time

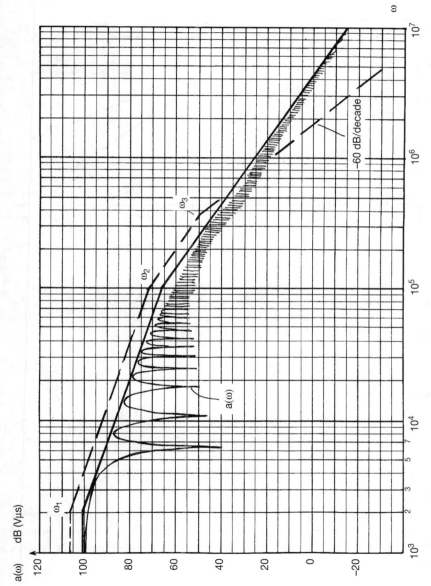

FIGURE 5.18 Exact and calculated envelopes of the waveform shown in Fig. 5.17

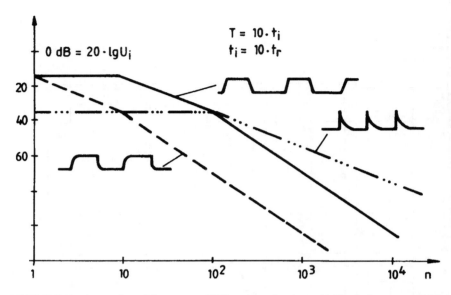

FIGURE 5.19   Comparison of the spectra of different time functions with identical parameters [87]

of the impulse-like current. These characteristics depend on the rectifier circuit, the value of the smoothing capacitance, and the load. The calculation task can be simplified using a chart cited by Edelman [94] and Rhoades [258]. To use this chart, the quantity M first should be calculated as follows:

$$M = \arctan\left[\omega \times RC \times 10^{-6}\right] \tag{5.11}$$

where

$\omega = 2\pi f$ = the angular frequency of the mains

C = capacitance of the filter capacitor in μF

R = load resistance of the rectifier, calculated from the
nominal voltage and current, in Ω

The conduction angle, θ_i, is shown in Fig. 5.20, and the current multiplying factor, P, is shown in Fig. 5.21 as a function of the quantity M. Knowing the multiplying factor, P, the peak value of the surge current flowing through the rectifying circuit is:

$$I_s = P\frac{U_s\sqrt{2}}{R} \tag{5.12}$$

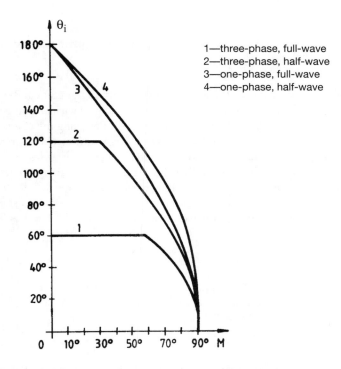

FIGURE 5.20   Rectifier conduction angle versus M (see Eq. [5.11] in text) [94, 258]

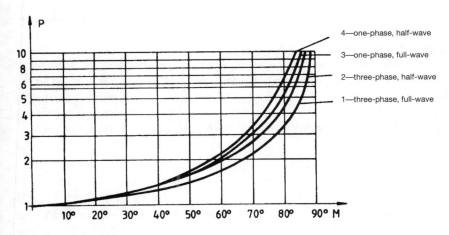

FIGURE 5.21   Multiplying factor versus M (see Eq. [5.11] in text) for rectifier circuits [94, 258]

where

U_s = the effective value of the ac voltage supplying the rectifier, in V_{rms}

For estimating EMI levels, the current of the rectifying circuit can be considered trapezoidal, with amplitude, I_s, and width, θ_i, in electrical degrees. This
means that the spectrum can be taken from Fig. 5.10. For using the chart, the rise
time of the surge current also should be known. The rise time can be approximated
with the assumption that, during rectifier commutation, the current is limited only
by the total circuit resistance (including the losses of the mains transformer, if
used), the internal resistance of the smoothing capacitor, and the differential resistance of the rectifiers. These losses can be represented by a total resistance, R_t.
The rise time (t_r) and width (T) of the surge current can be calculated as follows:

$$t_r = 2.2 \times R_1 \times C \quad \mu s$$

$$T = \frac{\Theta_i}{360°} \times \frac{10^6}{f} \quad \mu s$$

(5.13)

where

C = smoothing capacitor value in μF

R_t = total resistance in Ω

Θ_i = conducting angle of the rectifiers in electrical degrees

For estimating the EMI of power supplies, the chart of Fig. 5.10 can be used
where the unit of vertical axis is dBμA. To the values taken from the chart, we
must account for the value of A in amperes.

In most cases, the rectifier unit is connected to the mains via a step-down transformer. To estimate the emissions toward the mains, we must also consider the
transformer ratio. With the simplifying assumption that only magnetic coupling
exists (i.e., that there is no leakage inductance and no winding stray capacitance),
the calculated current emission of the rectifying circuit should be decreased by

$$20 \log \left(\frac{U_p}{U_s} \right)$$

This assumption requires an excellent electrostatic shield. It is also assumed that
the transformer ratio is unchanged throughout the frequency, which can be accepted for the first few tens harmonics.

It should be noted that calculated and measured EMI levels can be quite different. To account for the variation, one should look not to the mathematical relationships and assumptions but to the EMI measurement methodology. When measuring EMI voltage with a LISN, the impedance of the LISN can greatly influence the commutation of the rectifiers and thereby the shape of mains surge current [258]. The high serial impedance of the LISN has the most critical effect, as it can decrease the peak value of the current pulses drawn from the mains. Empirical data have shown that variations in LISNs and inconsistencies in current measurement methods can produce measurement anomalies of as much as 40 dB.

5.3.3 EMI Prediction for Switching Power Supplies

Since the early seventies, major changes have occurred in the power supply market—most notably a rapid shift from linear to switching power supply systems. Switching power supplies produce EMI that affects not only the mains but the dc load as well. If the designer has employed effective noise suppression methods (see Chapter 7, Sections 7.6 and 7.7), the EMI emitted into the mains can be accurately calculated as described in Section 5.3.2.

During power transistor switching, voltage surges are created that can appear on the output terminals and induce malfunctions in control units supplied by the switcher. For predicting the EMI problems generated by switching power supplies, we should study these disturbances that appear at the output terminals.

An investigation of HF disturbances at the output terminals of switching power supplies was conducted for a switching regulator with a two-winding configuration, as shown in Fig. 5.22 [129, 225, 228, 229]. This configuration has some excellent performance qualities and therefore is commonly used for ac-to-dc converters. During the analysis of this circuit, four assumptions were introduced

1. Both the transistor and diode are represented as short circuits at on-state and as small capacitances at off-state.

2. The magnetic inductance of the transformer is large in comparison with leakage inductances.

3. For studying surges, the impedance of the smoothing capacitor consists only of the stray inductance and winding resistances, and the stray capacitance between the two windings is neglected.

4. A voltage step-down regulator with high ratio of primary to secondary turns was studied.

Calculations showed that during the switching of the power transistor, a sinusoidal noise voltage arises at the output terminals. A higher noise voltage is expected at switch-on of the transistor. The frequency of this noise is well approximated as follows:

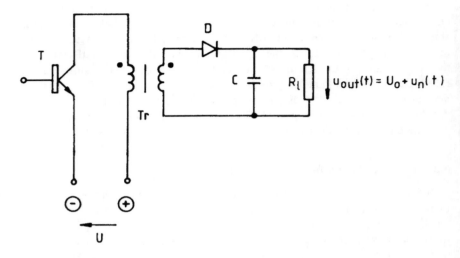

FIGURE 5.22   Dc-to-dc converter configuration

$$\omega_p = \frac{1}{\sqrt{(L_1 + L_2 + N^2 L_c) \times (C_c + C_s)}} \tag{5.14}$$

where

L_1 = stray inductance of the primary coil

L_2 = stray inductance of the secondary coil

L_c = parasitic inductance of the filter capacitor

C_c = collector-emitter capacitance of the transistor

C_s = stray capacitance between the transformer coils

N = turn ratio of the transformer

Additionally, the peak value of the noise signal is:

$$U_{n\,max} = N \times \omega_p L_c \times I_c(0) \tag{5.15}$$

where

$I_c(0)$ = the peak value of the transistor collector current

The above relationships show that the leakage inductances of the transformer and the parasitic inductance of the smoothing capacitor are the principal causes of increased output noise (for more details, see Chapter 7, section 7.7).

Note that this section has examined only the EMI generator and neglected the issue of differential- and common-mode noise reflected onto the mains by switch-mode power supplies. This subject is addressed in Chapter 7.

5.3.4 EMI Prediction for SCR AC Controllers

Standard ac choppers with SCRs and triacs are widely employed to control power flowing into a load from an ac supply. The most typical applications of ac choppers are lamp dimmers, heat regulators, and motor speed controls. The rms value of the supply voltage of the load can be controlled by the firing angle of SCRs or triacs. The ac choppers generate high EMI levels toward the power mains.

The EMI of ac choppers can be estimated by Fourier analysis. The typical voltage shape of the load can be seen in Fig. 5.23. The firing signal is imposed on the SCR or triac at a time, t_1. The switch-on time is labeled as $t_0 = t_2 - t_1$ in the figure. The Fourier series of the load voltage is:

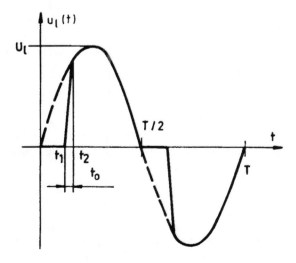

FIGURE 5.23   Time function of the voltage on the ac power control output terminals

$$u_1(t) = U_1 \times \sum_1^\infty [A_n \times \sin(n\omega t) + B_n \times \cos(n\omega t)] \qquad (5.16)$$

where

$$U_1 = \text{peak value of the load voltage}$$

The Fourier factors A_n and B_n can be calculated by means of Fourier integrals. For the time function shown in Fig. 5.23, and with the assumption that the voltage change is linear during the switch-on process, the Fourier-factors are as follows [37]:

$$A_n = \frac{2 \times U_1 \times \sin x}{n\pi t_o} \{t_o \times \sin(nx) + \frac{T}{2\pi n} \times [\cos(nx) + \cos(ny)]\}$$

$$B_n = \frac{2 \times U_1 \times \sin x}{n\pi t_o} \{t_o \times \cos(nx) + \frac{T}{2\pi n} \times [\sin(nx) - \sin(ny)]\}$$
$$\qquad (5.17)$$

where

$$n = 2 \times k - 1$$

$$k = 1, 2, 3, \ldots$$

$$x = \frac{2\pi \times t_1}{T}$$

$$y = \frac{2\pi \times t_2}{T}$$

$$T = \text{time period}$$

$$U_1 = \text{as in Eq. (5.16)}$$

For calculating the EMI, it is reasonable to use the complex form of the Fourier series:

$$C_n = \sqrt{A_n^2 + B_n^2} \qquad (5.18)$$

Making EMI measurements corresponding to CISPR recommendations, the estimated EMI can be calculated simply by adding the value C_n to the measuring bandwidth.

6

EMI Filter Elements

Because EMI produced by semiconductor equipments usually exceeds acceptable levels, the emissions must be reduced. In the majority of practical cases, noise suppression is accomplished using lowpass filters.

EMI filter elements are subject to two important requirements: they must safely tolerate the nominal operating voltage and current of the electrical equipment, and their HF characteristics must not vary with frequency. Understanding the HF characteristics of filter elements is of basic importance in EMI filter design and for obtaining pertinent data from catalogs and data sheets.

EMI filter elements cannot be considered ideal within their frequency reduction bands. In the frequency range of 0.15 to 30 MHz, the equivalent circuit of EMI filter elements becomes a two- or four-terminal network containing several components. Determining the HF characteristics of EMI filter elements requires analysis of the entire equivalent circuit.

6.1 MEASURING HF CHARACTERISTICS OF EMI FILTER ELEMENTS

The HF characteristics of non-ideal circuit elements can be described by means of the equivalent circuit. In widely accepted practice, EMI filter elements are characterized by insertion loss (I.L.) as well as their nominal ratings.

6.1.1 Definition of HF Characteristics

A four-terminal network can be described by impedance parameters (among others). Impedance parameters also can be properly used for describing the HF characteristics of EMI filter elements. In practice, the open-circuit and short-circuit transfer impedances are of primary importance. The open-circuit transfer impedance, Z_o, with voltage and current measuring directions as shown in Fig. 6.1, is calculated as follows:

$$Z_o = \frac{U_{2o}}{I_1} \tag{6.1}$$

The short-circuit transfer impedance can be determined by measuring the input voltage and the output short-circuit current:

$$Z_s = \frac{U_1}{I_{2s}} \tag{6.2}$$

The open-circuit transfer impedance is usually chosen for describing components or circuits of low impedance (capacitors, filter circuits), whereas the concept of short-circuit impedance is for characterizing components of high impedance (choke coils, resistors).

For practical measurements, however, a real open-circuit or short-circuit cannot always be realized. However, sufficiently accurate results can be obtained when $I_1 \gg I_2$ (for measuring the open-circuit transfer impedance) or $U_1 \gg U_2$ (for measuring the short-circuit transfer impedance).

The insertion loss of EMI filter elements or filter circuits is expressed by the ratio of two powers:

$$IL = 10\log\left(\frac{P_1}{P_2}\right) \tag{6.3}$$

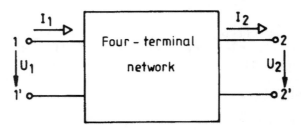

FIGURE 6.1   Voltage and current measurement directions

In the above formula, P_1 is the power delivered to resistor R_2 by the generator having a voltage of U_o and an internal resistance R_2, as shown in Fig. 6.2a. For measuring the insertion loss, the EMI filter element or circuit to be measured should be connected between the voltage generator and the measuring instrument, as shown in Fig. 6.2b. Then the power P_2 will be present on the terminator resistance R_2. The powers can be described by means of the generator voltage and the voltage shown by the measuring instrument. The power P_1 is:

$$P_1 = U_o^2 \times \frac{R_2}{(R_1 + R_2)^2} \tag{6.4}$$

Expressing the power P_2 across the termination resistor by means of the measured voltage U_m:

$$P_2 = \frac{U_m^2}{R_2} \tag{6.5}$$

and substituting it and the power P_1 into Eq. (6.3), the insertion loss can be expressed by the generator and the measured voltage:

$$IL = 20\log \left[\frac{U_o}{U_m} \times \frac{R_2}{R_1 + R_2} \right] \tag{6.6}$$

In practice, when measuring insertion loss, the condition of $R_1 = R_2 = R$ is usually met. In this case, Eq. (6.6) is simplified as follows:

$$IL = 20\log \left(\frac{U_o}{2 \times U_m} \right) \tag{6.7}$$

Insertion loss can be expressed by the impedance parameters, too. For measuring a serial component with an impedance $Z_s = R_s + jX_s$, the insertion loss (again assuming the condition of $R_1 = R_2 = R$) is as follows [10]:

$$IL_s = 20\log \left[\left\{ 1 + \frac{R_s}{R} \right\} \times \sqrt{1 + \left(\frac{X_s}{2R + R_s} \right)^2} \right] \tag{6.8}$$

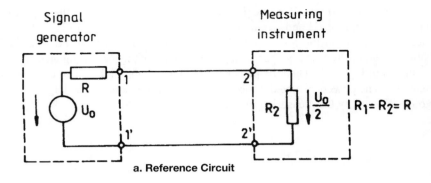

a. Reference Circuit

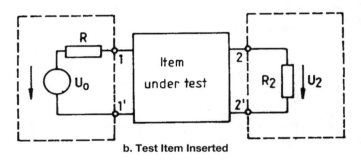

b. Test Item Inserted

FIGURE 6.2   Insertion loss definition

The value of the insertion loss of a parallel component with an impedance Z_p = $R_p + jX_p$ is [10]:

$$IL_p = 20\log\left[\left\{1 + \frac{R_p}{\frac{2}{R} \times (R_p^2 + X_p^2)}\right\} \times \sqrt{1 + \left\{\frac{X_p}{R_p + \frac{2}{R} \times (R_p^2 + X_p^2)}\right\}^2}\right] \quad (6.9)$$

Equations (6.8) and (6.9) become extremely simple if the real (resistive) part of the serial and parallel impedance is negligibly small; i.e., about zero. This approximation can be permitted on most EMI filter elements. The insertion loss of a serial circuit element having only reactive component (X'_s), is:

$$IL'_s = 20\log\sqrt{1 + \left(\frac{X'_s}{2R}\right)^2} \quad (6.10)$$

Equation (6.9), in the case of parallel impedance with clearly reactive component, is simplified to the form below:

$$IL'_p = 20\log\sqrt{1 + (\frac{R}{2 \times X'_p})^2} \tag{6.11}$$

By means of voltage measurements taken according to Fig. 6.2, the impedance of the examined two-terminal network can also be determined. Combining Eqs. (6.7) and (6.10), the serial reactance can be calculated as follows [10]:

$$X'_s = 2R\sqrt{\left(\frac{U_o}{2 \times 2U_2}\right)^2 - 1} \tag{6.12}$$

The parallel reactance, expressed with the measurable voltage, is [10]:

$$X'_p = \frac{R}{2} \times \frac{1}{\sqrt{\left(\frac{U_o}{2 \times U_2}\right)^2 - 1}} \tag{6.13}$$

If the voltage U_m, measured according to Fig. 6.2b, is much lower than the one measured according to Fig. 6.2a, the Eqs. (6.12) and (6.13) become simpler [10]:

$$X_s = 2 \times R \times \frac{U_o}{2 \times U_2} \tag{6.14}$$

$$X'_p = \frac{R}{2} \times 2 \times \frac{U_2}{U_o} \tag{6.15}$$

The above formulas are also valid for calculating the resistive part of the impedance, if the reactive part is negligible relative to the real part. It is to be noted that the simplification by two in Eqs. (6.14) and (6.15) is not reasonable, since only the voltage $U_o/2$ can be measured, as can be seen in Fig. 6.2a.

6.1.2 Measuring the HF Characteristics

The HF parameters of EMI filter elements must be known first of all in the frequency range of 0.15 to 30 MHz, and preferably beyond, for conducted noise suppression design. The measurement of HF parameters can be traced back to voltage and current measurements. The relationships for calculating the HF parameters become very simple if the internal resistance of the signal generator and the measuring instrument are selected to be equal (see Part 6.1.1). In HF instrument technology, signal generators and measuring sets of 50 or 60 Ω are widely used. Therefore, this measuring impedance is normal for evaluating the HF parameters of EMI filter elements.

The minimal and maximal value of the measurable impedance is limited by the power of signal generator, the sensitivity of the measuring instrument, and the value of the measuring impedance. A measuring set having, for example, a dynamic range of 0.5 to 100 dB can measure the serial impedance in the range of about 30 Ω to 10 MΩ, and the parallel impedance in the range of about 0.3 mΩ to 120 Ω.

Signal generators made for HF measurements usually require a constant load (termination). This requirement, however, is not fulfilled when measuring the HF characteristics of EMI filter elements, given that the impedance of the network to be tested can vary from zero to infinity. The signal generator load change can be reduced by attenuators, which are connected to output of the signal generator. The insertion loss of these attenuators is usually 10 dB; therefore, such attenuators are often referred to as *10 dB attenuators*. Using a 10 dB attenuator, the load change of the signal generator can be reduced to between 110 and 133 Ω, which is usually acceptable.

Insertion loss measurement with attenuators is shown in Fig. 6.3. The use of attenuators does not change the insertion loss calculation formulas given in Part

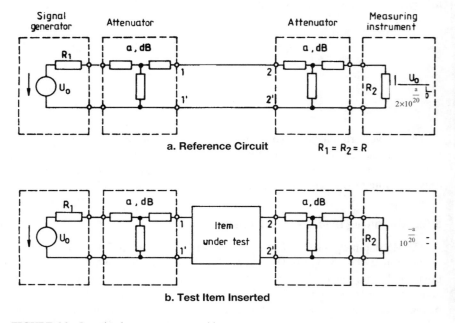

a. Reference Circuit $R_1 = R_2 = R$

b. Test Item Inserted

FIGURE 6.3   Insertion loss measurement with attenuators

6.1.1; it merely modifies the values of $U_o/2$ and U_m. In case of a 10 dB attenuator, we must substitute $U_o/20$ for $U_o/2$, and $U_m/10$ for U_m, in the relationships. Using attenuators, the accuracy of the insertion loss measurements can be increased but, at the same time, the sensitivity range of the measuring instrument will be decreased by twice the attenuation. Short of using up the entire sensitivity range, one should choose the highest possible attenuation to increase measurement accuracy.

The impedance of LISNs corresponding to CISPR requirements (see Chapter 4, Part 4.3) is 50 Ω, and measuring sets with other than 50 or 60 Ω impedance are often used (e.g., R = 600 Ω for measuring telecommunication equipment). Therefore, the HF parameters of EMI filter elements are also examined in measuring systems of R = 50 Ω and R = 600 Ω impedance. Such measurements can be made using 50 Ω instruments by inserting suitable impedance matching networks, as shown in Fig. 6.4. The 100 Ω impedance matching network increases the impedance of the signal generator and the measuring instrument to R = 150 Ω, and an R_i = 550 Ω instrument to R = 600 Ω.

When measuring the insertion loss with impedance matching networks, the relationships for calculating the impedances and insertion losses remain valid. However, their numerical values depend on the resistance of the measuring system. The numerical value of the insertion loss of a parallel connected circuit element increases with the impedance of the measuring system. For example, the insertion loss of a low-impedance parallel network can be 10 dB higher in a 150 Ω system, and 22 dB higher in a 600 Ω system, than in a 50 Ω measuring system. The measured value of the insertion loss of a serial connected network decreases as the impedance of the measuring set increases.

The relationship between the measured insertion loss and the impedance of the measuring system is shown in Fig. 6.5 [10], which gives the insertion loss results for an EMI filter choke coil. As this figure illustrates, it is not sufficient to consider only the value of the insertion loss of an EMI filter element; primarily for comparing various elements, the impedance of the measuring system also should be considered.

The layout of the measuring system (or, rather, the position of the element under test) can have an effect on the value of the insertion loss. Here, first of all, the large, surface-grounded parts should be considered. If such grounded surfaces get too near to the EMI filter element, the equivalent circuit of the element under test will be more complex because of stray capacitances toward the grounded parts. This phenomenon should be given more consideration when measuring the insertion loss of choke coils than when measuring that of capacitors. The effect of the grounded surface is well illustrated by Fig. 6.6 [10]. Curve 1 was measured with a choke coil installed far from grounded metal parts, and curve 2 was measured with the same choke coil installed near a large, grounded metal surface. The reverse would occur if an ungrounded metal part were close to the choke; i.e., the insertion loss would be decreased.

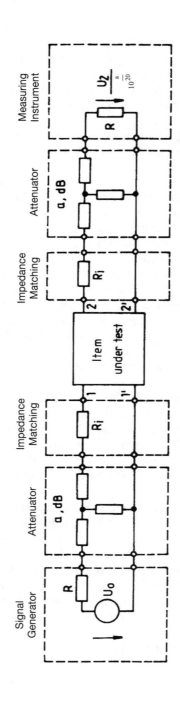

FIGURE 6.4 Insertion loss measurement with impedance matching

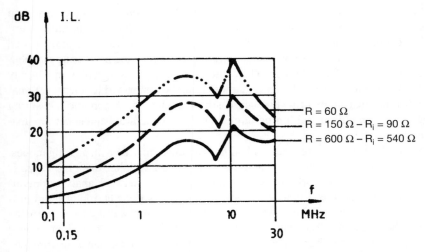

FIGURE 6.5 Insertion Loss of a choke coil as a function of measurement impedance

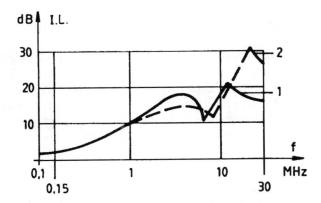

FIGURE 6.6 Insertion loss of a choke coil as a function of measurement layout

6.1.3 Automatic Instruments for Insertion Loss Measurement

Insertion loss measurements, especially for a wide frequency range, can be quite tiresome, given that two measurements must be carried out for every frequency. In addition, to draw the curve of the insertion loss as a function of frequency, the insertion loss measurements must be made for many frequencies. Such a measurement procedure, in the case of comparative investigations, is not highly desirable in practical engineering.

With network analyzers, however, it is possible to make repetitive insertion loss measurements with ease. With such instruments, the short-circuit transfer impedance, open-circuit transfer impedance, and the insertion loss can be measured comfortably. The network analyzer automatically sweeps the measuring frequency through a user-defined range. The results can be seen on a display in a highly informative manner. Usually, an X-Y plotter can be connected to a network analyzer so the measurement results can easily documented. Automatic measuring routines can be built up with programmable signal generators and spectrum analyzers, too.

Networks analyzers are made with high dynamic range. Two measuring modes, broadband and narrowband, are addressed. In general, higher dynamic range is achieved by going to a narrowband operational mode. A network analyzer operating in narrowband mode is usually more immune to source harmonics and provides a higher resolution. However, in most cases, the narrower the bandwidth, the slower the maximum frequency sweep. This limitation results from the response of the filter built into the analyzer unit. Network analyzers are linear instruments, so one must adjust the magnitude of the source signal, since operation in the nonlinear region is likely to give unreliable results.

6.2   CAPACITORS

A real capacitor is not a pure capacitance (even at low frequencies), since the leakage resistance of the isolation and equivalent series resistance (ESR) cannot be neglected in every case. At higher frequencies, the effect of stray inductances is also considered. The characteristics of a real capacitor can be properly discussed in a relatively wide frequency range by means of the equivalent circuit shown in Fig. 6.7. The impedance of the capacitor, Z_C, is as stated below:

$$Z_C = R_s + j\omega L + \frac{R_p}{1 + j\omega R_p C} = \frac{1 + \dfrac{R_s}{R_p} - \omega^2 LC}{\dfrac{1}{R_p} + j\omega LC + j\omega \left(\dfrac{L}{R_p} + R_s C\right)} \qquad (6.16)$$

At dc and low frequencies, the impedance of a real capacitor becomes equal to the leakage resistance, R_p. With increasing frequency, the reactance of the capacitor C will be lower than the resistance R_p. In this frequency range, Eq. (6.16) is simplified to the form below:

$$Z_C \approx R_s + j\omega L + \frac{1}{j\omega C} \qquad (6.17)$$

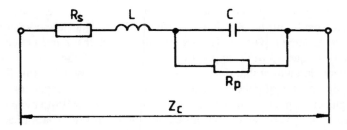

FIGURE 6.7   Equivalent circuit of capacitors

With a further increase in frequency, the impedance of real capacitors will be determined more and more by the parasitic inductance L instead of the capacitance. This means that the real capacitors behave more like an inductor than a capacitor at high frequencies. For real capacitors, a resonance frequency can be defined:

$$\omega_o^2 = \frac{1}{LC} \tag{6.18}$$

The impedance of real capacitor is just R_s (ESR) at resonance frequency. The impedance as a function of frequency, defined by Eq. (6.16), is shown in Fig. 6.8. As seen from the curve, the capacitor can be regarded as capacitance only in the frequency range labeled ω_c.

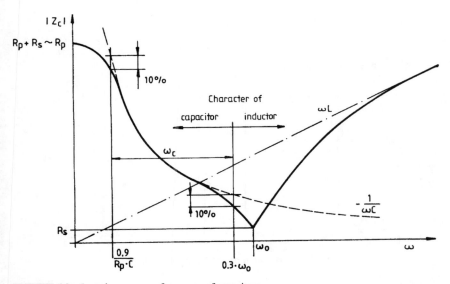

FIGURE 6.8   Impedance versus frequency of capacitors

For studying the HF characteristic of EMI filter capacitors, the series resistance, R_s is usually negligible because it has an effect only at dc and low frequencies. The value of the leakage resistance, R_p, is generally so high that it need not be considered. As the frequency of the applied voltage is increased, the value of the dielectric constant, and with it that of capacitance C, may decrease. But this fluctuation is negligible for studying the HF characteristics of EMI filter capacitors.

In engineering practice, the HF characteristics of real EMI filter capacitors are examined by means of the equivalent circuit shown in Fig. 6.9a, rather than the total equivalent circuit shown in Fig. 6.7. The serial parasitic inductance is principally the result of three partial inductances:

1. the inductance of the wound structure, L_s
2. the inductance of internal leads, L_l
3. the inductance of the connecting wires, L_w (see Fig. 6.9b).

The values of L_s and L_l (i.e., the internal parasitic inductance of the EMI filter capacitors) depend on their dimensions and structure. This inductance is generally about 5 to 50 nH.

EMI filter capacitors can be well characterized by the resonance frequency, f_o. To increase the effective frequency range where the real capacitors act like a capacitance (see Fig. 6.8), noise suppression capacitors with the highest possible resonance frequency should be applied. Because (1) the internal parasitic inductance, L_i, of the EMI filter capacitor cannot be changed, and (2) the connecting wires are

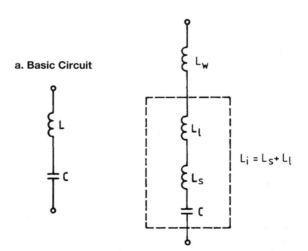

a. Basic Circuit

b. Parts of the Serial Inductance

FIGURE 6.9   Simplified equivalent circuit of capacitors

usually the major factors of inductance that determine the resonant frequency, connecting wire length should be kept to an absolute minimum for noise suppression. The inductance of the connecting leads, L_w, can be calculated with good approximation as follows:

$$L_w = 2 \times l_w \times \left(\ln \frac{4 \times l_w}{d_w} - 1 \right) \times 10^{-7} \quad H \qquad (6.19)$$

where

l_w = length of the connecting wires in meters

d_w = diameter of the connecting wires in meters

Figure 6.10 illustrates the effect of connection leads. The insertion loss measurements were made ob a 0.6 µF noise suppression capacitor [10]. The insertion loss curves in the figure well demonstrates that the highest (natural) resonance frequency of 1.7 MHz is reduced to 0.4 MHz by connection leads with 20 cm length. As a result of the lower resonance frequency, the insertion loss measured at 3 MHz has decreased even by 30 dB.

Figure 6.11 is helpful for a quick determination of the insertion loss curves of noise suppression capacitors. The value of the resonance frequency is determined by the intersection point of the proper capacitance and parasitic inductance lines. Figure 6.12 offers a brief survey for different types of capacitors, applied in various frequency ranges [1, 90].

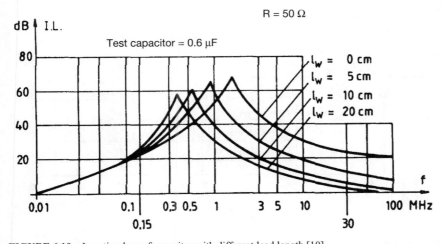

FIGURE 6.10   Insertion loss of capacitor with different lead length [10]

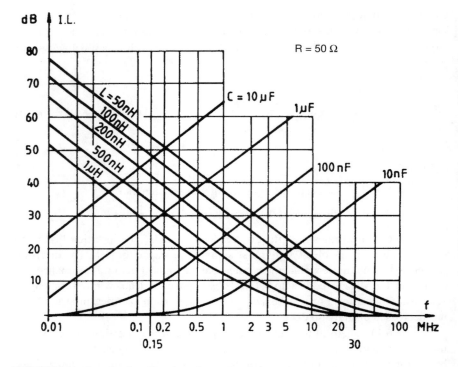

FIGURE 6.11 Insertion loss chart for noise suppression capacitors

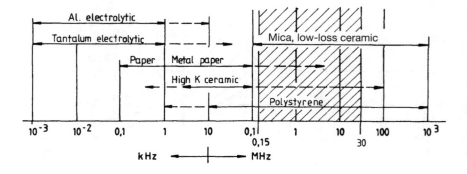

FIGURE 6.12 Approximate usable frequency range for various types of capacitors

6.2.1 Electrolytic Capacitors

The primary advantage of electrolytic capacitors is their high capacitance-to-volume ratio. But they also have some disadvantages. Electrolytics have high internal parasitic inductance due to the wound structure, and high serial resistance due to dielectric losses. The latter increases with frequency. There are two major types of electrolytic capacitors: aluminium and tantalum. Tantalum capacitors offer better performance, but they also can be regarded as a capacitance only up to 20 to 50 kHz. Because of these shortcomings, electrolytic capacitors are rarely used for noise suppression.

The HF characteristics of electrolytic capacitors may be improved by special internal structures. For reducing internal parasitic inductances, manufacturers place the leads in the middle of the wound structure or incorporate more lead points. The effects of these techniques are demonstrated in Fig. 6.13, which shows the insertion loss of a 4,700 µF, 63 V electrolytic capacitor [106]. The lead positions are marked by symbols.

6.2.2 Paper Capacitors

Various paper and metallized paper capacitors are widely employed in power electronics. Because of their high performance and reliability, they are also commonly used in power line filtering applications. Paper capacitors are produced

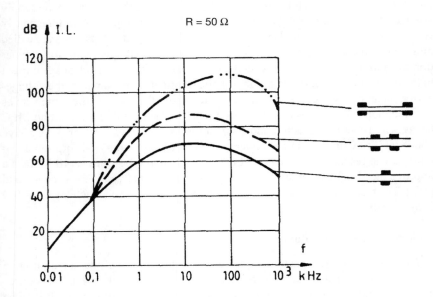

FIGURE 6.13 Approximate usable frequency range for various types of capacitors

with very wide capacitance and voltage ratings. The paper capacitors have considerably lower series resistances than electrolytics, but they still have moderately high parasitic inductance. The resonance frequency of noise suppression paper capacitors is usually in the range of 0.5 to 5 MHz. This is relatively low—at least compared to the 30 MHz upper frequency limit of noise suppression. To avoid further decreases in the resonance frequency, the wiring layout should be devised with special care (see Fig. 6.10).

6.2.3 Polystyrene and Ceramic Capacitors

Polystyrene capacitors have extremely low dielectric losses and very stable capacitance-to-frequency characteristics. They are manufactured in capacitance and voltage ratings that make them highly suitable for noise suppression in electronic units.

Ceramic capacitors are characterized by excellent HF performance. Until the eighties, ceramic capacitors were manufactured with ratings up to 100 V and a few nanofarads. With development in the technology, the voltage and capacitance ratings have been increased. The high-K (loss) ceramic capacitors are relatively unstable with respect to time, temperature, and voltage, and they may be damaged by voltage transients. The performance stability of low-loss ceramic capacitors has been improved considerably.

Because of their extremely small size, ceramic capacitors are widely used in printed wiring boards. Malfunctions and damage in electronic units are caused not only by HF disturbances but also by impulse-like signals that appear on the supply voltage. As a result, a new ceramic capacitor was developed to simultaneously suppress HF noise and impulse-like transients. These bypass capacitors are called multifunction ceramic (MFC) capacitors [270]. In normal operation, MFCs function as ceramic capacitors with excellent HF characteristics and a leakage current in the microamp range. But if a high-voltage transient is encountered, the voltage-current characteristic causes the capacitor to absorb the surge in the same way as a conventional ZnO varistor. These electrical characteristics mean that an MFC behaves as a ceramic capacitor and a varistor in parallel. As a varistor, MFCs have good stability with respect to temperature, plus excellent pulse absorption capabilities. The resonance frequency of the two-terminal type is about 10 MHz. To improve the HF characteristics of MFCs, a three-terminal version was developed [315]. The resonance frequency of these ceramic capacitor can reach levels of 100 to 500 MHz.

6.2.4 X- and Y-Capacitors

Noise suppression capacitors in power line filters are usually divided into two groups: X- and Y-capacitors. The terminology refers only to the electrical connection mode and not to insulation material, capacitance, or construction. It is to be noted that this categorization/designation is mainly used in Europe. Unlike Y-capacitors, X-capacitors do not pose a threat of electrical shock for personnel, even in the case of a breakdown. Because of possible injury to humans and electrical equipment, safety standards relating to the use of Y-capacitors are much stricter than those applied to X-capacitors. Y-capacitors provide CM attenuation in most powerline filters, but their maximum values are often limited because of leakage current (see Chapter 7, Section 7.5.1).

The noise suppression capacitors in power line filter not causing any electric shock for the personnel, even in case of a breakdown, are called X-capacitors.

6.2.5 Feedthrough Capacitors and Feedthrough Filters

EMI filter capacitors are often qualified by their resonance frequency. The HF performance of noise suppression capacitors can be improved by decreasing the parasitic inductances. Manufacturers strived to accomplish this task by reducing the impedance of internal leads, creating the so-called *feedthrough capacitors*. In feedthrough capacitors, the leads to the wound structure are not led out to the housing surface of the capacitor; rather, the wires connecting to external circuits are placed within the capacitor structure. This is the reason why, in contrast to capacitors with conventional structure, the nominal current should be considered in addition to capacitance and voltage ratings. The feedthrough structure makes it possible to connect the wound structure directly to the cable to be protected from HF disturbances. Two basic feedthrough capacitor structures exist: non-coaxial and coaxial.

In non-coaxial feedthrough capacitors, the leads pass close to the wound structure. Two types are defined, depending on internal structure, as shown in Fig. 6.14. In Fig. 6.14a, the leads pass through the wound structure, and in Fig. 6.14b, the of leads pass by it. The high performance characteristics of the non-coaxial structure can be utilized only if coupling between the leads in the capacitor hous-

a. Leads Passing through the Wound Structure	b. Leads Passing by the Wound Structure

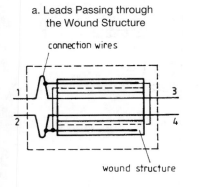

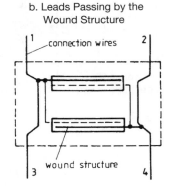

FIGURE 6.14 Internal structure of non-coaxial feedthrough capacitors

ing is low. With respect to this, the structure in Fig. 6.14b is preferable. The resonance frequency of a feedthrough capacitor of this type, in the range of hundreds of nanofarads capacitance, is about 10 MHz.

The goal of achieving higher resonance frequencies led to the development of coaxial feedthrough capacitors. The internal structure is shown in Fig. 6.15. The parasitic inductance of the leads is decreased by applying concentric contact elements. One contact element is connected to the lead passing along inside the capacitor, and the other contact element, generally grounded, is connected to the metal housing. The resonance frequency of coaxial feedthrough capacitors is much higher than that of non-coaxial types, and the contacts and soldering resistances can become I.L. limiting factors.

Coaxial feedthrough capacitors are most suitable for noise suppression above a few megahertz. As a comparison, Fig. 6.16 shows the insertion loss of capacitors of standard structure (curve 1) and that of a feedthrough capacitor. Both capacitors have the same value of 0.6 μF. The insertion loss of the standard type was measured without connection wires [10].

Although designers found it possible to minimize internal lead inductance by applying concentric contact elements in coaxial feedthrough capacitors, further reduction of internal parasitic inductances was blocked by the inductance of the wound structure. A method of increasing the resonance frequency was found by developing the so-called *cascaded feedthrough capacitors* [280]. Cascaded feedthrough capacitors, as shown in Fig. 6.17a, comprise several feedthrough capaci-

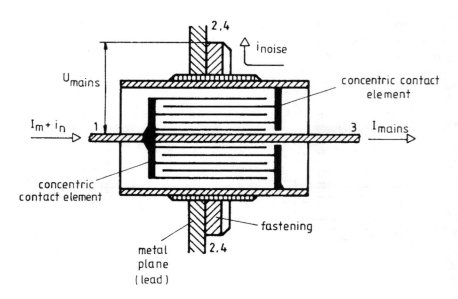

FIGURE 6.15 Internal structure of coaxial feedthrough capacitors

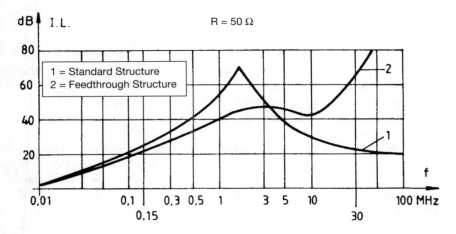

FIGURE 6.16 Insertion loss of noise suppression capacitors with different structures

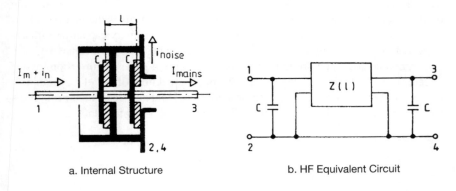

a. Internal Structure

b. HF Equivalent Circuit

FIGURE 6.17 Cascaded feedthrough capacitor

tor elements connected electrically parallel but geographically cascaded. The equivalent circuit of the cascaded feedthrough capacitor is shown in Fig. 6.17b. The four-terminal network $Z(l)$ represents the distributed impedance of the wire; i.e., the transmission line network with the length, l, between the capacitor discs. Because of their better HF parameters, ceramic capacitors are generally built into cascaded feedthrough capacitors. For increasing the insertion loss, ferrite beads (washers, rings) are placed on the wire. As a result of ferrite beads, the four-terminal network shown as $Z(l)$ in Fig. 6.17b can be replaced by the serial impedance of a ferrite core choke coil of one turn. Thus, the equivalent circuit is transformed into a common lowpass filter circuit. The character of the serial impedance depends on the complex permeability, μ^*, of the ferrite material (see Section 6.3.5).

Figure 6.18 shows a demonstrative example for the relation between the structure and the HF performance [280]. In this figure the insertion loss curves of 5.6 nF capacitance are shown. Curve 1 refers to a single ordinary (two-lead) ceramic capacitor of 5.6 nF. Curve 2 is the insertion loss of four elements of individual but electrically parallel connected ceramic capacitors of 1.4 nF. Curve 3 is the result of a cascaded feedthrough capacitor with ferrite beads. As seen from the curves, the cascaded feedthrough capacitor has the best HF performance.

The feedthrough filter can be listed among the feedthrough capacitor types, both in terms of their HF characteristics and internal structure. A feedthrough filter is, in fact, made up of electrically parallel feedthrough capacitors with the same serial impedance, constituting an n-stage Π-filter. Figure 6.19 shows the typical insertion loss curve of a feedthrough filter [10]. The feedthrough filter contains two feedthrough capacitors of 1 μF. The shape of insertion loss curve depends highly on the HF losses of the serial impedance. The solid line shows the typical I.L. curve of a feedthrough filter with higher quality factor (Q-factor), whereas the dashed line shows the curve of a feedthrough filter containing a damped inductor.

The HF parameters of various types of EMI filter capacitors can be compared by means of the insertion loss measurements shown in Fig. 6.20. In all cases, the total capacitance of paper capacitors was 0.6 μF. Curve 1 gives the theoretical insertion loss of an ideal capacitor. Curve 2 refers to a customary capacitor with approximately a zero-length connection to achieve the best HF performance. Curve 3 is from a coaxial feedthrough capacitor, and Curve 4 gives the insertion loss of a feedthrough filter containing two feedthrough capacitors of 0.3 μF capacitance. As seen in Fig. 6.20, the application of feedthrough capacitors or feedthrough filters can be effective for suppressing HF noises above a few megahertz.

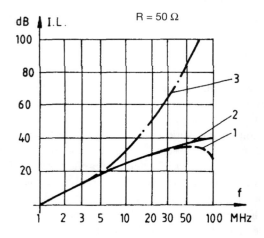

FIGURE 6.18   Insertion loss of various ceramic feedthrough capacitors

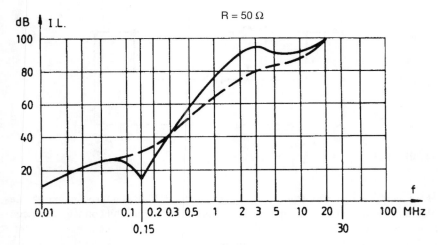

FIGURE 6.19 Typical insertion loss curves of feedthrough filters

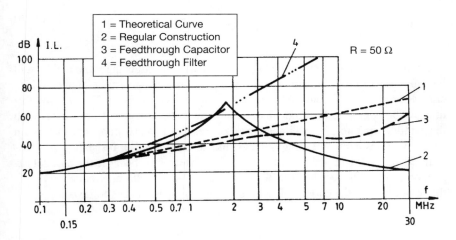

FIGURE 6.20 Insertion loss of noise suppression capacitors with various structures

6.3 CHOKE COILS

Noise suppression choke coils can be well characterized in a wide frequency range by the equivalent circuit seen in Fig. 6.21. This information is necessary to understand the terms that define the performance and the limitation of inductors in the frequency range of conducted noise suppression. The resistance in the equivalent circuit represents the losses of the choke coil. Parasitic effects on high-

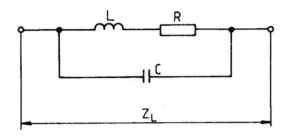

FIGURE 6.21 Equivalent circuit of a choke coil

er frequencies, resulting from the stray capacitances between turns, cannot be neglected. Although the turn-capacitance is distributed, a parallel connected concentrated capacitor provides a suitable approximation. The impedance of the choke coil according to the equivalent circuit is:

$$Z_L = \frac{R + j\omega L}{1 - \omega^2 LC + j\omega RC} \tag{6.20}$$

At low frequencies, impedance Z_L is dominated by inductance, and at dc it will be equal to R. In the frequency range of ω_L, the impedance of the choke coil increases proportionally with frequency, as can be seen in Fig. 6.22. At some frequency (belonging to the angular frequency, ω_0), the inductor, L, resonates with the parallel capacitor, C, and impedance, Z_L, reaches its maximum. The maximum value increases with increasing Q-factor and decreasing series resistance. At higher frequencies, the impedance of the choke coil decreases because the parallel capacitor dominates; i.e., the inductor acts like a capacitor. In this frequency range, the impedance of the choke coil with good approximation is as follows:

$$Z_L \approx \frac{j\omega L}{\omega^2 LC} \approx \frac{1}{j\omega C} \tag{6.21}$$

To examine the impedance function of the choke coil, Eq. (6.20) should be transformed into the root-factor form:

$$Z_L = \frac{\dfrac{1}{J\omega C} + \dfrac{R}{LC}}{(\omega - \omega_1) \times (\omega - \omega_2)} \tag{6.22}$$

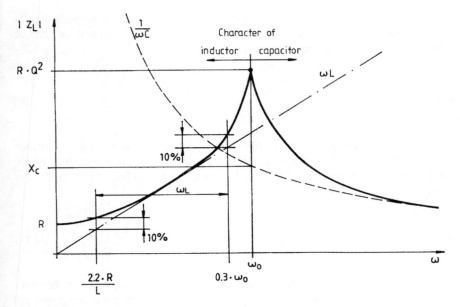

FIGURE 6.22 Impedance versus frequency of a choke coil [2]

The roots of the denominator in the above relationship are:

$$\omega_{1,2} = \frac{R}{2L} \pm \omega_o \sqrt{\frac{1}{4Q^2} - 1} \qquad (6.23)$$

To simplify the equation, defining the roots, it is reasonable to introduce two variables.

1. The self-resonance angular frequency of the choke coil:

$$\omega_o = \frac{1}{\sqrt{LC}} \qquad (6.24)$$

2. The Q-factor of the choke coil:

$$Q = \frac{\omega L}{R} \qquad (6.25)$$

Given Eq. (6.24), the resonance frequency commonly used in engineering practice can be calculated dividing by 2π.

The resonance frequency where the impedance is at maximum depends slightly on the resistance and capacitance. The resonance angular frequency also can be expressed by the self-resonance angular frequency:

$$\omega_r^2 = \omega_o^2 \times \frac{-R^2C + \sqrt{2 \times R^2LC + L^2}}{L} \tag{6.26}$$

As seen from the above relationship, the serial resistance reduces the resonance angular frequency (resonance frequency). A resistance value can be defined where a resonance first occurs, and above that value there is more resonance. This critical resistance is:

$$R_r \leq \sqrt{\frac{L}{C}} \times \sqrt{1 + \sqrt{2}} \tag{6.27}$$

Hence, in practice, the Q-factor of the choke coils is usually much higher than one; i.e., the value of the serial resistance is low, and the roots described by Eq. (6.23) can be rewritten in the following form:

$$\omega_{1,2} \approx -\frac{R}{2L} \pm j\omega_o \tag{6.28}$$

The impedance of the choke coil at low frequencies (more precisely, in the frequency range below the resonance angular frequency) with fair approximation is:

$$Z_L' \approx R + j\omega L \tag{6.29}$$

The impedance of the choke coil at resonance angular frequency can be determined by means of Eqs. (6.20) and (6.24). Supposing that the impedance of the choke coil at resonance is much higher than the serial resistance, the impedance of the choke coil at resonance is:

$$Z_L(f_o) \approx \frac{L}{RC} \tag{6.30}$$

Using Eq. (6.25), describing the Q-factor, the above equation also can be rewritten into the following form:

$$Z_L(f_o) \approx R \times Q^2 \tag{6.31}$$

Much as in the case of capacitors, we can use a graph (Fig. 6.23) to quickly determine the approximate insertion loss curve for choke coils. The value of the resonance frequency can be read where proper inductance and parasitic capacitance lines cross.

In the design of noise suppression choke coils, several aspects should be considered. In general, the primary goal is to push the resonance frequency to the highest possible value. This can be done in two ways: (1) by reaching the highest inductance for a given size or, (2) by minimizing the parasitic capacitance between the turns. For designing the form and structure of the choke coil, the inductance and parasitic capacitance of various coil arrangements should be known.

6.3.1 Calculation the Inductance of Various Coil Arrangements

Choke coils for EMI filtering are usually single-layer solenoid structures. The scheme of a solenoid-type choke coil is shown in Fig. 6.24. Suitable formulas are known for calculating the inductance of such coil arrangements, if the internal and

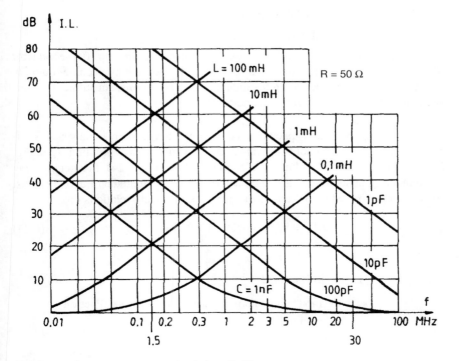

FIGURE 6.23 Insertion loss chart for choke coils [7]

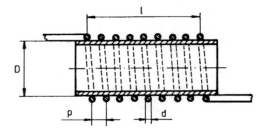

FIGURE 6.24   Structure of single-layer solenoid choke coil

external diameters are not very different, and if the coil length is much higher than its diameter (i.e., length > 0.3 D):[*]

$$L = \frac{(\pi ND)^2}{l + 0.45 \times D} \times 10^{-7} \quad H \qquad (6.32)$$

where

N = number of turns

D = internal diameter of the coil in meters

l = axial length of the coil in meters

In the case of multilayer coils or choke coils made from a strip (flat) conductor of large cross-section, the internal and external diameters may become mutually commensurable. For calculating the inductance of such coils, substitute the average coil diameter in place of the diameter, D, in Eq. (6.32).

For reducing the stray capacitance between the turns, often some space is left between the turns of solenoid-type choke coil; i.e., the pitch of windings (p in Fig. 6.24) is higher than the width of the conductor (d in Fig. 6.24). For determining the inductance of such a coil arrangement, the value calculated by Eq. (6.32) must be corrected. The reduction term is given by the following relationship:

$$\Delta L = k_m \times N \times D \times 10^{-7} \quad H \qquad (6.33)$$

where

N, D = as in Eq. (6.32)

k_m = the modifying factor, shown in Fig. 6.25 as a function of ratio p/d

[*]The basic formulas given hereafter are primarily for air coils. If the choke has a magnetic core, the value of L must be multiplied by the relative permeability, as will be seen in Eq. (6.42).

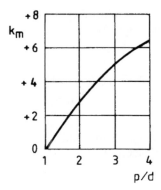

FIGURE 6.25 Modifying factor versus coil parameters a for loosely wound, single-layer coil arrangement

The inductance of short solenoid choke coil cannot be calculated by Eq. (6.32). A really good relationship for a relatively short solenoid-type choke coil is hard to establish because stray phenomena also should be considered. But the inductance of such a coil arrangement can be calculated with suitable approximation by the following formula:

$$L = \frac{(\pi ND)^2}{l + 0.45 \times D - \dfrac{0.01 \times D^2}{l}} \times 10^{-7} \quad H \tag{6.34}$$

Some EMI filter choke coils are made with two windings on a common cylindrical form. The windings can be connected to each other but also may belong to potentially independent circuits. The layout of such a choke coil is shown in Fig. 6.26. Calculating the inductance of a multiple-coil arrangement, the mutual inductance (M) should also be considered.

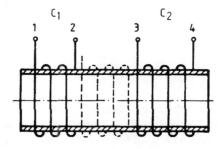

FIGURE 6.26 Two separate, identical coils arranged on the same coil form

Connecting the two windings in parallel and in a way that the magnetic fields add up, the resulting inductance is [5]:

$$L = \frac{1}{\dfrac{1}{L_1 + M} + \dfrac{1}{L_2 + M}} \qquad (6.35)$$

Connecting the two windings again parallel, but inversely to the above method (i.e., so that the magnetic fields of the two windings are opposing), the resulting inductance is:

$$L = \frac{1}{\dfrac{1}{L_1 - M} - \dfrac{1}{L_2 - M}} \qquad (6.36)$$

The mutual inductance, M, often is not calculated but determined by measurements. While there are impedance bridges that can measure the mutual inductance, a common method to determine M is measuring the inductance with different coil connections. If the two coils are connected in series so that the mutual inductance aids the net inductance, the value L' can be measured, which is:

$$L' = L_1 + L_2 + 2M \qquad (6.37)$$

Connecting the windings again serially, but in a way that their magnetic fields are in opposition (i.e., contrary to the way the inductance L' was measured), the inductance L'' can be measured, which is:

$$L'' = L_1 + L_2 - 2M \qquad (6.38)$$

The mutual inductance can be calculated, Possessing the two measured inductance values, we can calculate the mutual inductance:

$$M = \frac{L' - L''}{4} \qquad (6.39)$$

The inductance of the multiple-coil arrangement can be calculated with fair approximation by Eqs. (6.35) and (6.36), but only if the two winding are placed in close proximity. If there is an air gap between the coils as shown in Fig. 6.26, the mutual inductance is to be calculated by placing an assumed coil with the same construction between the points 2 and 3 [5, 95]. Under this assumption the mutual inductance can be calculated as follows:

$$M = \frac{1}{2}(L_{14} + L_{23} - L_{13} - L_{24}) \tag{6.40}$$

The above relationship also can be used for calculating the mutual inductance of tapped (autotransformer) solenoid choke coils [5, 95]. Using the Eq. (6.40) for autotransformer-type solenoid choke coils, $L_{23} = 0$ should be substituted. In this case, L_{14} means the inductance of the windings without a tap. Calculation of the inductance of various solenoid arrangements is aided by a chart provided by Edson [95].

To increase the inductance value, but partly to create HF losses, an open iron core is often placed in the solenoid choke coil. The effect of the iron core can be considered by introducing the equivalent permeability, μ_e. The inductance of a solenoid arrangement with an iron core is:

$$L_i = \mu_e L \tag{6.41}$$

The value of the equivalent permeability depends on the material used in the iron core and the dimensions of the coil. Using the symbols of Fig. 6.24, the equivalent permeability can be calculated when the cross-section of the iron core is at least 80 percent of the coil cross-section and the iron core is at least 20 percent longer than the coil. In the case of a silicon-iron transformer core, the equation is as follows:

$$\mu_e = \frac{0.45 \times D + 1}{0.45 \times D + 0.1 \times 1} \tag{6.42}$$

In the equivalent circuit of a choke coil with an iron core, the resistor, R, represents not only the losses of the winding but also that of iron core. The iron core losses are composed of several factors. The eddy current losses are proportional to the square of the frequency, whereas the hysteresis losses vary directly with the frequency.

To achieve a better inductance/volume ratio, EMI filter choke coils are sometimes made with multilayer windings. The scheme of a multilayer coil arrangement is shown in Fig. 6.27. The inductance of multilayer solenoid coil, wound closely, can be calculated by the following relationship:

$$L = \left[\frac{(\pi N D)^2}{1 + 0.45 \times D} - k_w \times \frac{2\pi N^2 D t}{1} \right] \times 10^{-7} \quad H \tag{6.43}$$

where

> D = average coil diameter (see Fig. 6.27) in meters
>
> l = axial length of the coil in meters
>
> t = thickness of the coil in meters
>
> k_w = modifying factor, shown in Fig. 6.28 as a function of length/thickness ratio

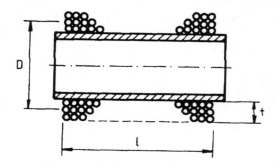

FIGURE 6.27 Multilayer coil arrangement

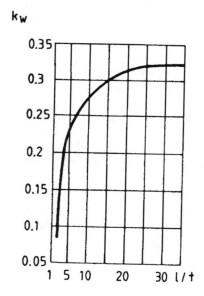

FIGURE 6.28 Modifying factor for multilayer coil arrangement versus coil parameters

For EMI filters, choke coils of toroid form are also available. The cross-section of the toroid coil can be circular or rectangular. The inductance of circular cross-section toroid coil, wound closely, is [5]:

$$L = 2\pi \times N^2 \times [D - \sqrt{D^2 - d^2}] \times 10^{-7} \quad H \qquad (6.44)$$

where

D = internal diameter of the coil in meters

d = diameter of the coil cross-section in meters

The inductance of a closely wounded toroid coil with rectangular cross-section:

$$L = 2\pi \times N^2 \times h \times \ln \left[\frac{D_o}{D_i} \right] \times 10^{-7} \quad H \qquad (6.45)$$

where

D_i = internal diameter of the coil in meters

D_o = external diameter of the coil in meters

h = height of the coil in m

Because of better cooling conditions and easier manufacturing, noise suppression choke coils for high current ratings are often made in spiral form. The inductance of a spiral coil arrangement can be calculated with good approximation as follows:

$$L = \frac{24.6 \times N^2 \times D}{1 + 2.75 \times \dfrac{t}{D}} \times 10^{-7} \quad H \qquad (6.46)$$

where

D = average diameter of the coil in meters

t = thickness of the coil ($v < 0.2 \times D$) in meters

6.3.2 Coil Capacitance

The impedance of choke coils above the resonance frequency is determined by coil parasitic capacitances developed between individual turns. Figure 6.29 shows the typical impedance curve of some noise suppression choke coils [10]. As seen in the figure, the resonance frequency of choke coils is in the range of 1

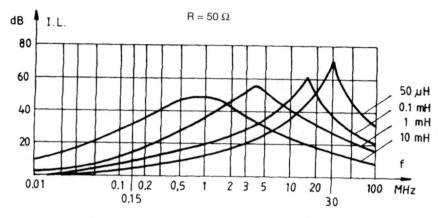

FIGURE 6.29 Insertion loss of various noise suppression choke coils [10]

to 10 MHz. Thus, choke coils of high inductance act like capacitors in the upper part of the 0.15 to 30 MHz frequency range, which is important for the suppression of conducted emissions from power electronic equipments.

The HF characteristics of noise suppression choke coils can be improved by decreasing their parasitic coil capacitances. Noise suppression choke coils are usually single-layer wound because, in this case, the parasitic capacitance consists only of turn capacitances, which are lower than the layer capacitance. The capacitance per turn of coils wound of round conductor can be taken from Fig. 6.30. For coils wound with strip conductor, Fig. 6.31 applies. The figures gives the value of stray capacitance between two adjacent turns. The value of the lumped capacitor C in the equivalent circuit of choke coils (see Fig. 6.21), representing the stray effects, can be calculated as follows:

$$C = \frac{C'}{N} \qquad (6.47)$$

Inductances in the order of millihenries can be realized as single-layer solenoid coil arrangement only in relatively large dimensions. For reducing the dimensions of noise suppression choke coils, multilayer windings are often applied. The rather high layer capacitance of the multilayer coil arrangement should be decreased to achieve better HF performance.

For single-layer construction (and, first of all, for coils wound with strip conductor), a question can be asked concerning which coil arrangement has the highest resonance frequency in a given size. If one chooses a winding pitch greater than the width of the conductor, the parasitic capacitance decreases, but so does the inductance. The extreme value calculations showed that the resonance fre-

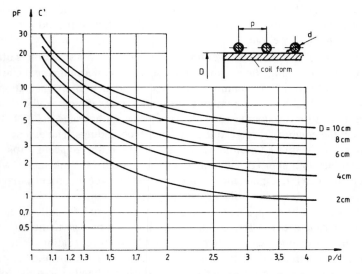

FIGURE 6.30   Stray capacitance of single-layer coil wound from round conductor (capacitance between two adjacent turns)

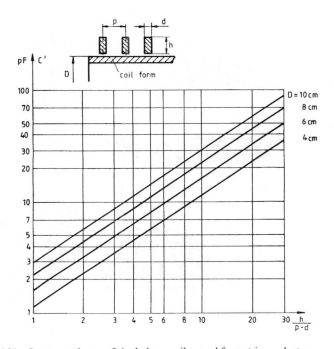

FIGURE 6.31   Stray capacitance of single-layer coil wound from strip conductor

quency increases if the coil is wound closely, using strip conductor having the highest possible height/width ratio. This means that a flat strip conductor, bent to the edge, is the proper choice.

For increasing the inductance/volume ratio, multilayer construction is often selected. The relatively high layer capacitance of such a coil arrangement can be decreased several ways. One commonly used technique is to place insulating rings as shown in Fig. 6.32. As the number of insulation rings increases, the stray capacitance attenuation also increases. Another solution is illustrated in Fig. 6.33. The numbers in the conductors refer to the layout of the windings. The essence of this solution is that, in spite of multilayer coil arrangement, the adjacent turns belong to coil sections with approximately identical turn numbers. With such a winding sequence, the parasitic capacitance can be reduced by even an order of magnitude compared to the standard multilayer winding arrangement.

The inductance of noise suppression choke coils is small, usually in the range of tens of microhenries to a few millihenries. The common way to measure the inductance of noise suppression choke coils with high current ratings is to form a parallel resonant circuit with a known capacitance and then determine the value of resonance frequency. Using this method, the parasitic capacitance of the coil cannot be neglected if the measuring and the parasitic capacitance differ from each other by less than one order of magnitude. This is because the parallel resonant

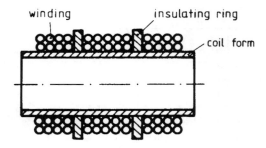

FIGURE 6.32 Reducing parasitic capacitance with insulation

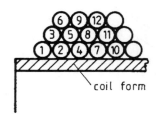

FIGURE 6.33 Reducing parasitic capacitance with special coil arrangement

circuit is formed not purely of the measuring capacitor, C_m, but the sum of it and the parasitic capacitance, C. In this case the resonance frequency is given by the following formula:

$$f_r = \frac{1}{2\pi \sqrt{L \, (C + C_m)}} \qquad (6.48)$$

For calculating the inductance of the noise suppression choke coil, the value of parasitic capacitance, C, must be also known:

$$L = \frac{1}{(2\pi \times f_r)^2 \times (C + C_m)} \qquad (6.49)$$

One can determine the parasitic stray capacitance experimentally by forming a parallel resonant circuit with known capacitance and then measuring the resonance frequency. To calculate the value C, two different resonant frequencies must be measured with two different capacitors. Assuming that the measuring capacitor, C_{m1}, results in the resonance frequency, f_{r1}, and C_{m2} results in the resonance frequency f_{r2}, the value of the unknown parasitic capacitance is [5]:

$$C = \frac{\left(\dfrac{f_{r1}}{f_{r2}}\right)^2 \times C_{m1} - C_{m2}}{1 - \left(\dfrac{f_{r1}}{f_{r2}}\right)^2} \qquad (6.50)$$

To simplify calculations using Eq. (6.50), the resonant frequencies $f_{r2} = 2 \times f_{r1}$ should be selected with an adjustable capacitor. In this situation, Eq. (6.50) becomes [5]:

$$C = \frac{C_{m1} - 4C_{m2}}{3} \qquad (6.51)$$

In practice, both the Q-factor and the serial resistance of EMI filter choke coils are often measured by resonating the choke coil. The calculated Q-factor value, and the value of the serial resistance representing the losses, must be corrected if the ratio of the measuring capacitance, C_m, and the parasitic capacitance, C, are less than 10 to 20. The correction formulas are as given below [5]:

$$Q = Q_m \left(1 + \frac{C}{C_m}\right) \qquad (6.52)$$

$$R = R_m \times \frac{(C_m^2)}{(C + C_m)^2} \tag{6.53}$$

6.3.3 Noise Suppression Choke Coils for High Current Ratings

The inductance of noise suppression choke coils in powerline filters rated for high currents cannot be very high, given that a voltage drop at mains frequency occurs across the choke coil. At low frequencies (i.e., at mains frequencies), the absolute value of the impedance of the choke coil with good approximation is as follows:

$$Z_f = \sqrt{R^2 + (\omega_m L)^2} \tag{6.54}$$

The mains frequency current, I_m, generates a voltage drop across the noise suppression choke coil equal to:

$$\Delta U = I_m \times Z_f \tag{6.55}$$

The allowable voltage drop at mains frequency is limited by the equivalent voltage regulation of the power mains and filter combination, and also by transient coupling when loads are switched on or off. Setting up limitations for voltage drop under full-load conditions, the transient coupling should be checked often, since these two limitations are closely related. Having already established a limitation from full-load condition, then considered the calculated transient coupling, we must take into account inrush currents that can be many times greater than nominal. For these reasons, the allowable voltage drop is usually limited to some percentage of the mains (supply) voltage. Neglecting the real component of the voltage drop (besides the inductive one), a simple relationship can be given for calculating the maximum inductance of noise suppression choke coils. Assuming that the acceptable voltage drop is ΔU_{max}, the allowable inductance with good approximation is:

$$L_{max} = \frac{\Delta U_{max}}{2\pi f_m \times I_m} \tag{6.56}$$

Designing solenoid-type noise suppression choke coils for a given inductance is seldom a simple job. In particular, the inductance value depends implicitly on geometric dimensions, as was shown by Eq. (6.32). The calculations become even more complex if an open iron core is placed in the coil to increase the inductance/

volume ratio. This is because the effect of the iron core also depends implicitly on geometry, as shown by Eq. (6.42). The next problem is the temperature rise of the choke coil, which is also an implicit function of geometry. The above indicates that the design of solenoid-type noise suppression choke coils requires multi-stage iteration.

Measurements have indicated that Eqs. (6.32) and (6.42) are not accurate enough for a length/diameter ratio of 3 to 7, which is normal for noise suppression choke coils of high current rating. The inductance of solenoid choke coils with this length/diameter ratio can be more accurately calculated as:

$$L_o = \frac{D_i \times 4N^2 (D^* + 1)^2}{13D^* + 18K - 7} \tag{6.57}$$

where

N = number of turns

D_i = internal diameter of the coil in meters

D^* = ratio of the external and internal diameters of the coil:

$$D^* = \frac{D_o}{D_i} \tag{6.58}$$

K = ratio of the length and the internal diameter:

$$K = \frac{1}{D_i} \tag{6.59}$$

As an alternative to using Eq. (6.42), the equivalent permeability, representing the effect of the open iron core, can be calculated for length/diameter ratio of 3 to 7 as follows:

$$\mu_e = 1 + \frac{36K}{(4.5 + K) \times (D^* + 1)^2} = f_2 (D^*, K) \tag{6.60}$$

Equations (6.57) through (6.60) are for designing solenoid-type choke coils, with open iron cores, for high current rating. In practice, the design procedure can follow any of several paths, but the following one may be advisable.

Prescribing the form factor of the coil by an appropriately wide range, calculate the actual equivalent permeability, then the coil data. As the last step, keeping in

mind the pertinent coil data, check the temperature rise. The allowable temperature rise can be determined within a reasonable range.

To facilitate the calculation and to decrease the number of iterations in the recommended design method, the formulas giving the equivalent permeability and
inductance must be transformed somewhat.

Choke coils for high current ratings are usually made with strip conductors. Let
the form of the strip conductor be characterized by the height/width ratio (see
Fig. 6.34):

$$M = \frac{y}{x} \tag{6.61}$$

Supposing that the current density, s, is known or set, the required copper
cross-section of the conductor can be determined. The width of the strip conductor
can be calculated as follows:

$$x = \sqrt{\frac{A_{cu}}{M}} \tag{6.62}$$

where

$$A_{cu} = \text{copper cross-section in } mm^2$$

The height of the strip conductor is as follows:

$$y = M \times x \tag{6.63}$$

Having the height of the strip conductor, the ratio of diameters as defined by
Eq. (6.58) takes the following form:

$$D^* = 1 + \frac{2y}{D_i} \tag{6.64}$$

The number of turns, using Eqs. (6.64) and (6.62), can be also given by means
of coil diameters and cross-section of the copper conductor. Introducing the filling
factor, k_w, as the ratio of actual coil length and the pure axial copper length ($N \times x$), the number of turns we can wind on the coil form of length l can be calculated
as follows:

$$N = D_i^2 \times \frac{k_w (D^* - 1)}{2 \times A_{cu}} = f_3 (D_i, D^*, K, A_{cu}) \tag{6.65}$$

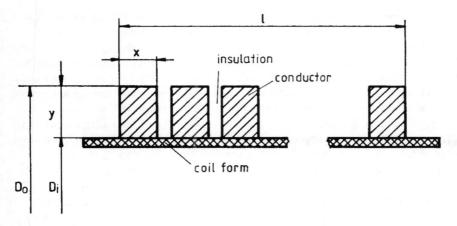

FIGURE 6.34 Structure of noise suppression choke coil for high current ratings

Substituting this expression in Eq. (6.57), the inductance of a solenoid coil without an iron core is:

$$L_o = D_i^2 \times \frac{k_w \times K \times (D^* - 1)}{A_{cu}^2 \times (13 \times D^* + 18 \times K - 7)} \times 10^{-6} \quad H \qquad (6.66)$$

Regarding K as unknown in the above relationship, the ratio K, belonging to known inductance L_o, can be calculated. Using notation as defined for Eq. (6.57), we can determine that:

$$K = \frac{9 \times L_o \times A_{cu}^2}{C} \times \left[1 + \sqrt{1 + \frac{C \times (13D^* - 7)}{81 \times L_o \times A_{cu}^2}} \right] = f_1 (L_o, D_i, D^*, A_{cu}) \qquad (6.67)$$

where

$$C = D_i^5 \times k_w^2 \times (D^* - 1)^2 \times (D^* + 1)^2 \times 10^{-6}$$

L_o = inductance of the air core solenoid choke coil in henries

It should be noted that, in the above formula, L_o represents the inductance of the air-core choke coil but, in fact, only the value of the iron-core choke coil is known. The value of L_o can be calculated, according to Eq. (6.41), by dividing the inductance of the iron-core choke coil by the value of equivalent permeability.

The design process should start with the current rating, inductance, allowable temperature rise, and form factor of the solenoid type choke coil (see Fig. 6.35). After entering the value of current density, s; the initial value of the internal coil diameter, D_i; the axial filling factor, k_w; and the form factor of the strip conductor, M, the initial value of μ_e still must be given. With these data and using Eqs. (6.63), (6.64), and (6.67), first calculate the length of the choke coil needed to produce the required inductance. Having established the coil length, the actual value of μ_e can be calculated using Eq. (6.60). If the calculated and the entered value of μ_e does not fall within the appropriate small error limit, the inductance of the air core arrangement, L_o, must be figured again with the calculated μ_e. This procedure should be continued until the two values of μ_e are equal (see Fig. 6.35). After this iteration, the calculated value of K must be checked to be sure it is in the prescribed range; i.e., between K_{min} and K_{max}. If not, the value of the internal coil diameter should be changed and the above calculations repeated with the new D_i. These calculations require two-stage iterations as seen from Fig. 6.35.

Having established the coil data, the actual temperature rise of the choke coil can be calculated. For calculating the temperature rise, one may suppose that the iron losses, compared to the copper loss, are negligible. This assumption is reasonable if the magnetic force remains under 1,000 A-turn. With the allowable approximation of $D_i = D_o = D$, the copper loss of the coil is as follows:

$$P = I_2 R = I^2 \times \frac{ND\pi}{57 A_{cu}} \quad W \tag{6.68}$$

where

$$I = \text{nominal supply (mains) current in amps}$$

$$D = \text{average diameter of the coil in meters}$$

$$A_{cu} = \text{pure copper cross-section in square millimeters}$$

$$N = \text{number of turns calculated by Eq. (6.65)}$$

The copper loss increases with the temperature; hence, the resistance of the conductor depends on the temperature. The total copper loss is:

$$P_t = P (1 + \alpha \times \Delta T) \tag{6.69}$$

where

$$\Delta T = \text{temperature rise in degrees Celsius}$$

$$\alpha = 4 \times 10^{-3}/\,^\circ C \text{ for copper wire}$$

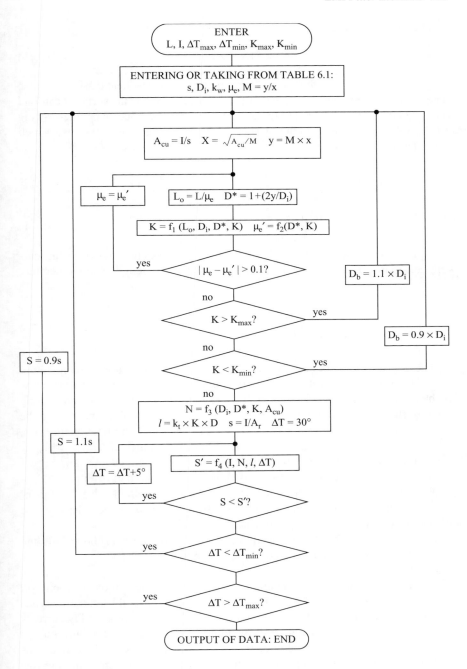

FIGURE 6.35 Flowchart for noise suppression choke coil design

The surface area responsible for the loss is approximately:

$$S = D \times \pi \times l \tag{6.70}$$

The loss occurs via radiation and heat transmission. The heat dissipation, also function of the temperature rise, can best be characterized by the surface heat-load (i.e., the ratio of the power loss to the heat dissipating surface). For free-flowing air, the relationship between the surface heat load and the temperature rise is well approached in the range of 30 to 150 °C by the following power series:

$$Q(\Delta T) = 4.56 \times 10^{-7} \times \Delta T^4 - 1.43 \times 10^{-4} \times \Delta T^3$$

$$+ 0.0167 \times \Delta T^2 - (0.665 \times \Delta T + 12.75) \tag{6.71}$$

Using Eqs. (6.68) through (6.71), a relationship can be created between the current density and the current rating, number of turns, coil length, and temperature rise:

$$s = \frac{57 \times l \times Q(\Delta T)}{I \times N \times (1 + \alpha \times \Delta T)} = f_4(I, N, l, \Delta T) \tag{6.72}$$

The above relationship can be used for the indirect calculation of temperature rise by determining the value of ΔT corresponding to the entered current density. Although this is most precisely done by iteration, we can shorten the computing time by stepping the value of ΔT by 5 °C (see Fig. 6.34). After determining the current density related to the calculated surface heat load, the temperature rise, ΔT, must be checked to determine whether it falls in the range of ΔT_{min} to ΔT_{max}. If not, the entire design procedure must be repeated, beginning with a new current density. The recommended procedure is shown in the flowchart of Fig. 6.35.

We can reduce the number of iteration steps using suggested values for the internal coil diameter, height/width ratio of the strip conductor, current density, and fill factor, k_w. These values are summarized in Table 6.1.

For designing solenoid-form, open iron core noise suppression choke coils, the following aspects should be considered (see also Section 9.7):

- The calculations give the length (l) of the coil. The length of the open iron core should be at least 20 to 40 percent longer than the coil length; otherwise, the stray effects on the ends of the iron core induce additional eddy currents in the conductor at the end of the winding. This can lead to coil burn-out.
- The current flowing through the noise suppression choke coils of semiconductor equipment usually has a significant harmonic content. The low-frequency harmonics can generate quite a bit of acoustical noise. This should be

TABLE 6.1 **Recommended Initial Data for Designing a Solenoid Noise Filter Choke Coil with High Current Rating**

Current Rating (A)	D_i (cm)	$M = y/x$	Current Density, s (A/mm²)
100 to 250	4	3	3
250 to 400	6	2	2.5
400 to 639	10	1.2	2
630 to 1,000	12	0.8	1.5

reduced, usually by special methods such as impregnating of windings or filling the entire choke coil with an appropriate material.

- A noise suppression choke coil for current ratings above 100 to 200 A must be made of parallel strip conductors. If such a coil is wound simply as single layers which are connected in parallel at the ends of the coil, the internal layer will burn out. This is caused by the unequal current distribution that arises because of the difference in resistance and inductance of the mutually isolated windings. Two solutions are commonly used to ensure an equal current distribution between the windings. One is to wind the coil of edge-bent, parallel-lead strip conductors; however, this is sometimes difficult to accomplish. The other solution is simply to place the single layers upon each other and change the single layers' symmetry at the halfway point in the coil; i.e., connect the bottom layer with the upper layer, and so on.

- It is advisable to calculate the magnetizing force (amp-turns) of a high-current, open-core noise suppression choke coil. If it is above approximately 1,000 A-turns, even the open iron core can saturate, causing an extreme increase in the iron loss. The iron core loss in saturation can be many times higher than the loss for copper. For reducing the magnetizing force, two noise suppression choke coils should be connected in series.

6.3.4 Common-Mode EMI Filter Choke Coils

The inductance of solenoid choke coils, even if they have an iron core, is rather small relative to their size. In EMI filters, high inductance is often required since the value of Y-capacitors is limited in many equipment to the nanofarad level; i.e., "low-leakage" power line filters are often needed (see Section 7.5.1).

The goal of creating a choke coil that displays a high impedance for common-mode noise components but a low impedance against differential-mode noise components was accomplished with the development of common-mode choke coils. Common-mode choke coils are often called *common-core* or *longitudinal* chokes. Mainly in Europe, they are also referred to as *balanced* or *current compensated* chokes. The common-mode choke coil consists of identical windings placed on a closed iron core. The windings are connected to the circuit so that, as seen in the case of the two-winding structure of Fig. 6.36, the magnetizing force of the sum of mains current (Φ_m) and the differential-mode noise current (Φ_d) cancel each other. That is why they are called *balanced* choke coils. Because the net mains frequency magnetizing force is very low, the iron core may be of high magnetic permeability, allowing a large common-mode inductance with only few turns. The iron core can be made without an air gap since the magnetizing force of common-mode current (i.e., the common-mode noise components $\Phi_{c1} + \Phi_{c2}$) is usually very low. Since the common-mode choke coil acts as a closed-core choke coil for common-mode components, the common-mode inductance can be calculated as follows:

$$L = \mu_o \times \mu_r \times N^2 \times A_{fe} l_{fe} \quad H \qquad (6.73)$$

where

A_{fe} = cross-section of the iron core in m^2

l_{fe} = average length of the induction lines in meters

μ_r = relative permeability of the iron core ($\mu_o = 4\pi \times 10^{-7}$ V/m^2)

The relative permeability of iron cores is not independent of the frequency; thus, the inductance of the common-mode choke coil also changes as a function of frequency. The relative permeability decreases with increasing frequency because of the HF electromagnetic fields which are excited by internal eddy current loops.

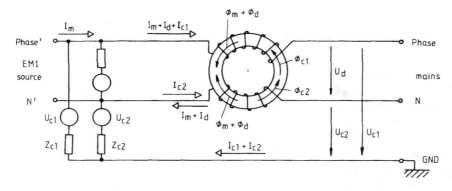

FIGURE 6.36 Magnetizing forces in common-mode choke coil

For normal silicon-iron core materials, the so-called eddy current frequency limit can be defined. At this frequency limit, the inductance of common-mode choke coil falls back to about 75 percent of the initial value. The eddy current frequency limit of transformer and other similar laminated silicon-iron core materials can be calculated with fair approximation as follows:

$$f_1 = \frac{4\rho}{\pi \times \mu_o \times \mu_i \times d^2} \tag{6.74}$$

where

ρ = specific resistivity of the iron core in Ω-m

μ_i = initial permeability of the iron core = $\mu_r(0)$

d = thickness of the plates in meters

The frequency limit of even the best quality transformer coil material is very low—generally not exceeding 50 kHz. Above this frequency, the inductance usually decreases very rapidly. Figure 6.37 shows the typical change of relative permeability of a normal SiFe transformer iron core material as a function of frequency. The dashed line refers to the real part of the permeability (see Section 6.3.5).

To increase the frequency limit, common-mode choke coils are usually made with ferrite cores. The permeability of ferrite cores also decreases with an increase in frequency, but for different reasons. For ferrite cores, a so-called *gyromagnetic* frequency limit can be defined (see Section 6.3.5). With fair approximation, it is:

$$f_g = \frac{5000}{\mu_i} \quad \text{MHz} \tag{6.75}$$

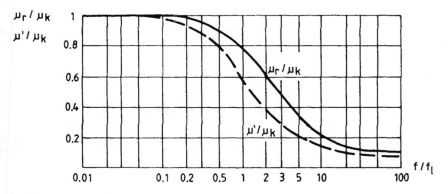

FIGURE 6.37 Relative permeability of laminated iron core vs. frequency

The gyromagnetic frequency limit of ferrite cores is in the megahertz order of magnitude. However, the inductance of ferrite common-mode choke coils decreases at gyromagnetic frequency to about the half of initial value. The frequency limit of ferrite cores is very high and can even exceed the megahertz level.

Common-mode choke coils are also produced for three-phase powerline filters. Figure 6.38 shows the scheme of common-mode choke coil for filtering the emission of an asymmetrical load. Mains frequency magnetizing force balancing is ensured by the winding being connected in the neutral conductor (N – N′).

To achieve a better HF performance, soft ferrite cores are used for low mains currents. Such ferrite cores are available in E-, U-, and spot core form. According to measurements [10], the common-mode choke coils with E- and U-cores show a very sharp resonance. The characteristic frequency functions of common-mode choke coils of various construction are shown in Fig. 6.39.

In case of perfect compensation, the inductance of common-mode choke coils might not depend on mains frequency currents, but perfection is not always achieved in practice. Figure 6.40 [10] shows the characteristic inductance change of identical common-mode choke coils as a function of mains frequency current.

Because common-mode choke coils are sensitive to external magnetic fields, they should be placed in a shielding cover whenever this is possible. The effect of external disturbing fields has been measured [10]. A common-mode choke coil with spot core was exposed to external magnetic field in the directions as shown in Fig. 6.41. The external magnetic field was excited by means of an open-core solenoid coil. The frequency of the excited magnetic field was changed in the range of 10 kHz to 1 MHz. The highest noise voltage (U_n in the figure) was received when the axis of the open core solenoid coil and the common-mode choke coil was the same (direction 1 in the figure). The measured noise voltage increased proportionally with frequency. Referring to this value as base, the noise voltage

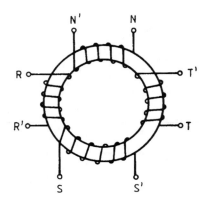

FIGURE 6.38 Common-mode choke coil for three-phase, four-conductor ac system

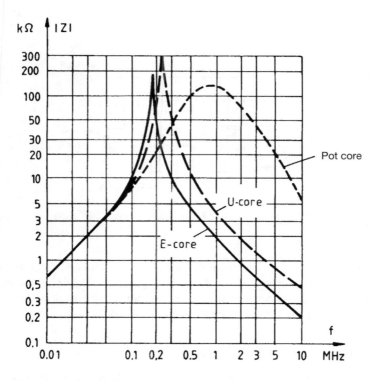

FIGURE 6.39 Impedance vs. frequency of common-mode choke coils with various ferrite iron cores

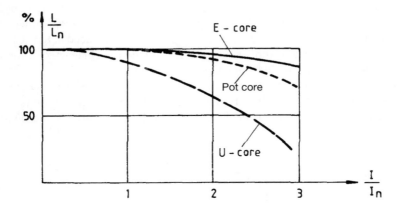

FIGURE 6.40 Impedance variation vs. load current of common-mode choke coils with various ferrite iron cores

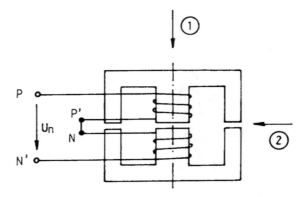

FIGURE 6.41 Measuring the sensitivity of a common-mode choke coil to disturbing magnetic fields [10]

became 20 dB lower when the axis of the open core solenoid coil faced in direction 2. Placing the common-mode choke coil in a grounded aluminium cover, the open core solenoid coil was adjusted again in direction 1. As a result of the shielding, the measured noise voltage decreased by about 30 dB.

6.3.5 Ferrite Cores and Ferrite Beads

To suppress noise signals above a few megahertz, ferrite beads are generally used. Ferrite beads have been produced for noise suppression since the middle 1950s. Small ferrite beads appeared first on the market, but by the 1980s they were available with cross-sections up to several square centimeters.

The ferrite beads used for common-mode choke coils are of toroid form but often placed only on the wire to be HF suppressed. In effect, a choke coil with one turn is formed. It is noteworthy that more wires can be also placed in ferrite beads. If a common-mode attenuation is desired, the directions of magnetizing force of the operating current should be in opposition and cancel each other.

The best way to study the characteristics of ferrite material is to analyze the impedance of a wire section with a ferrite bead. The HF equivalent circuit is shown in Fig. 6.42, where the inductor, L_f, represents the inductance of the one-turn

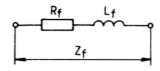

FIGURE 6.42 Equivalent circuit of ferrite bead

choke coil, and the resistor, R_f, represents the HF loss of the ferrite material. The impedance of the wire section with ferrite bead can be given as follows:

$$Z_f = \sqrt{R_f^2 + (2\pi f \times L_f)^2} \tag{6.76}$$

The inductance, L_f, of a wire section with a ferrite bead can be calculated using Eq. (6.73) by substituting $N = 1$.

The relative permeability of ferrite materials is dependent on frequency. The relative permeability of ferrite materials is complex and can be given in the following form:

$$\mu_r(f) = \mu^* = \mu'(f) - j \times \mu''(f) \tag{6.77}$$

In the above relationship, the real component (μ') represents the permeability of crystals, and the imaginary component (μ'') is in connection with the losses of magnetic rotation. The characteristic frequency function of the components μ' and μ'' is shown in Fig. 6.43. At low frequencies, μ'' is near zero; thus, the relative permeability is mainly determined by the component μ'. At higher frequencies, the

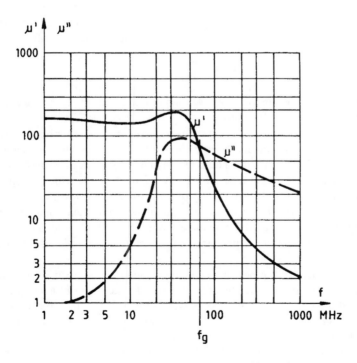

FIGURE 6.43 Real and virtual component of complex permeability vs. frequency

component μ'' becomes dominant and increases the value of resistor R_f, representing the magnetic losses of the ferrite material. The impedance, represented by ferrite beads, can also be expressed by the complex permeability:

$$Z_f = \mu''(f) \times 2\pi f \times L_f + J \times \mu'(f) \times 2\pi f \times L_f \qquad (6.78)$$

For choke coils made with ferrite beads, the Q-factor, defined by Eq. (6.25), is as follows:

$$Q_f = \frac{\mu'(f)}{\mu''f} \qquad (6.79)$$

The frequency at which the real and the imaginary components of the complex permeability are equal is called the *gyromagnetic frequency limit*. The Q-factor of ferrite choke coils is just unity ($Q = 1$) at gyromagnetic frequency. Above this frequency, the Q-factor decreases further. That is why the ferrite beads result in good HF damping (see Section 9.5.1) near and above the gyromagnetic frequency, and the choke coil will not resonate. The value of the gyromagnetic frequency depends on the ferrite material.

Rather than specifying the real and the imaginary components separately, catalogs and data sheets often give the absolute value of the complex permeability as a function of frequency. The absolute value of complex permeability is:

$$|\mu^*| = \sqrt{\mu'^2 + \mu''^2} \qquad (6.80)$$

Such curves are shown in Fig. 6.44. At the point marked $Q = 1$, the value of gyromagnetic frequency can be read.

There are basically two different types of ferrite materials for beads and choke coils: manganese-zinc and nickel-zinc alloys [242]. The former ones feature relatively high initial permeability. These ferrite materials generally can be used up to about 10 MHz. The nickel-zinc alloys have gained rather wider acceptance in EMC practice. The initial permeability of these materials is less than that of manganese-zinc alloys, but they can be used up to higher frequencies of approximately 100 MHz. The parameters of ferrite materials are rather highly dependent on temperature. The temperature dependence of nickel-zinc alloys is less than that of manganese-zinc alloys.

Two ferrite materials types can be distinguished as being resistive or inductive in character. The impedance characteristic of a ferrite material is determined by the ratio of the permeability components, $\mu'(f)$ and $\mu''(f)$. The typical impedance function of choke coils fabricated with resistive and inductive ferrite cores is shown in Fig. 6.45. Choke coils with inductive ferrite beads, completed with capacitors, are generally used for LC-filter circuits. Resistive ferrite beads are a reasonable choice for creating HF damping.

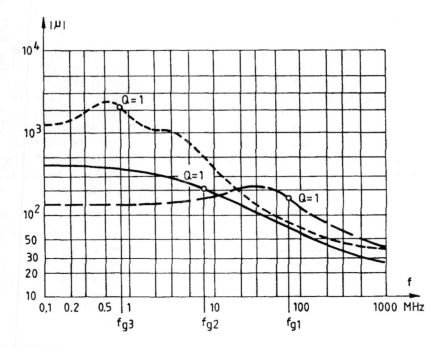

FIGURE 6.44 Typical complex permeability vs. frequency curves

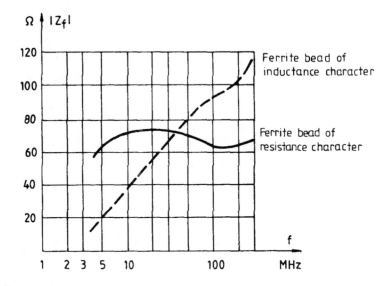

FIGURE 6.45 Typical impedance vs. frequency curve of inductive and resistive ferrite beads

6.4 RESISTORS

EMI filters contain not only capacitors and inductors but also resistors for HF damping. Depending on the loss power, metal film or wound resistors are used. Although the HF parameters of various types of resistors differ greatly, the equivalent circuit shown in Fig. 6.46 can be given for studying the HF characteristics. The value of capacitor C is in the range of 0.1 to 10 pF (μμF). The value of the serial inductance depends on the construction of the resistors. For metal film resistors, the parasitic inductance is limited practically by the parasitic inductance of the leads. The parasitic inductance of a wound resistor of conventional construction can be very high; therefore, only specially constructed resistors of low parasitic inductance should be used in EMI filters for HF damping.

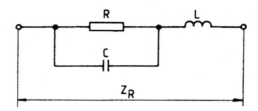

FIGURE 6.46 Resistor equivalent circuit

7

Noise Suppression

For most electrical equipment employing semiconductors, including the power electronic equipment, EMI must not exceed specific acceptable levels. For economic reasons, EMC should be considered early in the equipment design stage. As product development progresses from design to test and production, the range of available noise suppression techniques decreases steadily. As a first step, one should analyze the equipment's noise generation characteristics to determine the required level of EMI suppression. Because EMI filters are not cheap, it is best to avoid the practice of automatically designing and building an EMI filter into a system without any previous attention to its EMI characteristics.

Semiconductor circuits usually produce differential-mode noise components. Common-mode EMI can be very effectively controlled by reducing stray capacitances between the circuit and the grounded parts. First, reduce the stray capacitances between heat sinks and ground. In addition, EMI generated by switching processes can be reduced by proper design and also via the use of RC-snubbers. In many cases, suitable noise suppression can be achieved by applying EMI filter capacitors. The capacitor filter effectively short circuits the differential- and common-mode noise components close to their generation points.

7.1 NOISE SUPPRESSION IN RELAY SYSTEMS

It is well known that solenoid-operated control devices (i.e., relays) produce control and operational transients. Capacitor switching transients are primarily leading-edge phenomena, occurring when a switch is closed between a capacitor and a low-impedance voltage source. During the switching transient, a high inrush current can flow and a significant voltage drop will appear simultaneously across the wiring. It should be mentioned that inrush current may also be observed in noncapacitive devices such as incandescent lamps, motors, and ac contactors.

On the other hand, inductor switching transients are trailing-edge effects, occurring when a switch is opened to interrupt the inductance current. A very rapid voltage buildup occurs across the inductance in a direction that tends to prevent the interruption of the current. The peak voltage is limited only by the breakdown of either the air gap between the contacts or the inductor insulation.

HF disturbances produced by relay operation and coupled into the electronic unit can cause operational problems ranging from trifling annoyances to catastrophic malfunction and failure. Analysis of the problem is usually difficult, given that the switch noise characteristics are not precisely known. In this text, the detailed analysis is based on an ideal switch. The results are quite useful when properly applied to real switching situations. The examination showed that some basic rules can be formulated [58, 147, 198, 224, 331, 338]. However, the HF noise produced depends on the switching conditions and load characteristics, and repeated spark-over can be expected. The repetition frequency of the spark-over is in the range 10^4 to 10^7 Hz and continues up to approximately 0.1 ms. The impulse train is of sawtooth shape, of which the slope is determined by the distributed and lumped inductances and capacitances of the circuit and load.

In the moment of every spark-over, a damped oscillatory wave starts at a frequency determined by the inductance of the wiring and the stray capacitance of the load. The oscillation frequency can be in the range of 1 to 100 MHz, and the first amplitude can even reach a few kilovolts. Also, at the moment of interruption, the spark between the contacts appears as a damped oscillatory wave. The frequency of this wave depends on the parasitic inductance and capacitance of the wiring, and it can reach 10 to 1,000 MHz. The first amplitude is in the range of hundreds of volts at the relay contacts, and it decreases along the wire.

Noise suppression in relay systems is usually limited to reduction of the transient amplitude. The inrush current generally is controllable by serial impedance or other current-limiting methods. Increasing the internal impedance of the voltage source is not the best solution, since it can disturb the operation of other associated circuits. Figure 7.1 shows a solution for capacitive load switching in dc circuits without increasing the inrush current. The current peak is determined by the basic resistance and the current gain of the transistor.

Transients caused by switching inductive loads off are generally attenuated by reducing the decay rate of the magnetic field associated with the switched inductance. HF disturbances generated by switching dc inductive loads off may be suppressed by methods such as shown in Fig. 7.2 [338]. The decay time of the load current is longer with suppression than without. It may be seen that highly effective control of transient peak voltage is achieved with no major difficulties, so long as the load terminals are available. The transient voltage protection should be

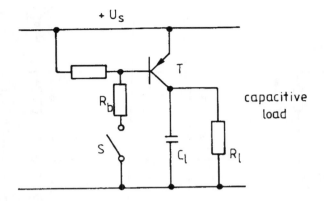

FIGURE 7.1   Switching transient reduction via control of the capacitive load inrush current

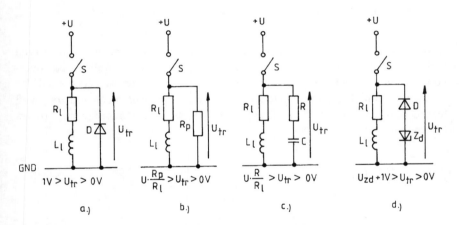

FIGURE 7.2   Inductive switching transient suppression in dc circuits

built up, as far as possible, on the terminals of the switched inductive load. If this is not possible, the protection can be placed on the relay contacts.

Transient overvoltage will be generated also by switching inductive loads in ac circuits. Overvoltage protection in ac circuits is a somewhat more difficult problem. Fig. 7.3 shows three ac circuit solutions that are essentially identical to those obtained for dc circuits. Effective transient suppression can be achieved by the application of solid-state relays (see Section 7.2).

For analyzing EMI from relay systems, it is not enough to study only the transient amplitude; the frequency spectra of the generated HF disturbances also should be examined [58, 198, 224]. A typical transient spectrum with various ov-

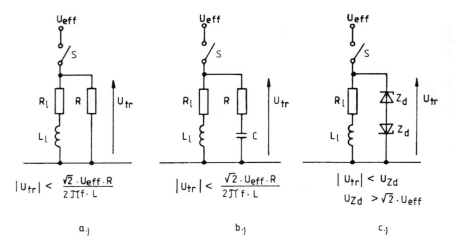

FIGURE 7.3   Inductive switching transient suppression in ac circuits

ervoltage protection schemes applied is shown in Fig. 7.4. Curve 1 was measured without any transient protection. Curve 2 was gained with a parallel diode overvoltage protection such as shown in Fig. 7.2a, and curve 3 uses a zener and diode connected parallel with the contacts as Fig. 7.2d. The measurements illustrate well that these overvoltage protections do not drastically reduce the frequency spectra of generated EMI signals.

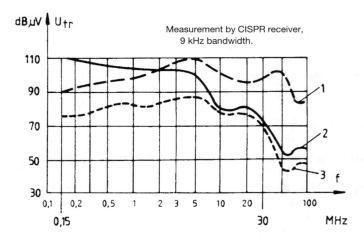

FIGURE 7.4   Spectrum of a switching transient (1) with no protection, (2) with protection as shown in Fig. 7.2a, and (3) with protection as shown in Fig. 7.2d

The overvoltage protections studied above did not drastically reduce the frequency span of the generated transients because they allowed rapid voltage changes on the load terminals. Two solutions for reducing not only the amplitude but also the frequency content of switching transients are shown in Fig. 7.5. The diode, D, provides overvoltage protection, and rapid voltage changes are prevented by the capacitor, C, which is connected in parallel with the switching contact. The spectra of the transient, measured after applying the two protections shown in Fig. 7.5, are shown in Fig. 7.6 [58]. Curve 1 was measured without any overvoltage protection, curve 2 with the circuit shown in Fig. 7.5a, and curve 3 with

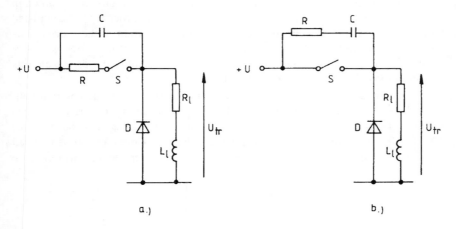

FIGURE 7.5 Transient protection on relay contacts (a) for amplitude and spectrum reduction and (b) for amplitude reduction only

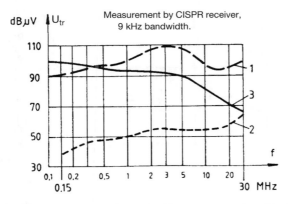

FIGURE 7.6 Spectra of switching transients (1) without any protection, (2) with protection as shown in Fig. 7.5a, and (3) with protection as shown in Fig. 7.5b

the solution as in Fig. 7.5b. As seen from the measured spectra, the protection shown in Fig. 7.5a is good for controlling overvoltage and frequency content. However, this solution is rarely used in practice since the load current flows though the protection resistor, which causes an additional loss. The commonly used parallel RC protection, shown in Fig. 7.5b, does not significantly change the frequency content of the generated EMI below 10 MHz because rapid voltage changes can appear on the contacts as a result of the serially connected resistor, R.

The amplitude and frequency content of switching transients can be reduced effectively by applying transistor circuits for controlling inductive dc loads; however, a proper driver circuit should be chosen for slowing down the switching procedure. Figure 7.7 shows the basic principle of two solutions, well-known in practice. In the driver circuits, the speed of the switching process can be adjusted by the capacitor, C_b. The spectra measured after application of these driver circuits are shown in Fig. 7.8 [58]. As a reference, curve 1 was measured with mechanical contact and no protection. Curves 2 and 3 employ the driver circuits shown in Figs. 7.7a and 7.7b, respectively.

We also must not neglect noise coupling between the contact pairs of multicontact relays [231]. Measurements have shown that, depending on the contact arrangement and impedance conditions, the coupled noise voltage is in the range of 10^{-1} to 10^{-5} times lower than the switching transients. Coupling can be reduced effectively by grounding the unused relay contacts between the one switching the

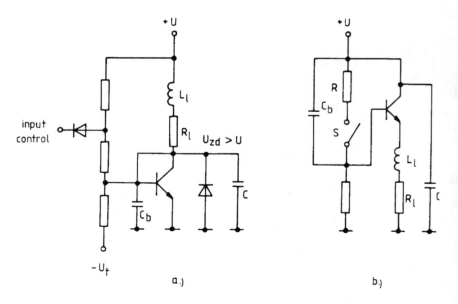

FIGURE 7.7   Inductive transient suppression by application of a transistor switch

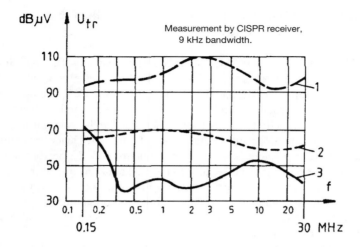

FIGURE 7.8 Inductive transient caused by transistor switching in dc circuits by (1) mechanical contacts, (2) a transistor switch as shown in Fig. 7.7a, and (3) a transistor switch as shown in Fig. 7.7b.

inductive load and the one associated with sensitive circuits. Measurements also indicate that, via the employment of proper RC networks for damping the oscillations, noise coupling between the contacts pairs is greatly decreased.

7.2 APPLICATION OF AC SWITCHING RELAYS

Since the contacts of electromechanical switches close and open stochastically in relation to the period of the ac supply voltage, worst-case conditions should be presumed. When switching an inductive load, the highest EMI is produced at switch-off at maximum current. When switching resistive and capacitive loads, the worst case is the switch-on at maximum voltage.

HF noise generation by solid-state relays in ac circuits can be much lower than that of electromechanical switches, if a suitable control method is applied. One of the worst switching cases is automatically precluded, since the SCRs and triacs interrupt the current only at zero-crossover.

Another useful EMI reduction method is the zero-crossover technique [224]. The zero-crossover switching technique means that, independent of the input control signal, relay turn-on occurs only near the zero crossing of the supply voltage sine wave. The switching transients of solid-state relays, controlled with zero-crossover technique, can be even 50 to 80 dB lower than those of electromechanical relays. Solid-state relays with zero-crossover control units are marketed in wide range of styles.

7.3 APPLICATION OF RC-SNUBBERS TO POWER SEMICONDUCTORS

Rectifier circuits produce EMI primarily at diode switch-off, while SCR control circuits do it at switch-on (see Section 5.1). EMI generation can be reduced by connecting a capacitor in parallel with the rectifier. Figure 7.9 demonstrates the effect of a parallel capacitor. The EMI measurements were made with a half-wave circuit connected directly to the mains [10]. In practice, however, it is not advisable to connect only a capacitor parallel with the rectifier since, in conjunction with the leakage inductances, oscillations can be formed. To prevent oscillation, a resistor should be connected serially to the capacitor. Although the noise suppression effect of the RC-snubber is inferior to that of a capacitor, the EMI of a rectifier with an RC-snubber will be much lower than without. Suggested values for RC networks that are also suitable for noise suppression are included in semiconductor catalogues and data sheets.

RC-snubbers are less effective in SCR circuits than in rectifier circuits. Specifically, SCRs produce higher EMI at switch-on, and RC networks serve mainly for suppressing HF noises generated at switch-off. Instead, RC-snubbers reduce EMI in the range of a small firing angle. Figure 7.10 shows the measured spectra of a full-wave, half-controlled SCR circuit with and without an RC network. The measurements were made with firing angle of 90°. It is to be noted that the EMI of an SCR control circuit with RC-snubbers can even be slightly higher in the frequency range above a few megahertz than without snubbers.

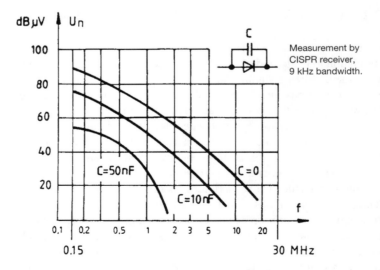

FIGURE 7.9 Spectra of rectifier disturbances as a function of parallel capacitor value

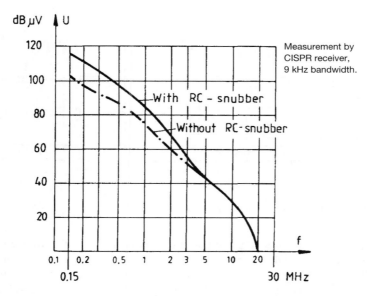

FIGURE 7.10   Spectra of an SCR control as a function of RC-snubber application

7.4   SHIELDED TRANSFORMERS

Generally, the rectifier and converter circuit is supplied through a transformer. The HF impedance of the transformer can be utilized for noise suppression as a serial/parallel impedance, since it is in the equivalent circuit between the EMI source and the mains. The HF impedance, and so the insertion loss of the transformer, depends on the winding structure and (eventually) applied shieldings.

The equivalent circuit of a transformer, reduced to the secondary side, is shown in Fig. 7.11, where r represents the ratio of primary and secondary turns. This equivalent circuit is valid in a very wide frequency range. The resistor, R_m, representing the iron losses, is not constant, and its frequency function can be described after further approximation. For studying the insertion loss of the transformer, the variation of this impedance in frequency dependence usually can be neglected.

At lower frequencies (up to about 10 kHz), the leakage inductances and parasitic capacitances of the windings are negligible. In this frequency range, the insertion loss of the transformer is determined by the winding and iron core losses. In most practical cases, the insertion loss is so small that it can be neglected, since the internal impedance of the noise source (Z_n) and the HF impedance of the mains (Z_m) are significantly higher than the serial transformer impedance.

With increasing frequency, the effects of leakage inductances and parasitic capacitances grow. In the middle frequency range, the insertion loss of the transformer is determined by the primary and secondary leakage inductances and wind-

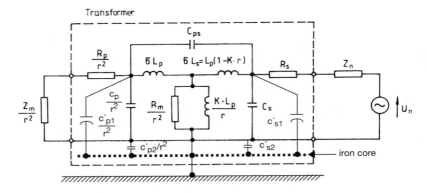

FIGURE 7.11 Transformer equivalent circuit

ing losses. The correct description of the insertion loss in this frequency range is very difficult and needs more approximation. The reason for this is that the ratio of the transformer can no longer be calculated as the ratio of the turns, because the flux of the iron core is decreased by eddy current and hysteresis losses. In this frequency range, the ratio of the transformer can be approximated as follows:

$$r = \sqrt{\frac{L_p}{L_s}} \tag{7.1}$$

With a further increase in frequency, the capacitive coupling between the windings continues to grow. As we approach megahertz frequencies, it becomes determinant. In the high frequency range, the ratio of the transformer can be expressed approximately as follows:

$$r = \frac{U_p}{U_s} = \frac{C_{ps} + C_s}{C_{ps}} \tag{7.2}$$

The parasitic capacitances ($C_{p1'}$, $C_{p2'}$, $C_{s1'}$, and $C_{s2'}$) of the winding to the core also play a role in the common-mode insertion loss, but they are too small to be significant up to approximately 10 MHz.

In the HF range, the equivalent circuit of Fig. 7.11 can be simplified into the one shown in Fig. 7.12. The insertion loss in this frequency range is basically determinant by the parasitic capacitances of the winding and the capacitive coupling between the windings. This parasitic capacitance shunts out the leakage inductances; however, for good noise suppression, a high serial impedance would be needed.

The insertion loss of transformers in the higher frequency range can be increased by a winding structure with higher leakage inductance and parasitic coil capacitances, but with smaller coupling capacitance between the windings. Sev-

Transformer

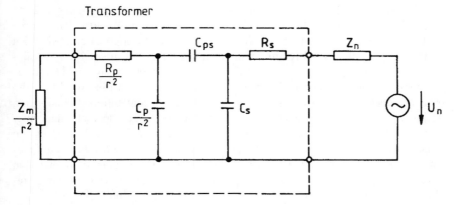

FIGURE 7.12   HF transformer equivalent circuit

eral solutions are well known for increasing leakage inductance and stray capacitance. However, it may be worthwhile to examine methods for decreasing the coupling capacitance.

The capacitive coupling between the primary and secondary windings can be best reduced via a shield between them. The equivalent circuit of a transformer with a shielding foil is shown in Fig. 7.13 [174]. The coupling capacitance, C_{ps}, theoretically could be decreased to zero by means of a grounded shield. In practice, this parasitic capacitance cannot be totally eliminated. The shielding between the primary and secondary windings is usually made of a one-layer winding or of an isolating sheet covered on one side by a conducting layer. This layer is generally takes the form of a coat of a paint with good conductivity, or a copper or aluminum film. Remember that the shielding coil should not forme a closed loop.

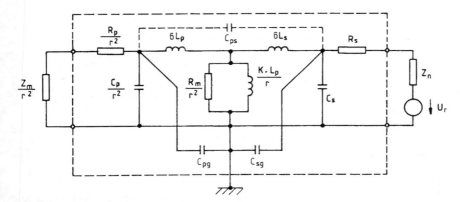

FIGURE 7.13   Shielded transformer equivalent circuit

After applying shielded transformers in power electronic equipments, a 20 to 40 dB EMI reduction can be expected. The shielded transformer also offers some protection to semiconductors against mains transients. The application of shielded transformers to meet EMI requirements is also very useful in small power supplies. The noise suppression effect of shielded transformers is illustrated by insertion loss measurements (per MIL-Std. 461 and 462), as shown in Fig. 7.14. Two transformers of about 1.5 kW/400 Hz were under test: a conventional transformer, designed only to provide the output dc voltage (curve 1), and a special transformer with shielding and increased reactances for providing the best EMI suppression effect (curve 2). The insertion loss measurements clearly show the benefit of shielded transformers. As the cost of EMI transformers are not much higher than of standard transformers, considerable savings in costs, size, and weight can be realized, and EMI requirements can be met without the use of powerline filters.

7.5   CAPACITOR FILTERS

In many cases, EMI can be suppressed to acceptable levels through the use of capacitive filters only. Filter capacitors, also called *noise suppression capacitors*,

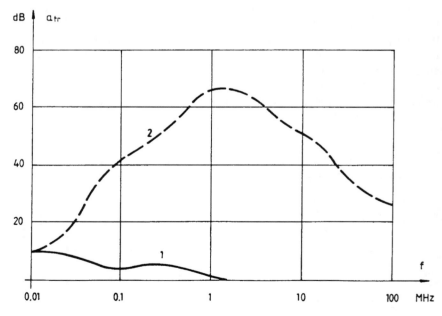

FIGURE 7.14 Insertion loss of (1) conventional transformer and (2) special transformer for differential-mode EMI reduction

are connected to various strategically chosen points in the electrical circuit. They are used in semiconductor equipments and other high-power electrical systems for which a serial noise suppression choke coil would be difficult to build and install. For economic reasons, capacitive filtering is advisable when a high level of EMI suppression is not required. Noise suppression up to about 30 to 40 dB can be achieved with capacitive filters.

Capacitive filtering will be discussed on the basis of the equivalent circuit shown in Fig. 7.15. In this figure, Z_g represents the impedance between the noise source and the capacitor filter, and Z_1 represents the HF impedance to be measured from the capacitor filter to the mains ground. Capacitive filtering is effective only if the following condition exists in the frequency range (0.15 to 30 MHz) to be suppressed:

$$Z_g \gg Z_C \ll Z_1 \qquad (7.3)$$

Since the impedances Z_g and Z_1, across the frequency of interest, are often unknown to the designer, in most cases, the proper capacitor filtering approach is derived from measurements. But a basic philosophy can be given for choosing the connection points for noise suppression capacitors.

Capacitor filters are most frequently connected to the input and output of power electronic equipment. These capacitor filters are star-connected and delta-connected noise suppression capacitors. When choosing the value of Y-capacitors (capacitors that are mounted line-to-earth), special attention should be paid to ground currents to avoid electric shocks to personnel (see Section 7.5.1). The noise suppression capacitor should be placed near the point to be filtered and connected with the shortest wires possible (see Section 6.2).

Certain elements of the electrical circuit (e.g., transformer, choke coil, and longer wires) represent a high enough serial impedance for noise suppression in the frequency range of 0.15 to 30 MHz. For selecting the points for capacitor filtering, it is reasonable to consider and utilize this impedance.

Noise suppression capacitors in conjunction with circuit inductances can form high-quality oscillators. As an effect of these oscillations, EMI in some narrow

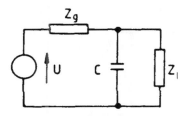

FIGURE 7.15   Capacitor filter equivalent circuit

frequency ranges can even be increased. To avoid oscillation, HF damping is advisable (see Section 9.5).

7.5.1 Noise Suppression with Y-Capacitors

Common-mode EMI components of the power electronic equipments are usually suppressed by so-called Y-capacitors, which are connected between phase/neutral and earth. Figure 7.16 shows the problem of connecting high value Y-capacitors for EMI suppression of electrical equipment supplied from the mains. The resistor R_i represents the leakage resistance of the insulation. Capacitor C_i represents the stray capacitances. In the case of a good insulation, the leakage currents flowing through the parallel impedance of R_i and C_i are inconsiderately small. The Y-capacitor is connecting parallel to C_i. As a result, a much higher current than the leakage current of the insulation flows through the ground wire. This ground current can be calculated as follows:

$$I_g = \sqrt{I_R^2 + (I_C^2 + I_y^2)^2} \approx I_y \tag{7.4}$$

Since the current and value of the Y-capacitor are determined by the mains voltage and frequency, the ground current can be calculated with good approximation as follows:

$$I_g = U_m \times 2\pi f_m \times C_y \times 10^{-6} \quad \text{mA} \tag{7.5}$$

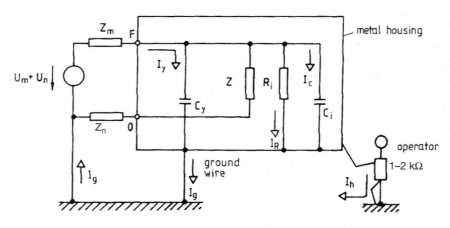

FIGURE 7.16 Equivalent circuit for determining the capacitive (leakage) current of noise suppression

where

U_m = mains voltage in volts

f_m = mains frequency in hertz

C_y = capacitance of the Y-capacitor in nanofarads

For many electrical equipment, such as portable electrical tools and other consumer appliances connected to the mains, an eventual break in the ground wire can be presumed. Touching such a faulty electrical equipment may cause current to flow through the human body rather than the ground wire as intended. In terms of impedance, the human body is regarded as a 1 to 2 kΩ resistor. Since this resistance is smaller than the reactance of the Y-capacitor, the current flowing through the human body (I_h in Fig. 7.16) is, with good approximation, equal to the current of Y-capacitor. Human health is endangered by a relatively small current intensity. Above 30 mA, the probability of mortal electrical shock must be considered, but even a current of about 10 mA can cause serious spasms. The current level that constitutes a health threat is a matter of disagreement, since its value depends on many factors, including the physical condition and age of the recipient, and other circumstances.

The acceptable capacitive current limits for various grounded devices are recommended by national standards. A short survey of acceptable capacitive current (I_g) in some countries is given in Table 7.1. (Class II equipment, without an earth wire, is not included in this table but can be found in various sources listed in the bibliography.) The limits depend on the type of equipment and operating condi-

TABLE 7.1   Acceptable Leakage Currents in Various Countries

Country	Standard	Leakage Current (mA)			
		Equipment		Components	
		Portable	*Fixed*	*Touchable*	*Not Touchable*
U.S.A.	UL 478	Max. 5 mA @ 120 V/60 Hz			
	UL 1283	0.5 to 3.5 MA @ 120V/60 Hz			
Canada	CSA 22.2 No. 1	Max. 5 mA @ 120V/60 Hz			
International	CEE 10 (household				
	motor-driven devices)	0.75	3.5	0.25	5.0
	CEE 11 (household heating)	0.75	5/kW	—	—
	CEE 12 (portable machine				
	tools)	0.5	—	0.5	3.5
Germany	VDE 0720 (household				
	heating)	3.0	0.75/kW	0.5	—
	VDE 0730 (household		3.5		
	appliances)	0.5	—	0.5	—
	VDE 0740 (portable tools)	0.5	3.5	0.5	3.5
	VDE 0875 (noise filtering)	0.5		0.1	3.5

tions. The most of these standards classify electrical equipment in three groups. For portable tools, the capacitive current usually cannot be higher than 0.75 mA. For consumer devices that are fixed and supplied through a connector, the capacitive current can be as high as 3 to 5 mA. In some cases, such as for electrical heating and electric stoves, a leakage current of 10 mA is also acceptable.

The above-mentioned restrictions for capacitive current maximize the value of Y-capacitors. For quick estimation, the maximum value of Y-capacitors can be approximately calculated as follows:

$$C_s = \frac{I_g}{U_m \times 2\pi f_m} \times 10^6 \quad \text{nf} \tag{7.6}$$

where

U_m = the mains voltage in volts

f_m = the mains frequency in hertz

I_g = the acceptable leakage current in milliamps

By the above relationship, we can determine the maximum allowable value of Y-capacitors in portable electrical equipment, which belongs to the strictest safety class. For a 127 V, 60 Hz ac power mains, the limit is about 5 nF. For a 220 V, 50 Hz ac power mains, it cannot exceed about 3 nF.

7.5.2 Designing Capacitor Filters

EMI generated by power electronic equipments is differential-mode in nature, with common-mode components originating as a result of secondary, usually parasitic, effects. Calculating common-mode EMI values is not so straightforward as with differential-mode components (see Section 5.3), but common-mode component generation modes are worth studying for correct EMI suppression design. In many cases, it is more economical and equally effective to use capacitor filters or to make small layout modifications, as opposed to applying EMI filters.

First, we will study how common-mode noise components originate from an EMI source producing differential-mode emissions [9]. In the equivalent circuit of Fig. 7.17, Z_g is the internal impedance of the noise generator, Z_{p1} and Z_{p2} are the lumped representations of distributed stray capacitances between the circuit and the metal housing, and Z_1 and Z_2 are serial impedances between the noise source and the output terminal of the electrical equipment. These impedances also include the serial impedance of EMI filter, which is occasionally applied. Z_x represents the impedance of the X-capacitor, and Z_y is that of the Y-capacitor. The impedance Z_y also includes the grounding impedance, but this is usually negligible in comparison to Z_y. The common-mode noise components can be calculated

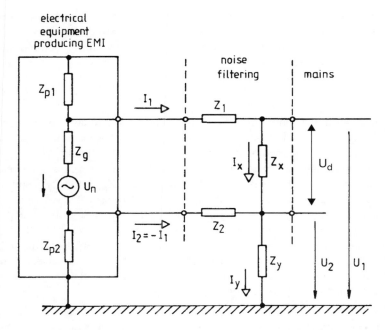

FIGURE 7.17 Origin of the common-mode noise component from a differential-mode noise source

by neglecting the mains impedances, which is a fair approximation in the frequency range of 0.15 to 30 MHz. Using the definition shown in Eq.(4.2), the common-mode noise voltage is as follows:

$$U_c = I_y \times Z_y + \frac{U_d}{2} = I_y \times Z_y + \frac{I_x \times Z_x}{2}$$

$$= U_n Z_y \times \frac{Z_2 \times Z_{p1} - (Z_1 + Z_2) \times Z_{p2}}{Z_g Z_2 (Z_1 + Z_x + Z_y + Z_{p1}) + (Z_1 + Z_x)(Z_g + Z_2)(Z_{p2} + Z_y)}$$

$$\times \frac{1}{(Z_1 + Z_g + Z_2 + Z_x)(Z_{p2} Z_y + Z_{p1} Z_{p2} + Z_{p1} Z_y)} + \frac{U_d}{2} \qquad (7.7)$$

The above relationship is too cumbersome to be useful for estimating the effect of the impedance change or to draw any conclusions for designing the EMI suppression. In most practical cases, it can be further supposed that the differential-mode noise voltage is not the dominant concern (relative to the common-mode value) since there are generally no barriers to increasing the value of X-capacitor (within reasonable limits, of course). With this assumption, the common-mode noise voltage will be zero if:

$$Z_2 \times Z_{p1} = (Z_1 + Z_x) Z_{p2} \qquad (7.8)$$

Assuming the above condition, the common-mode noise voltage component can be made equal to zero, independent of the value of the Y-capacitor. This fact indicates that a well designed X-capacitor can be more effective in noise suppression than a high-value Y-capacitor. In practice, of course, meeting the above condition can present difficulties, since the impedances Z_{p1} and Z_{p2} result from distributed capacitances. Compensation can be made most effective when these impedances (or at least one of them) are determined by a discrete capacitor or some characteristic parasitic capacitance. It is reasonable to try to meet the condition, expressed by Eq. (7.8), by designing a proper serial impedance in the EMI filter to force Z_1 and Z_2 to increase and help the equality. This means that, *in some cases,* noise suppression devices should connect to the neutral conductor and not the phase, contrary to usual practice.

7.6 EMI GENERATION AND REDUCTION AT ITS SOURCE

As described in Section 7.5.2, common-mode EMI is usually the result of parasitic effects. In semiconductor circuits, there are two major circuit elements that presumably are responsible for common-mode EMI: the heat sinks and the transformer. The use of proper assembly layout techniques can be very effective in controlling common-mode noise components.

We will discuss the relationship between assembly technology and common-mode EMI production mainly in terms of transistorized circuits, but the result can be used for diodes and SCR circuit design as well.

7.6.1 The Role of Heat Sinks in EMI Generation

Power semiconductors are usually mounted on heat sinks. A significant capacitance can be formed between the heat sink and a metal base (ground). This stray capacitance, in general, is especially high and plays a major role in production of common-mode noise components.

Study the common-mode EMI in a transistorized power supply as shown in Fig. 7.18. The metal case of the power transistor is usually connected to the collector. For mounting such a power transistor onto a heat sink, the metal case should be insulated from the grounded heat sink by a thin mica layer. Because of the thin insulating layer and the relatively large surface area, the stray capacitance

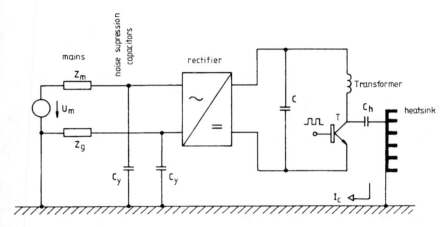

FIGURE 7.18 Origin of common-mode current through heat sink parasitic capacitance in an A/D converter

toward the ground (C_h) can reach approximately 100 pF. In switched-mode power supplies, this capacitance couples the harmonics of the switching waveform into the ground line as common-mode noise current, I_c. This common-mode current can cause a relatively high common-mode noise voltage across the suppression capacitors, C_y. The approximate value of the common-mode noise voltage can be simply calculated whereby the emitter-collector voltage is divided between the stray capacitance and the sum (parallel resultant) of noise-suppression capacitors. As the value of Y-capacitors is often limited to some nanofarads, the common-mode noise voltage can be unacceptable high (see Section 8.2.2 for more details).

One solution for suppressing the common-mode noise voltage is to use a transistor whose emitter is connected to the metal case. This ensures that the stray capacitance between the collector and ground, which changes with high amplitude and slope, is significantly reduced. Another solution is the use of a pnp-type power transistor. With this approach, the voltage of the switching waveform can be considerably reduced on the high side of the stray capacitance versus earth.

One of the most efficient suppression solutions is achieved by reducing the value of the stray capacitance. Two methods are known. One method is to avoid grounding the heat sink and, alternatively, to connect it to the transistor emitter as shown in Fig. 7.19. This ensures that the noise current, I_n, flowing through the capacitance, C_{h1}, remains mostly in the primary circuit and is kept from flowing into the mains. The common-mode noise current flowing through the capacitance C_{h2} is much smaller than the current shown in Fig. 7.18, because capacitance C_{h2} in Fig. 7.19 is much smaller than capacitance C_h in Fig. 7.18. The heat sink could be connected to one of the supply lines that produce the above-described effect, but technical and safety considerations usually prevent the application of these solutions.

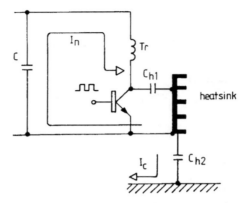

FIGURE 7.19 Reducing common-mode current by isolating the heat sink

The stray capacitance between the transistor collector and the ground can be re-duced by applying a screen between the metal case of the power transistor and the heat sink. The screen is to be insulated both from the transistor case and heat sink, and it must be made of material with good thermal conductivity. The screen is usu-ally connected to the transistor emitter as shown in Fig. 7.20. This solution is the subject of an Advance Electronics, Ltd., patent. This screening has the same effect as illustrated in Fig. 7.19. The high-pass circuit formed by C_{h1}/Z_s is very efficient in bypassing the capacitive current back to the emitter, provided Z_s is as small as possible.

The above stated principles are also valid for SCR circuits. As a general rule, it can be stated that the stray capacitance between (a) the circuit elements of which potential changes with high amplitude and high slope and (b) ground should be minimized. With proper design of the assembly, EMI can reduced by about 20 to 40 dB.

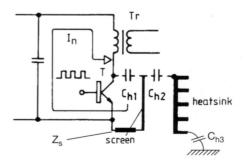

FIGURE 7.20 Reducing common-mode current by screening the power transistor

7.6.2 The Role of Transformers in EMI Production

In switched-mode power supplies and similar transistorized circuits, aside from heat sink stray capacitance, transformer parasitic capacitance is the main common-mode EMI source. In most practical cases, the iron core of the transformer and the secondary circuit are grounded. The layout of such a power supply is shown in Fig. 7.21. The parasitic stray capacitances between the transformer windings, C_w, as well as capacitances between the primary winding and the iron core, C_i, run in parallel with the heat sink stray capacitance. That means that the transformer parasitic capacitances have the same effect on common-mode EMI production as the heat sink stray capacitance, but this EMI now appears on the secondary (dc) side.

The unwanted inter-winding capacitance (C_w) in the output transformer can be reduced by placing a screen between the windings. The screen, for example, may consist of a thin layer copper sheet [140, 255, 308]. The screen causes the capacitive current from the primary to return harmlessly to the supply lines. For low output voltages, this is an adequate solution, but for higher output voltages, the stray capacitance between the screen and the secondary winding can produce common-mode EMI currents. To circumvent this current path, a second screen becomes necessary. The usual connection of the screens is shown in Fig. 7.21. In this manner, capacitive currents generated in the primary by the switching waveform are returned to the primary, and similar currents generated in the secondary are returned to the secondary. Notice that, dynamically, the hot side on the primary is the transistor end of the winding.

The stray capacitance between the transformer primary and its iron core can also produce excessive common-mode EMI in the manner previously described,

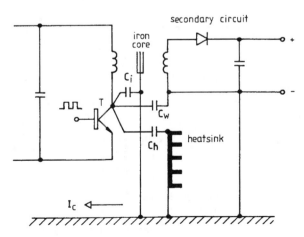

FIGURE 7.21 Origin of common-mode current through the transformer parasitic capacitance in a switch-mode converter

if the iron core is grounded. Although the stray capacitance, C_w, is usually much higher than C_i, common-mode EMI can be suppressed by not grounding but connecting the iron core electrically to the positive supply line as shown in Fig. 7.22. It should be noted that the insulation of the iron core and connection to a potential, which is in some cases even higher, can cause technical and safety problems. It is to be noted that, because of the ratio of stray capacitances, a relatively small increase in common-mode EMI should be expected by grounding the iron core of the output transformer.

The same suppression technique needs to be applied to every base-drive and feedback transformer that carries switching waveform.

7.7 INFLUENCE OF LAYOUT AND CONTROL OF PARASITICS

Transistorized switched-mode power supplies, commonly used in SCR power electronic equipments as well, can produce higher EMI levels on the mains terminals than the SCR circuits—even those with high power ratings. It is not unusual to find that noise suppression must be improved or modified solely because of the EMI from switched-mode power supplies. In such cases, rather than increasing the insertion loss of the powerline filter, the generation of common-mode noise components should be analyzed and prevented. In addition to the noise suppression methods detailed in Sections 7.6.1 and 7.6.2, some further techniques should be considered.

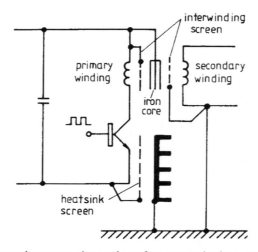

FIGURE 7.22 Proposed power transistor and transformer screening in an ac-to-dc converter

The EMI of switched-mode power supplies often exhibits an increase above a few megahertz. This occurs because of the very high slope of the recovery current of the rectifiers (see Section 5.1.1) in the primary and secondary circuits. If the EMI of switched-mode supplies exceeds the acceptable limit only beyond the megahertz frequency range, it is reasonable to begin our noise suppression efforts by trading the rectifiers for soft-recovery types.

The EMI of switched-mode power supplies also can be influenced by designing the circuit elements to meet EMI requirements. By means of theoretical calculations and detailed measurements, the relationship between the generated EMI and the characteristics of certain circuit elements has been studied [129, 225, 228, 229]. Although the subject of these examinations was the noise appearing on the output terminals, a brief review of the results can be useful because there is a close relationship between differential- and common-mode noise.

The analyses and measurements of noise generation were made using the scheme of the switching regulator shown in Fig. 7.23, and also using a forward dc/dc converter [225]. These commonly used configurations are excellent choices for analysis because of characteristics such as their high voltage ratio and isolation of the primary and the secondary circuits. Capacitor C_g represents the stray capacitance between the primary circuit and ground. The parasitic capacitance between the transformer windings is represented by C_w, drawn with a dashed line. The choke coil, L_w, also drawn with a dashed line, represents the inductance of wiring to the load, and choke coil L_g is the parasitic inductance of the grounding.

First of all, the analyses and measurements showed that, for the two configuration, higher output noise voltage will be generated at transistor turn-on; thus the

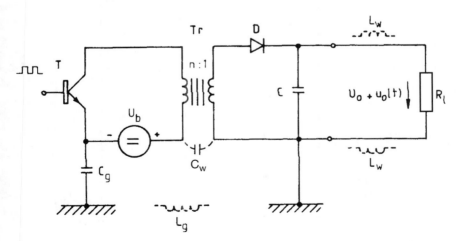

FIGURE 7.23 Forward dc-to-dc circuit scheme for examining output noise voltage vs. circuit parameters

spectrum of the output noise is also determined by the turn-on noise. This means that the switching speed of the transistor has a severe influence on the output noise voltage, but no obvious connection exists between the storage capacitance of the transistor and the output noise voltage.

The measurements were made using a 20 kHz converter configured as shown in Fig. 7.23 and supplied from 6 Vdc. The power rating was a few watts. The secondary circuit was grounded near to the load. The parameters of the measured dc/dc converter were $L_w = 0.35 \mu H$, $L_g = 2 \mu H$, $C_w = 0.6$ nF, and the stray inductance of the transformer $(L_s) = 6.5 \mu H$. The storage (depletion) capacitance (C_d) of the applied rectifiers was 0.2 nF. These values can be considered typical. Figure 7.24 gives the measured peak-to-peak output noise voltage as a function of the stray capacitance toward the ground (C_g). This measurement was made with two different transformer constructions; i.e., with two different transformer parasitic capacitances. The curves clearly show that the output noise increases rapidly with C_g, if it is small.

Figure 7.25 shows the relationship between the parasitic ground inductance and the peak-to-peak output noise voltage. The measurements illustrate well that the inductance in the ground path decreases the output noise voltage; however, if the depletion capacitance of the rectifiers is higher than the stray capacitance between the transformer windings (i.e., $C_w < C_d$), a peak appears for a certain value of the ground path inductance.

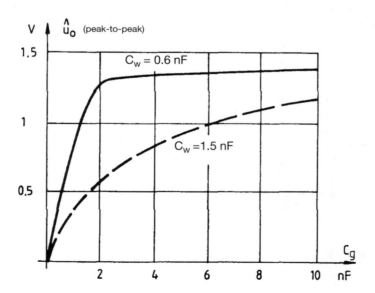

FIGURE 7.24 Output noise voltage amplitude vs. parasitic grounding capacitance; parameter C_w = parasitic capacitance between the transformer coils

In Fig. 7.26, the effect of stray capacitance between the transformer windings is shown. As seen from the figure, the relationship between the windings' stray capacitance and the output noise depends on the depletion capacitance of the rectifiers. The effect of stray capacitance C_w is weak for a rectifier with small depletion capacitance (i.e., for $C_w > C_d$) and is strong in the opposite situation (i.e., $C_w < C_d$). It was demonstrated that the increase in depletion capacitance of the rectifiers is reduced but, conversely, the increase in the stored charge of the rectifiers also produces an increase in output noise voltage. These two effects are opposite and nearly equal; therefore, the output noise voltage is only slightly affected by the depletion capacitance of the rectifiers [225]. The rectifiers have an additional effect on the output noise voltage. The output noise voltage increases with the output voltage, so the reverse current of the rectifiers will be also increased.

The measured results (Fig. 7.27) clearly indicate that the parasitic wiring inductance, L_w, increases the output noise voltage. In real circuits, this parasitic inductance cannot be reduced below a given limit. Here, we note that the output noise voltage can be effectively reduced by a common-mode choke coil connected as shown in Fig. 7.28 (see Section 9.4). Noise suppression can be improved by applying an iron core with high eddy current loss and closely coupled windings [228].

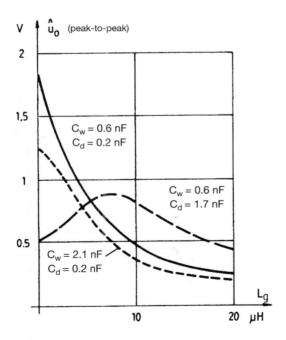

FIGURE 7.25 Output noise voltage amplitude vs. parasitic grounding path. inductance; $C_w =$ parasitic capacitance between transformer coils, and $C_d =$ depletion capacitance of the rectifiers

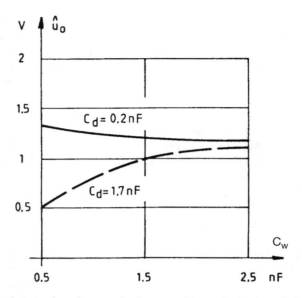

FIGURE 7.26 Output noise voltage amplitude vs. parasitic transformer interwinding capacitance

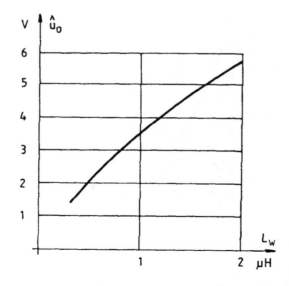

FIGURE 7.27 Output noise voltage amplitude vs. parasitic inductance of the wires to the load

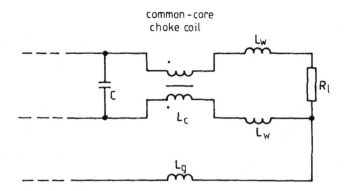

FIGURE 7.28 Noise suppression in a dc-to-dc converter with a common-mode choke coil; L_c = inductance of the common-mode choke coil, L_w = parasitic inductance of the wires to the load, and L_g = parasitic inductance of the ground path

For the relationship between the leakage inductance of the transformer (L_s) and the output noise voltage, only theoretical results are presented in Fig. 7.29, because it was impossible to separate the effects of leakage inductance and stray capacitance in the measurements. It will be predicted from the mathematical analysis that the increase in leakage inductance results in the decrease of output noise voltage as a whole, and that the worst case is the small leakage inductance combined with relatively low stray capacitance between the windings and relatively high depletion capacitance of the rectifiers.

Although the above detailed result were obtained by measurement with a dc/dc converter as shown in Fig. 7.23, the characteristic of relationships between the parameters of certain circuit elements and the peak-to-peak value of the output noise voltage is the same for forward dc/dc converters [225].

In summary, it can be stated that several parasitic phenomena, and the parameters of certain circuit elements, play an important role in the generation of noise voltage on the output of switched-mode power supplies. In general, it is worthwhile to reduce the stray capacitance between the transformer windings and to select a rectifier type with large depletion capacitance and small stored charge.

The HF disturbance on the output and mains terminals can be very effectively reduced by a well designed wiring arrangement and layout [140, 308]. The principles described in Section 7.6 should be applied to every part of a practical circuit that carries a switching waveform. These techniques include minimizing inter-wiring capacitances and any stray capacitances toward the ground. In addition to capacitive coupling, all unwanted inductive coupling between the circuit carrying switched waveforms and connected sensitive units must be minimized (see Chapter 11). Simple examples for bad and good wiring layouts are shown in Fig. 7.30.

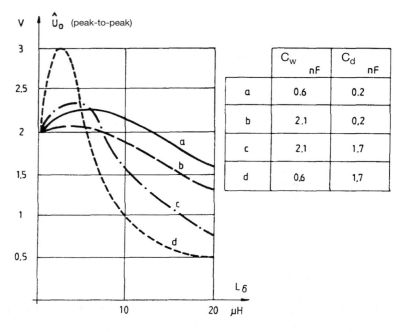

FIGURE 7.29 Output noise voltage amplitude vs. leakage inductance of the transformer; $C_w =$ parasitic capacitance between the transformer coils, and $C_d =$ depletion capacitance of the rectifiers

Reducing the noise voltage on the output and mains terminals via well designed layout produces savings in cost, size, weight, and (not least of all) annoyance. The first step for designing the layout of a circuit that employs a switched-mode supply or similar transistorized circuit is to carefully select the location of EMI-sensitive components and circuits that carry switching waveforms. The primary goal is to prevent electromagnetic coupling between these circuits, but one should also remember to minimize coupling to ground. Circuit elements carrying the switching currents should be placed as near to each other as possible. In addition, circuit elements producing the highest HF disturbances should be electromagnetically shielded (see Section 11.5).

It will be predicted from the analytical results made for the given simplified dc-to-dc circuit that the leakage inductance of the transformer L_σ increases as the output peak-to-peak voltage as a whole decreases, and that a peak in peak-to-peak voltage can appear for a certain value of leakage inductance and parasitic capacitance.

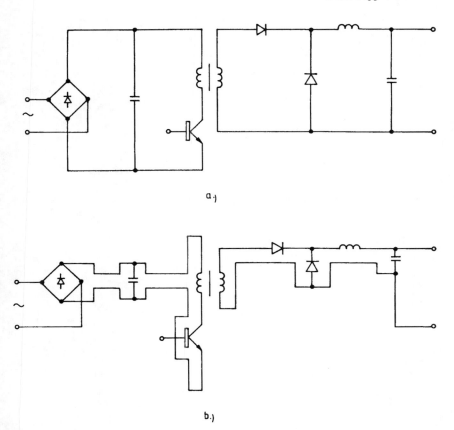

FIGURE 7.30   Wiring layout in dc-to-dc converters: (a) bad layout and (b) proposed layout for reducing electromagnetic coupling

8

EMI Filter Circuit
Selection and Measurement

In practice, EMI generated by power electronic equipments can be reduced to acceptable levels using filter circuits, usually referred to simply as EMI or RFI filters. Standard EMI filters are lowpass filter circuits with serial choke coils and parallel capacitors. Within the EMI filter category, to meet special requirements, the powerline filters may be differentiated.

A major problem in designing EMI filters is determining the most economical circuit configuration that gives a high insertion loss in the frequency range of 0.15 to 30 MHz and is also capable of withstanding substantial low-frequency (operational) voltage and current ratings. The decision in favor of a particular filter circuit is also influenced by the input and output impedances and the occasional limitations on the serial and parallel impedances. EMI filters for power electronic equipments cannot be discussed properly in the same terms as filters for telecommunications. The situation is different not only because of the high voltage and current ratings, but due also to the mismatched impedance conditions.

EMI filters are usually designed to provide a specified insertion loss. The difference between predicted levels of conducted EMI and the target level obviously equals the minimum attenuation performance required of the EMI filter. To avoid overkill, the amount of insertion loss should be calculated with special care. Although there are known methods for determining the differential-mode noise components produced by semiconductor circuits (see Chapter 5), the common-mode noise components can be calculated only with more or less approximation. To give the final requirement for EMI filter, a safety margin of approximately 10 dB should be added to the calculated insertion loss.

The EMI filter may alter the electrical characteristics of electrical equipment whose output is to be filtered. Undesired effects can include a voltage drop or rise at the power frequency or excessive leakage current to ground. These characteristics are often specified for powerline filters.

The specified insertion loss value is in many respects a function of the measurement technique employed. For precise specification, one must consider and analyze the various measurement methods.

It is important here to define EMI filter input and output from a noise suppression point of view. In this text, the terminals connected to the noise source are referred to as *input*, and the ones connected to the power mains to be protected from conducted EMI constitute the filter *output*. One should not confuse these designations with those of powerline filters, which are exactly the opposite.[*]

8.1 DEFINITION OF EMI FILTER PARAMETERS

EMI filters are characterized by insertion loss (see Section 6.1.1) rather than voltage attenuation. Insertion loss, in contrast to voltage attenuation, is not a transfer function. An insertion loss measurement method is shown in Fig. 8.1. Assuming that the generator and load impedances are identical, the value of the insertion loss can be expressed according to Eq. (6.3). The difference between the measured voltages appearing beyond the insertion point before and after the insertion is as follows:

$$IL = 20 \log\left(\frac{U_1}{U_2}\right) \tag{8.1}$$

Like other filter circuits, EMI filters are discussed in terms of four-terminal network theory. For EMI filter design practice, impedance and cascade parameters are most often used.

For network analysis, the voltage and current measurement directions will be as shown in Fig. 8.2. The EMI filter, using the impedance parameters, is described by the following equations:

$$U_1 = Z_{11} \times I_1 + Z_{12} \times I_2$$

$$U_2 = Z_{21} \times I_1 + Z_{22} \times I_2 \tag{8.2}$$

The open-circuit input impedance, Z_{11}, and the open-circuit output transfer impedance, Z_{21}, can be measured with unloaded (open) output; i.e., $I_2 = 0$:

$$Z_{11} = \frac{U_1}{I_1} \tag{8.3}$$

---

[*]In practice, vendors do not refer to the power mains side of a filter as "output," presumably because most filters are bidirectional (LC circuits notwithstanding). The author's intent here is to avoid any ambiguity about how the filter should be connected. —Ed.

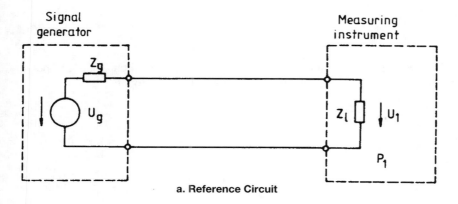

a. Reference Circuit

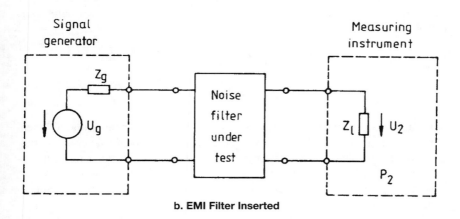

b. EMI Filter Inserted

FIGURE 8.1   Measuring the insertion loss of EMI filters

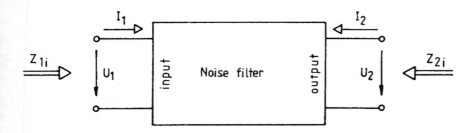

FIGURE 8.2   Measurement directions and input impedances of EMI filters

$$Z_{21} = \frac{U_2}{I_1} \qquad (8.4)$$

The open-circuit output impedance, Z_{22}, and the open-circuit transfer input impedance, Z_{12}, can be determined with unloaded ($I_1 = 0$) input:

$$Z_{22} = \frac{U_2}{I_2} \qquad (8.5)$$

$$Z_{12} = \frac{U_1}{I_2} \qquad (8.6)$$

For calculating the insertion loss of EMI filters, the cascade parameters are used more often than the impedance parameters, since they create a relationship between I/O voltage and current. The equation routine with cascade parameters yields:

$$U_1 = A_{11} \times U_2 - A_{12} \times I_2$$

$$I_1 = A_{21} \times U_2 - A_{22} \times I_2 \qquad (8.7)$$

In the above equations, the parameters A_{11} and A_{22} are quantities without dimension. The former has the character of voltage attenuation, and the latter that of current attenuation. The cascade parameter A_{12} has impedance dimension, and A_{21} has admittance dimension. To avoid the subscripts, the designations $A_{11} = A$, $A_{12} = B$, $A_{21} = C$ and $A_{22} = D$ are also common. Sometimes the inverse of chain parameters are used.

The chain parameters can be expressed by impedance parameters as follows:

$$A_{11} = \frac{Z_{11}}{Z_{21}}$$

$$A_{12} = \frac{Z_{11} \times Z_{22}}{Z_{21}} - Z_{12} = A_{11} \times Z_{22} - Z_{12}$$

$$A_{21} = \frac{1}{Z_{21}}$$

$$A_{22} = \frac{Z_{22}}{Z_{21}} \qquad (8.8)$$

Input and output impedances are important parameters for both the study of EMI filter performance and the design of filter element values. Loading the output of the four-terminal network with impedance Z_L, the input impedance seen from primary (input) side, expressed by the impedance and cascade parameters, is as follows:

$$Z_{1i} = Z_{11} - \frac{Z_{12} \times Z_{21}}{Z_{22} + Z_L} = \frac{A_{12} + A_{11} \times Z_L}{A_{22} + A_{21} \times Z_L} \qquad (8.9)$$

Loading the input of the four-terminal network with impedance Z_g, the input impedance seen from the secondary (output) side can be received by appropriate index changes in the above relationship:

$$Z_{2i} = Z_{22} - \frac{Z_{12} \times Z_{21}}{Z_{11} + Z_g} = \frac{A_{12} + A_{11} \times Z_g}{A_{22} + A_{21} \times Z_g} \qquad (8.10)$$

In four-terminal network theory, the insertion loss obtained by measurements as in Fig. 8.1 is often called *insertion factor* and marked with a T. The insertion loss by means of impedance parameters is as follows:

$$IL = \frac{(Z_{11} + Z_g) \times (Z_L + Z_{22}) - Z_{12} \times Z_{21}}{(Z_g + Z_L) \times Z_{21}} \qquad (8.11)$$

The insertion loss can be also expressed by the cascade parameters:

$$IL = \frac{A_{11} \times Z_L + A_{12} + A_{21} \times Z_g \times Z_L + A_{22} \times Z_g}{Z_g + Z_L} \qquad (8.12)$$

The insertion loss of EMI filters is usually given by curves plotted in a dB-logω coordinate system.

8.2 EMI FILTER CIRCUITS

After determining the required insertion loss, the next step is to choose a circuit configuration for the EMI filter. For this design, it is often necessary to consider not only the insertion loss value but some other specifications as well. Pertinent factors may include a limitation on capacitive current for grounded equipments or the acceptable voltage drop across powerline filters. For very stringent noise suppression standards, one should also consider the mismatched impedance conditions.

8.2.1 Basic Circuit Configurations

The basic EMI filter circuit configurations are shown in Fig. 8.3. The π-circuits are the most common, but multistage (ladder type) filters are also commonly used for high-performance applications. Multistage LC-circuits are often realized by adding an input capacitor to a multistage π-circuit. In engineering practice, multistage filters having more than four stages are not very common.

The EMI filters cannot be treated as the simple circuits shown in Fig. 8.3. Even for two-wire supply, the interference of both wires must be suppressed, since EMI can be generated across the relatively high HF impedance between the electrical equipment (EMI source) and the grounding point. To suppress EMI on all wires, filter circuits as shown in Fig. 8.3 must be inserted in every wire. This is why a two-wire EMI filter should be studied as a six-terminal network, as shown in Fig. 8.4. The network becomes more complex with an increase in the number of wires to be filtered.

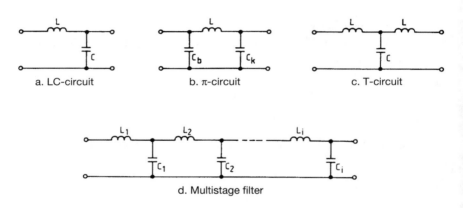

a. LC-circuit   b. π-circuit   c. T-circuit

d. Multistage filter

FIGURE 8.3   Basic EMI filter circuit configurations

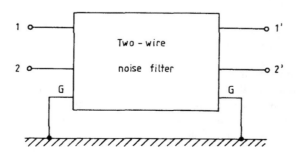

FIGURE 8.4   Two-wire EMI filter as six-terminal network

The filter circuits shown in Fig. 8.3 are for suppressing single-wire EMI components, but differential-mode emissions must be reduced as well. Some basic circuit configurations for reducing both common-mode and differential-mode EMI components are given in Fig. 8.5. Although this figure shows two-wire configurations, EMI filter circuits for more than two wires can be easily set up on the basis of Fig. 8.5 by multiplying the circuits of Fig. 8.3.

For EMI filters, one can interpret the insertion loss differently, depending on whether it is considered from a common-mode or differential-mode point of view.

8.2.2 EMI Filters under Mismatched Impedance Conditions

One major problem in designing EMI filter circuits for power electronic equipment is caused by the arbitrary generator and load impedances. Unlike filter circuits in telecommunications engineering, EMI filters operate under mismatched impedance conditions. The mismatched impedance condition (or, for short, *mismatch*) means that we have no assurances with regard to optimal, nor even known, source and load impedances. These impedances are really arbitrary because neither their value nor character can be known, given that a typical filter is to be installed in different equipments and supply networks. The HF impedance of

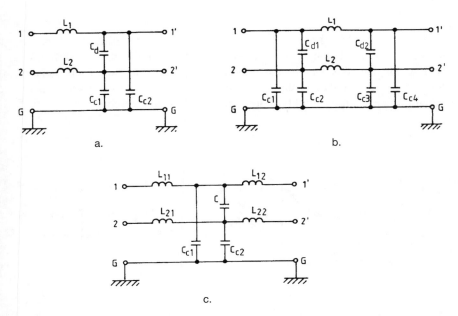

FIGURE 8.5 Basic EMI filter configurations for common-mode and differential-mode EMI components

electrical equipments as noise sources varies widely, because it depends on the circuit configuration, the layout technology, the applied semiconductor and, of course, varies with frequency. The load impedance of EMI filters (i.e., the HF impedance of supplies) can be deemed even less knowable than the noise source impedance, since this is the function of the connection point of the electrical equipment. It also cannot be regarded as constant in time, since it depends on the number of users supplied by this branch. Power EMI filtering methods must be discussed in terms of impedance matching, which is why techniques known to the telecommunications community are not directly applicable.

Problems related to the mismatched impedance condition can be well illustrated by studying the insertion loss of a single LC-circuit at various load impedances. For the sake of simplicity, we will consider a negligibly small generator impedance because in such a case the insertion loss is equal to the voltage attenuation of the circuit.

The general requirement for an EMI filter is to provide an insertion loss higher than a given IL_m value in the frequency range above an upper limit, f_1. Using the symbol ω_o as the self-resonant angular frequency of the LC-circuit, the insertion loss for a resistive load can be given as:

$$IL_R(\omega) = \sqrt{\left(1 - \frac{\omega^2}{\omega_o^2}\right)^2 + \frac{(\omega L)^2}{R_l^2}} \tag{8.13}$$

The insertion loss as a function of frequency for different resistive loads is shown in Fig. 8.6. It can be seen from the figure that, although the insertion loss will not change considerably above the frequency f_1, a large dip in the negative loss side (i.e., gain) can occur. As a result of resonances, noise components in a certain frequency range will be amplified instead of suppressed.

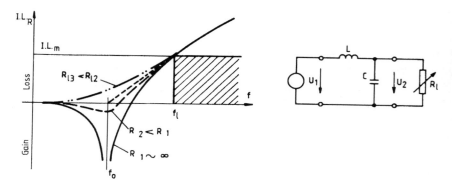

FIGURE 8.6 Insertion loss curves vs. frequency for different resistive loads

Let us study the insertion loss for inductive load, represented by a resistor and inductor connected in parallel. If we define the symbol for resonance angular frequency of the circuit as

$$\omega_L^2 = \omega_o^2 \left[1 + \frac{L}{L_1} \right]$$
(8.14)

the insertion loss of the circuit is as follows:

$$IL_L(\omega) = \left(\frac{\omega_L}{\omega_o} \right)^2 \sqrt{ \left(1 - \frac{\omega^2}{\omega_L^2} \right)^2 + \frac{(\omega L)^2}{R_1^2} \times \frac{\omega_o^2}{\omega_L^2} }$$
(8.15)

The function $IL_L(\omega)$ is plotted for different inductive loads in Fig. 8.7. As the effect of the inductive load, the resonance frequency of the circuit is increased. For extreme case, the voltage gain can occur in the cutoff frequency range.

Finally, look the insertion loss as a function of frequency for a capacitive load. Defining the symbol for the resonance angular frequency of the circuit as

$$\omega_c^2 = \frac{\omega_o^2}{1 + \dfrac{C_1}{C}}$$
(8.16)

the insertion loss of an LC-circuit is described by the following relationship:

$$IL_c(\omega) = \sqrt{ \left(1 - \frac{\omega^2}{\omega_c^2} \right)^2 + \frac{(\omega L)^2}{R_1^2} }$$
(8.17)

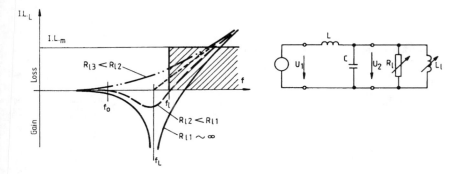

FIGURE 8.7 Insertion loss curves vs. frequency for different inductive loads

The insertion loss as a function of frequency is shown in Fig. 8.8. As a result of the load being capacitive in nature, the resonance frequency is decreased; thus, the insertion loss in the cutoff frequency range has increased.

The above calculations illustrate the effect of load impedance mismatch only. It is not necessary to perform detailed examinations to see that the problem of mismatched impedance conditions is further complicated by the source impedance.

Let us examine two other examples to make clear the problem with mismatched impedance conditions. The insertion loss curve of an EMI filter, in consideration of CISPR-defined LISNs, should be measured in a 50 Ω system. The actual insertion loss of such an EMI filter, because of mismatched impedances, can differ by as much as 40 dB/decade from the calculated results for matched conditions. Figure 8.9 shows an LC-circuit inserted between the noise source and the supply network. In ideal impedance conditions or in a 50 Ω measuring system, the insertion loss increases by 40 dB/decade slope as a function of the frequency, but if we have high source and low load impedances, it may even be reduced to zero, as indicated in the figure.

Figure 8.10 shows a π configuration. Although for matched impedance the slope of the insertion loss curve is 60 dB/decade, for low source and load impedances, the slope of the insertion loss curve as a function of frequency will be no more than 20 dB/decade.

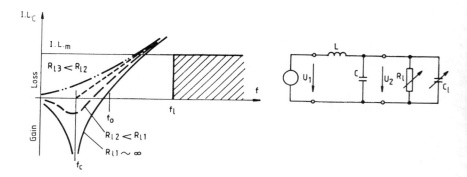

FIGURE 8.8 Insertion loss curves vs. frequency for different capacitive loads

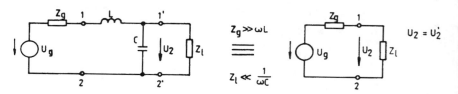

FIGURE 8.9 LC-configuration EMI filter under extreme impedance mismatch conditions

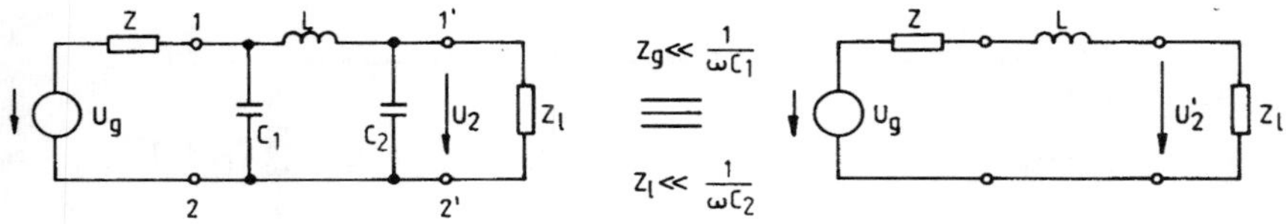

FIGURE 8.10 π-configuration EMI filter under extreme impedance mismatch conditions

Knowledge of the source and load impedances, or at least of their magnitudes, is important for selecting the most appropriate filter configuration. With a fairly good design, the expected and actual insertion loss curves will agree. Figure 8.11 shows the advisable filter configurations for extreme source and load combinations. These configurations provide an insertion loss curve of 60 dB/decade slope as a function of frequency. The philosophy, then, is to connect either a series choke coil to a low-impedance source or a parallel capacitor to a high-impedance load. Similarly, a series choke coil should face a low-impedance load, and a parallel capacitor should look to a high-impedance source [1]. This philosophy ensures the optimal use of filter elements as well as partly compensating for the variation in source and load impedance. The input and output filter elements should be designed so that the impedances of the source, load, and filter elements are about equal at the desired cutoff frequency. The above considerations should be kept in mind when the source and load impedances differ greatly.

The relationship between the mismatched impedance conditions and the shape of insertion loss curve for different filter configurations has been studied [282]. The measurements were made with one-stage, two-stage, and four-stage LC configurations under four extreme source and load impedance combinations. The filter circuits were designed to provide a 60 dB insertion loss at 150 kHz for very low source and very high load impedances. The total capacitance of each filter was the same. The following filter circuits were compared:

1. One-stage LC-filter (n = 1) with 1 mH inductance and 1 μF capacitance
2. Two-stage LC-filter (n = 2) with 63 μH inductances and 0.5 μF capacitances
3. Four-stage LC-filter (n = 4) with 22.5 μH inductances and 0.25 μF capacitances

The measured insertion loss curves are shown in Fig. 8.12. The plotted curves clearly show that the character of the insertion loss as a function of the frequency is essentially effected by the mismatched impedance conditions. The worst mismatch that can happen is the high source and low load impedance for a one-stage LC-filter; i.e., the reverse of the impedance condition taken as a basis for designing its element for providing the required insertion loss. The response of multistage LC-filter is also poor given high source and low load impedances, but the

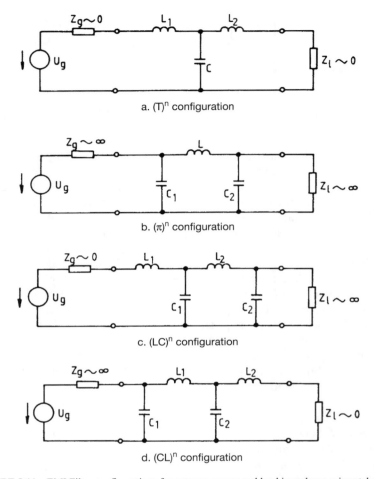

FIGURE 8.11 EMI Filter configurations for extreme source and load impedance mismatches

performance improves rapidly with increasing n. The measured insertion loss curves illustrate that resonances in the passband may be expected only in case of very different source and load impedances.

To improve EMI filter performance under mismatched impedance conditions, multistage filter configurations should be applied. When the absolute value of the source and load impedances can be approximated, the use of the following configurations, analogous to Fig. 8.11a through 8.11d, is recommended:

   a. For low source and load impedance, a $(T)^n$ filter configuration
   b. For high source and load impedance, a $(\pi)^n$ filter configuration.

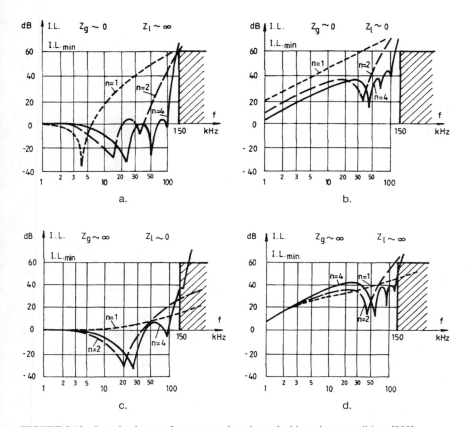

FIGURE 8.12   Insertion loss vs. frequency under mismatched impedance conditions [282]

  c. For low source and high load impedance, an $(LC)^n$ filter configuration
  d. For high source and low load impedance, a $(CL)^n$ filter configuration

Until now, EMI filters have been examined as real four-terminal networks, but this idealistic case does not exist in engineering practice. Common-mode EMI usually has to be suppressed; therefore, for designing the best circuit configuration, the equivalent four-terminal network should be set up for the common-mode components. To draw the equivalent circuit, the origin of common-mode noise components should be studied. For a better understanding of the above statement, review the problem of noise suppression for switched-mode power supplies as detailed in Section 7.7. If we provide a power supply with an EMI filter as shown in Fig. 8.5a, the simplified circuit shown in Fig. 8.13 can be realized. Note that the noisy input of the EMI filter (i.e., the serial choke coil) faces the power supply as the noise source, and the cleaned-up output of the EMI filter is connected to the mains.

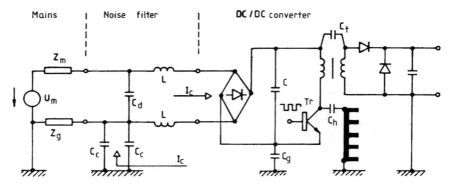

FIGURE 8.13 Noise suppression for switched-mode power supplies

The switching waveform of the power transistor is coupled through parasitic capacitances (C_g, C_t, C_h) into the ground as common-mode EMI current (I_c). The coupling capacitance is the sum of heatsinks, transformer, and circuit stray capacitances. In practice, the value of X-capacitors is much higher than that of Y-capacitors; thus, for differential mode, the X-capacitor (C_d) can be regarded as a short circuit beside the reactance of Y-capacitors in the frequency range of 0.15 to 30 MHz. With this approximation, the equivalent circuit shown in Fig. 8.14 can be useful for examining common-mode EMI. In this circuit, capacitor C_r is the total parasitic coupling capacitor. Its capacitance can even reach some nanofarads since more power semiconductors usually are applied in switched-mode power supplies. The load inductance of the EMI filter is the parallel result of the mains inductances, Z_g and Z_m.

Theoretically, the EMI filter element could be well designed on the basis of the equivalent circuit. But, as was demonstrated above, the actual insertion loss depends largely on the value as well as the type of the load impedance. The common-mode source impedance is very high when the assembly technology meets the EMI requirements discussed in Sections 7.6 and 7.7. For high source and low load impedance conditions, the insertion loss of a one-stage LC-circuit can offer insufficient performance, depending on the impedance mismatch (see Fig. 8.12c). EMI filtering performance can be improved by using a multistage filter configuration.

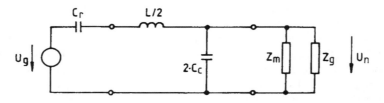

FIGURE 8.14 Equivalent circuit for examining common-mode noise suppression

EMI from low-power equipments is usually measured with a LISN, but in Fig. 8.14, the HF inductances of the mains, Z_g and Z_m, are substituted for the measuring resistance of the LISN. This results in a more exact situation for designing the EMI filter, but its performance can be considerably changed under actual mains impedance conditions. This is one example of the problems involved in EMI measurement with LISNs (see Section 4.3.3).

8.2.3 Powerline Filters

In addition to impedance mismatch, one more very important condition usually should be considered for designing powerline filters. In such filters, there are restrictions for both the serial and the parallel inductances. The value of the serial inductance is limited by the operating frequency voltage drop (see Section 6.3.3), and the value of Y-capacitors is limited by the acceptable ground leakage current for safety and shock hazards (see Section 7.5.1). These limitations are usually very strict, and it can be very difficult to meet them while achieving the insertion loss objectives. For the differential (X) capacitor, although the differential-mode capacitive current flowing through it cannot be as high as one might like, this restriction does not make the powerline filter design difficult.

At the same time, the limitations on the serial and parallel inductances mean that the sum of the reactances in powerline filters seen by the normal differential current (50/60 or 400 Hz) cannot be unrestrictedly high. For powerline filter design, this limitation can be best expressed by the LC product. For powerline filter characterization, the maximum LC product can be calculated from Eqs. (6.56) and (7.6) as follows:

$$L_m \times C_m = \frac{\Delta U \times I_n}{I_g \times U_m} \times \frac{1}{\omega_m^2} \tag{8.18}$$

with $\qquad \Delta U$ = maximum permissible voltage drop at mains frequency ω_m

The value of the LC product depends on the type as well as the power rating of the equipment. As an example, for low-power appliances, this product is about 100 μHμF. It is a very low value which, in fact, can be best illustrated if we assume a one-stage LC-filter configuration. For sake of simplicity, let only the voltage attenuation be calculated. In the attenuation frequency range, the voltage attenuation as a function of the frequency is, with good approximation, as follows:

$$K(\omega) = \omega^2 \times LC \tag{8.19}$$

According to the above relationship, for $L_m C_m = 100$ μHμF, the voltage attenuation barely reaches 40 dB at 150 kHz. This value is too small for most practical cases, since the EMI filters must provide 60 to 80 dB insertion loss at this frequency (see Chapter 5). Recall that, because of the inductance mismatch, the actual insertion loss of a one-stage LC-filter may be much lower than the calculated one. The above train of thought clearly shows the difficulty in meeting the limitations represented by the LC product and the noise suppression requirements.

Let us examine the above mentioned noise suppression problem one more time, and let us assume that there is no limitation on the capacitive current. As an example, for stationary equipments, shock hazards do not exist. Let us suppose further that for proper noise suppression we must provide 80 dB insertion loss at 150 kHz. Since the voltage attenuation of the one-stage LC configuration changes by 40 dB/decade slope in frequency, the cutoff frequency of the filter circuit should be designed to about 1 kHz. Although this is realizable, resonances may develop in the vicinity of the cutoff frequency, especially in a case of high Q-factor. In this frequency range, the harmonics produced by semiconductor equipments are usually high, and they may even be amplified.

The problem caused by the impedance mismatch is to stay within the reactive voltage drop limitations, but at the same time to provide the required insertion loss. This difficulty can be solved by using a multistage filter configuration. For mathematical simplicity, we will examine the multistage filter's voltage attenuation, since there is a very close connection between voltage attenuation and insertion loss for this configuration.

Introducing the symbol k, where

$$k = \omega^2 \times LC \tag{8.20}$$

the voltage attenuation of one-stage LC configuration is as follows:

$$K_1(\omega) = 1 - k = 1 - \alpha_{11} \times k \tag{8.21}$$

Two series-connected LC configurations consisting of identical inductances and capacitances is referred to as a two-stage filter configuration. The voltage attenuation in this case is as follows:

$$K_2(\omega) = 1 - 3 \times k + k^2 = 1 - \alpha_{21} \times k + \alpha_{22} \times k^2 \tag{8.22}$$

The voltage attenuation of an n-stage filter consisting of identical inductances and capacitances can be given in the form below [10]:

$$K_n(\omega) = \quad = 1 - \alpha_{n1} k + \alpha_{n2} k^2 - \ldots \pm \alpha_{nn} k^n \tag{8.23}$$

For $n < 7$, the values of coefficient α are included in Table 8.1.

From the above equations, we see that any relationship between the voltage attenuation (insertion loss) and the number of filter stages is difficult to establish;

TABLE 8.1   Coefficients for Calculating the Voltage Attenuation of N-Stage LC-Filters

n	α_{n1}	α_{n2}	α_{n3}	α_{n4}	α_{n5}	α_{n6}
1	1					
2	3	1				
3	6	5	1			
4	10	15	7	1		
5	15	35	20	9	1	
6	21	70	84	45	11	1

therefore, it is useful to plot these relationships in a chart. Figure 8.15 is a voltage attenuation chart of one-stage (continuous line), two-stage (dashed line), and three-stage (dotted line) filter configurations for some typical LC products.

Two important conclusion can be drawn from the chart. First, with an increase in the number of stages, the voltage attenuation improves; however, the gain shows a reduction trend. The other conclusion is that, for high insertion loss, a multistage configuration is better than a single-stage one. To support this, take a numerical example. The voltage attenuation is 80 dB at 160 kHz for a one-stage filter with LC = 10^4 µHµF. For the same voltage attenuation with a two-stage filter, LC = 10^2 µHµF at this frequency. This means that the same voltage attenuation can be realized by a two-stage filter with only 10^2 µHµF reactance as by a one-stage filter with 10^4 µHµF reactance. By decreasing the reactance, we may achieve a considerable saving in cost and size.

Equation (8.23) and its chart do not tell us how to design an EMI filter to meet the reactance limitations and insertion loss requirements. The limitations made for serial and parallel impedances are independent of the number of stages; i.e.:

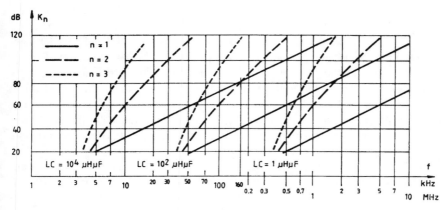

FIGURE 8.15   Voltage attenuation vs. frequency of multistage filter configurations (one LC cell is counted as n = 1)

$$L_m C_m = \Sigma L_i \times \Sigma C_i \qquad (8.24)$$

Supposing that the stages consist of identical elements, the inductance and capacitance of single elements can be calculated easily as follows:

$$L_i = \frac{L_m}{n} \quad \text{and} \quad C_i = \frac{C_m}{n} \qquad (8.25)$$

The symbol k, defined by Eq. (8.20), can be rewritten using the reactance limitations in the following form:

$$k = \frac{\omega^2 \times L_m C_m}{n^2} = \frac{k_m}{n^2} \qquad (8.26)$$

Equations (8.23) and (8.26) yield the voltage attenuation as a function of angular frequency as follows:

$$K_n(\omega) = 1 - \alpha_{n1}\left(\frac{k_m}{n^2}\right) + a_{n2}\left(\frac{k_m}{n^2}\right)^2 - \alpha_{n3}\left(\frac{k_m}{n^2}\right)^3 + \ldots \pm \alpha_{nn}\left(\frac{k_m}{n^2}\right)^n \qquad (8.27)$$

The voltage attenuation as a function of frequency, given by the above expression, can be plotted for given LC products. As an example, Fig. 8.16 shows such a chart for $L_m C_m = 100\ \mu H\mu F$. The parameter of the chart is the stage number. The one-stage filter provides a higher voltage attenuation under 80 kHz than the two-stage filter for $L_m C_m = 100\ \mu H\mu F$.

The LC product value defines a maximum available voltage attenuation curve. This envelope curve consists of sections belonging to various stage numbers (n). In Fig. 8.16, the envelope curve above 220 kHz is indicated by $n = \infty$. This envelope curve means that while keeping the limitations on serial and parallel reactances, a higher voltage attenuation (given by this curve) cannot be provided.

Going back to the example, one can read from Fig. 8.16 that for $L_m C_m = 100\ \mu H\mu F$ reactances, the voltage attenuation of a one-stage filter is 40 dB at 150 kHz. This can be increased to 52 dB by applying a two-stage filter configuration. The envelope curve also shows that, because of reactance limitations, a voltage attenuation higher than 56 dB cannot be provided at 150 kHz; therefore, this value belongs to a three-stage filter configuration.

It should be noted that the use of multistage instead of one-stage filters is advantageous not only to meet reactance limitations and achieve a sufficiently high insertion loss. An additional benefit is that the impedance mismatch will have considerably less effect on the filter's noise suppression performance (see Section 8.2.2).

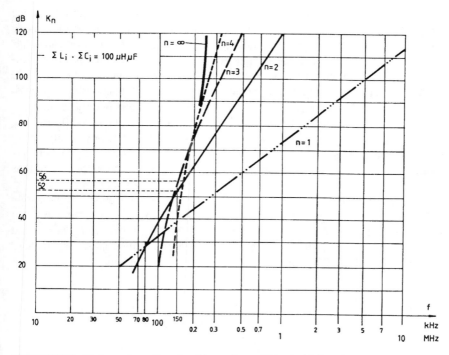

FIGURE 8.16   Voltage attenuation of multistage filters of identical LC product

8.2.4 EMI Filter Configurations with Common-Mode Choke Coils

One of the most difficult problems in powerline EMI filter design is staying within reactance limitations while providing a high insertion loss. To illustrate this difficulty, let us return to the example of noise suppression in switched-mode power supplies and study Fig. 8.14, also from a powerline point of view. Because of size limitations, the inductance of an air-core solenoid coil arrangement is usually not more than about 100 µH. Supposing that the capacitance of Y-capacitors is limited by safety and shock hazards to a few nanofarads, the resonance frequency of the one-stage LC configuration is in the order of 100 kHz. This means that the insertion loss at 150 kHz (the lower frequency limit to be filtered) will be negligibly small. Worse still, in the frequency range of 0.15 to 30 MHz, voltage gain can occur. The insertion loss cannot be increased by multistage filters, because for LC products lower than 10, the highest voltage attenuation is produced by the one-stage filter configuration (see Fig. 8.15).

To overcome the above mentioned design problem, common-mode choke coils with magnetic cores were developed that exhibit a very small impedance for differential-mode (operational frequency) components and a relatively high impedance for common-mode components (see Section 6.3.4). Applying a common-mode choke coil in an EMI filter configuration, the limitation on voltage drop can be ignored since $\Delta U = 0$ for the operational frequency. Because of the high common-mode inductance of common-mode choke coils, the LC product can be increased even if the capacitance of the Y-capacitors is limited by safety and shock hazards.

Common-mode choke coils are used in noise suppression not only for increasing the LC product but also for decreasing the size and weight of EMI filters. The common-mode choke coils in powerline filters of intermediate power ratings are generally more economical and provide the same insertion loss as solenoid coil arrangements, even when the common-mode choke coils use more noble (and thus more expensive) iron materials.

The basic filter circuit configuration with a common-mode choke coil is shown in Fig. 8.17. The configurations of Figs. 8.17a and 8.17b essentially correspond to Figs. 8.5a and 8.5b, respectively. A T configuration as shown in Fig. 8.5c is not common with common-mode choke coils. The configurations shown in Fig. 8.17 are for suppression of the common-mode noise components only.

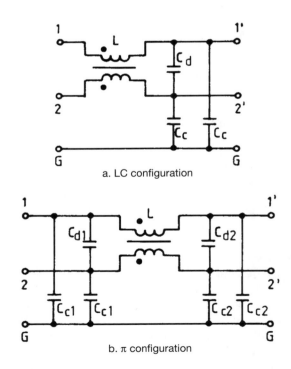

a. LC configuration

b. π configuration

FIGURE 8.17   Filter configurations with common-mode choke coil

There is considerable variation in the differential-mode and common-mode insertion losses of EMI filter configurations with common-mode choke coils. The two insertion loss values can be calculated by drawing up separate equivalent circuits for differential-mode and common-mode components. The common-mode choke coil acts as a choke coil of N number of turns with a closed iron core for common-mode current components (I_{c1} and I_{c2} in Fig. 8.18a), as shown in Fig. 8.18b. The inductance, L, is usually quite high. The impedance for differential-mode components is proportional only to the leakage between the two coils. The equivalent circuit of the common-mode choke coil for differential-mode components is also a choke coil circuit, shown in Fig. 8.18c. Inductance l_d is very small relative to L_c of Fig. 8.18b. The differential-mode inductance can be expressed by the mutual inductance, M, and the coupling coefficient, K, as follows:

$$L_d = 2(L_c - M) = (2) L_c (1 - K) \tag{8.28}$$

Since their differential-mode inductance is only a few percent of LC, the configurations shown in Fig. 8.17 are mostly used with low-power electronic equipments that produce a moderate amount of differential-mode noise. EMI filters for larger equipments must also provide significant insertion loss for differential-mode

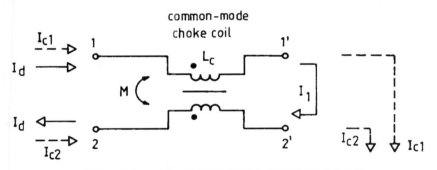

a. Differential-mode and common-mode current components

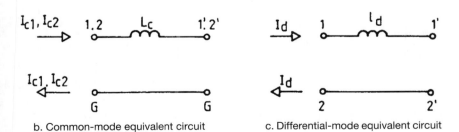

b. Common-mode equivalent circuit   c. Differential-mode equivalent circuit

FIGURE 8.18   Equivalent circuits of common-mode choke coils

noise components. To improve differential-mode noise suppression, the filter configurations of Fig. 8.17 are generally completed with regular noise suppression choke coils. Figure 8.19a shows the commonly used filter configuration for suppressing both the differential-mode and common-mode noise components. These filter circuits can be designed by setting up the differential-mode and common-mode equivalent circuit separately. Figure 8.19b shows the equivalent circuit for differential-mode noise source, and Fig. 8.19c shows that for common-mode noise source. As seen from the figures, the EMI filter of Fig. 8.19a acts as two-stage LC-configuration for both noise components types, only the values of circuit elements vary. The common-mode insertion loss is considerably higher than the differential-mode insertion loss because the first filter stage has a much higher voltage attenuation for common-mode noise voltages than for differential-mode noise.

For improved performance, the filter configuration of Fig. 8.20a is advisable, although many other configurations are known. The Y-capacitors on the output are feedthrough type for improving the HF performance. The differential-mode equivalent circuit is of a two-stage π-filter configuration (Fig. 8.20b). In practice, inductance l_c (the leakage inductance of the common-mode choke coil in the differential-mode equivalent circuit) is much less than that of the regular noise suppression choke coils (L_d); therefore, this circuit can be simplified to a one-stage π-filter configuration where the capacitance of the first capacitor is the sum of C_i and C. The common-mode equivalent circuit is a two-stage π-filter (Fig. 8.20c). This filter configuration provides quite a high insertion loss, not only because of the three stages but also because of the high serial inductance. The three-stage configuration has another advantage in that it offers excellent performance even under impedance mismatch conditions.

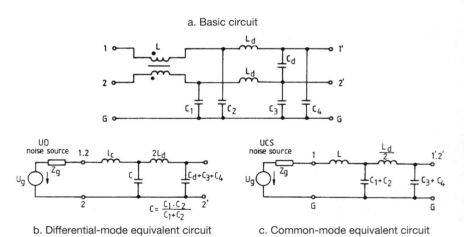

a. Basic circuit

b. Differential-mode equivalent circuit   c. Common-mode equivalent circuit

FIGURE 8.19 EMI filter configuration for suppressing both differential- and common-mode noise components

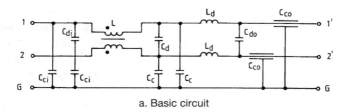

a. Basic circuit

b. Differential-mode equivalent   c. Common-mode equivalent

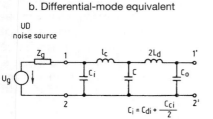

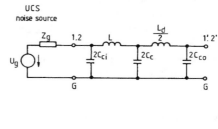

FIGURE 8.20 High-performance EMI filter configuration for suppressing both differential- and common-mode noise components

8.3 INSERTION LOSS TEST METHODS

The basic circuit scheme for measuring the insertion loss of EMI filters is shown in Fig. 8.21. For determining insertion loss as a function of frequency in a given range, the insertion loss measurement must be made at several frequencies. Switch position 1 is for calibration, and insertion loss is measured in position 2.

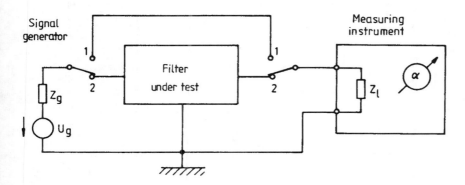

FIGURE 8.21 Basic circuit for measuring the insertion loss of EMI filters

The internal impedance of most signal generators and measuring instruments is 50 Ω. This is why most national and international standards suggest testing the insertion loss in a 50 Ω measurement system. Some specifications, oriented toward the impedance of certain CISPR-defined LISN, accept measurements from a 150 Ω system. Attenuators are often used for impedance matching and stabilizing the load of the signal generator (see Section 6.1.2).

According to specifications, insertion loss is usually measured under clearly defined impedance conditions. However, it is well known that the insertion loss is highly dependent on the load and source impedances (see Section 8.2.2). In determining EMI filter insertion loss characteristics, it is of primary importance to utilize laboratory measurement techniques whose results are indicative of actual real-world performance. To get realistic data regarding EMI filter performance under operational conditions, and to avoid overspecification of the insertion loss as a safeguard, new laboratory measurement techniques have been developed and suggested for general use.

8.3.1 Insertion Loss Test in Various Measuring Impedances

The numerical value of the insertion loss depends on the source and load impedance; i.e., on that of the measuring set (see Sections 6.1.1 and 6.1.2). This dependence is well illustrated by the measured insertion loss curves shown in Fig. 8.22. The same EMI filter was measured in both low-impedance and 50 Ω measuring systems. The resulting values differ by 20 dB at 150 kHz. The other im-

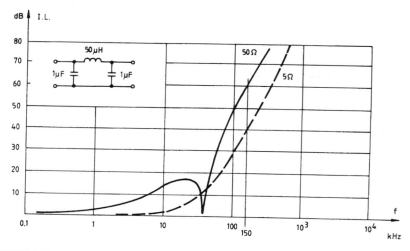

FIGURE 8.22 Insertion loss as a function of the measurement system impedance

portant difference is that the resonance is not indicated by the low-impedance measuring set.

To allow the laboratory test to better approach operational conditions, a recommendation was elaborated to specify the absolute value of the source and load impedance [40]. Impedance levels encountered in power mains may be considerably lower than 50 Ω, particularly below 1 MHz; therefore, proposals have been offered whereby the insertion loss of powerline filters should be measured using a low impedance system or, better still, a few selected worst cases of low/high impedance combinations for input and output of the tested filter. The source impedance might be calculated from the mains drop, and the load impedance from the nominal voltage and current of the EMI filter. Although this laboratory measurement does not adequately characterize EMI filter performances in situ, it does provide more realistic data than that of a 50 Ω measuring set.

8.3.2 Insertion Loss Test under Heavy Load Currents

EMI filters insertion loss depends more or less on power bias [121], which can be explained by the inductance variations of EMI filter choke coils. The insertion loss degradation in heavy load current may be quite high for solenoid choke coils with iron cores, but that of common-mode choke coils is usually negligible (see Section 6.3.4). For a detailed performance study of EMI filters, the test setup shown in Fig. 8.23 might be suggested for measuring EMI filter insertion loss. In the U.S.A., this corresponds to MIL-STD-220A. Performing insertion loss measurements under various load conditions is intended to increase the usefulness of the insertion loss data because it will reveal the existence of an incorrectly designed choke coil.

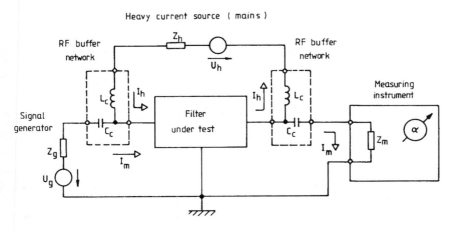

FIGURE 8.23   Basic circuit for insertion loss measurement under heavy current load (ac or dc)

The purpose of the RF buffer network in Fig. 8.23 is to provide RF isolation between the low frequency (operational) terminals and the measurement system; i.e., to present a high-impedance path (L_c) and a low-impedance path (C_c) at measurement frequencies for the injected HF signal. The impedance of the isolation choke coil must be much higher in the measurement frequency range so as not to influence measurement accuracy. The choke coil, L_c, must have excellent HF performance.

Figure 8.24 shows typical insertion loss variation under power bias. The measurements were made in a 50 Ω setup using a 100 A rated powerline filter [70].

8.3.3 Worst-Case Insertion Loss Test Method

EMI filters insertion loss depends not only on the magnitude but also the character of the source and load impedances (see Section 8.2.2). Whether the goal is to design an EMI filter or simply select one from a catalog, it is essential to obtain realistic data under all possible source and load impedances. To study the effect of impedance mismatch conditions on the insertion loss, the worst-case test method has been developed [96, 284, 335].

It can be shown that the insertion loss of a filter reaches a minimum for a given frequency when its input and output are matched with the complex conjugate of their respective impedances. Supposing a high Q-factor, this is caused by the development of resonances. Conventional powerline filters are of one-stage or multistage π configuration; thus, capacitors are placed across the input and output. In

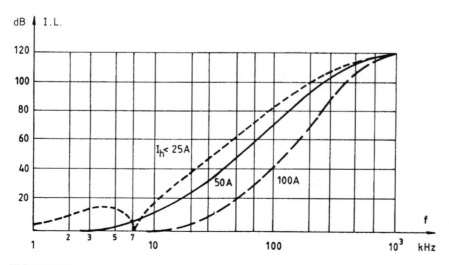

FIGURE 8.24 Insertion loss vs. frequency of a 100 A rated filter for various loads

order to match such filter configurations with the complex conjugate impedance, a choke coil of appropriate value must be inserted in series with the input and output. The inductance value to be used at a given measuring frequency (f_x) is given by

$$L = \frac{1}{4\pi^2 \times f_x^2 \times C}$$ (8.29)

For calculating the inductance of the input and output matching choke coil, one must use the capacitance of the proper side.

The worst-case insertion loss of LC configurations can be measured by placing adjustable matching capacitors on the input and an adjustable matching inductor on the output. The value of the matching capacitance can be calculated, after some transformation, also by Eq. (8.29).

The matching choke coil must have a high Q-factor to develop some clear resonance. The required value for the Q-factor of the resonant circuit is at ≥ 10, and the losses are generated more in the inductor than in the capacitor. Figure 8.25 shows the worst-case insertion loss curves at a given frequency, measured with the same EMI filter but with matching inductors of three different Q-factors [334].

Due to the capacitances possessed by most powerline filters, it is impractical to obtain choke coils of the required value above a few megahertz. But given the usual power mains impedance, laboratory measurements with a 50 Ω system provide quick test data to approximate real worst-case insertion loss values.

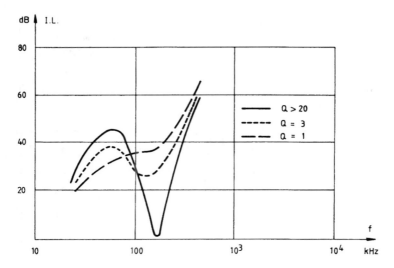

FIGURE 8.25 Worst-case insertion loss curves measured with matching inductors of various Q-factors [334]

For determining the worst-case insertion loss curve as a function of frequency, the minimum insertion loss must be measured at many frequencies with properly adjusted, matching source and load impedances. With automatic, swept-frequency measuring instruments, the worst-case insertion loss curve also can be determined by stepping the values of the input and output matching impedances and then plotting the insertion loss curve in a common coordinate system. The worst-case insertion loss curve is the minimum envelope of the insertion loss curves, as shown in Fig. 8.26. The figure shows the insertion loss curve measured in a 50 Ω system, and the plotted worst-case curve. The figure demonstrates the considerable difference between insertion loss curves measured by conventional laboratory test methods and by the worst-case test method.

For better characterizing the EMI filters, worst-case insertion loss measurements are usually made under power bias. Two test methods types are known: the serial and parallel injection methods.

A schematic diagram of series injection worst-case measuring setup is shown in Fig. 8.27. The given measuring setup serves to determine the minimum insertion loss of π-type filter configuration, since two matching inductors are used (capacitor input and output). The insertion loss can be measured in two ways, depending on the setup of the measuring system. Adjusting the output voltage of the signal generator to a constant level, the insertion loss can be determined by

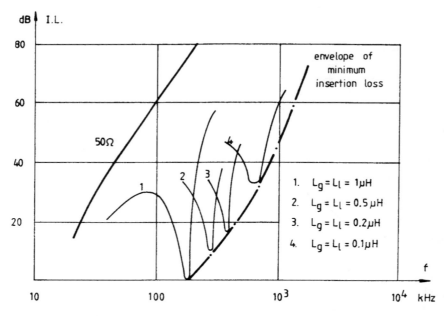

FIGURE 8.26 Plotting the worst-case insertion loss curve

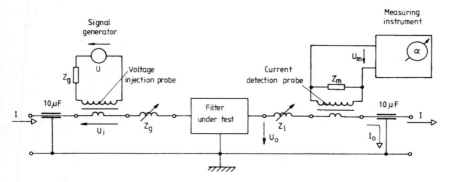

FIGURE 8.27   Schematic circuit of series injection worst-case test system

voltage or current measurements on the output. The insertion loss with the notations of Fig. 8.27 is defined as:

$$IL = 20 \, \log \left(\frac{U_{g2}}{U_{g1}} \right) \Bigg|_{I_{o1} = I_{o2}} = 20 \, \log \left(\frac{I_{o1}}{I_{o2}} \right) \Bigg|_{I_{g1} = I_{g2}} \tag{8.30}$$

The subscript 1 indicates measurements made without the filter, and the subscript 2 refers to measurements with the filter. These subscripts correspond to the switch position as shown in Fig. 8.21.

Insertion loss can be measured by adjusting the output voltage of the signal generator to get a constant measuring instrument indication. The insertion loss expressed by the voltage of the signal generator, with the above notation of the second subscript, is as follows:

$$IL = 20 \, \log \left(\frac{U_{g2}}{U_{g1}} \right) \Bigg|_{I_{o1} = I_{o2}} \tag{8.31}$$

The 10 μF capacitors are needed to bypass the mains impedance and to form a HF short-circuit for the measurement signal. For series injection, an ideal voltage source would be needed. The impedance associated with any realizable voltage generator must be small in relation to the equivalent series resistance of the resonant circuit at resonance. If this condition is not met, a degradation in Q-factor occurs, and the worst-case insertion loss cannot be determined. Usually, a few milliohms of equivalent series resistance is required. To obtain such a low value, a voltage injection probe must be used. The voltage injection probe must have a large turn ratio and often must be terminated by low resistance. Furthermore, the inductance inserted by the voltage injection probe (as well as current detection probe) must be minute, since it limits the highest interface resonance frequency that can be attained with a given EMI filter. This condition requires very tight coupling between the two probe windings. With specially wound probes, about 50 nH equivalent series inductance can be achieved.

Because of parasitic inductances, the worst-case insertion loss can be measured only up to the megahertz range with serial injection. The upper limit of the measurement frequency range can be increased by parallel injection. The schematic circuit of a parallel injection worst-case insertion loss test setup is shown in Fig. 8.28 [334]. Similarly to serial injection, the insertion loss can be measured in two ways. Adjusting the output voltage of the signal generator to a constant value, the insertion loss can be calculated by Eq. (8.30). The other possibility is to adjust the output of the signal generator so that the measured voltage or current is constant. Applying this second adjusting method, the insertion loss, expressed by the current of the signal generator measured without the filter (I_{g1}) and with the filter (I_{g2}), is as follows:

$$IL = 20 \log\left(\frac{I_{g2}}{I_{g1}}\right) \tag{8.32}$$

The most important benefit of the parallel injection method is that no inductive injection probe is needed, since the injection is made by a highpass network that keeps the low-frequency biasing voltage off the current source and detector. This network is represented by the coupling capacitor, C, in Fig. 8.28. To establish a characteristic interfacial resonance, it is essential that the current source and the detector have a large resistance compared to the resistance of resonant circuits on the input and output at resonance. Typically, a 50 Ω current source and detector is large enough for parallel injection worst-case insertion loss measurement.

The upper measurement frequency limit will be determined by various parasitic inductances. In general, an upper limit of about 10 MHz can be expected.

The lower measurement frequency limit for both serial and parallel injection will be affected by the size of the matching choke coil and the 10 µF capacitor

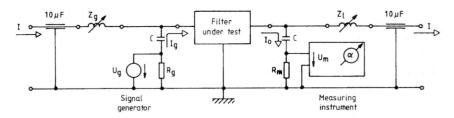

FIGURE 8.28   Schematic circuit of parallel injection worst-case test system

needed for HF isolation. The lower frequency limit can be calculated from the relationship as follows [96]:

$$Q = 2\pi \times f_u \times R \times C \tag{8.33}$$

As an example, the lower measurement frequency limit does not exceed the 20 kHz when the measurement is made with instruments of $R = 50\ \Omega$, assuming an average filter capacitance of 2 µF and a Q-factor of 10 for the matching choke coil.

8.3.4 Insertion Loss Test Methods for Multi-Terminal EMI filters

Up to this point, we have examined insertion loss measurements only for four-terminal networks (see Fig. 8.21), as described by most specifications. EMI filters, however, are seldom four-terminal networks; usually, they are designed to suppress HF noises on all lines including neutral. Many national standards do not specify an insertion loss test method for EMI filters with more than four terminals. But when this test method is described, it usually differs markedly. In general, measurements must be made separately for all terminal pairs, and the unused terminal pairs must be connected together to obtain the lowest insertion loss value. Some specifications require the unused terminals to be grounded, ungrounded, or linked to ground through a specific impedance. In the latter case, the impedance is usually the resistance of LISN used. In any case, this choice has an impact on the insertion loss values obtained by the measurement process. Therefore, it may be helpful to describe applicable test methods.

8.3.5 Insertion Loss Test Methods for EMI filters with Common-Mode Choke Coils

The original insertion loss test method was elaborated for EMI filters without any mutual inductance. But as common-mode choke coils gained wide application in engineering practice to provide a large common-mode insertion loss (although only a moderate one for differential-mode components), modification of insertion loss test methods became necessary.

An EMI filter that contains a common-mode choke coil provides an insertion loss for common-mode noise components that differs considerably from its insertion loss for differential-mode components. The main problem relates to the different test methods mentioned in Section 8.3.4. The measured value of the insertion loss can also change considerably under various test condition [146]. To demonstrate the problem, consider the insertion loss of the π-configuration shown in Fig. 8.29 under various test conditions.

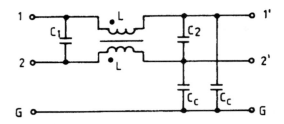

FIGURE 8.29 EMI filter configuration with common-mode choke coil

Measuring the insertion loss with the unused terminals open circuited (see Fig. 8.30a), we derive the equivalent circuit of Fig. 8.30b. Grounding the unused terminals for insertion loss measurement as shown in Fig. 8.31a, the equivalent circuit becomes as shown in Fig. 8.31b. Moreover, consider the insertion loss test method with connecting the unused terminals together as shown in Fig. 8.32a. For

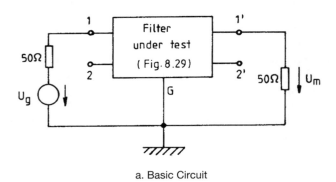

a. Basic Circuit

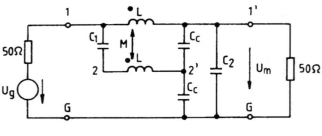

b. Equivalent Circuit

FIGURE 8.30 Insertion loss test method with open circuited unused terminals

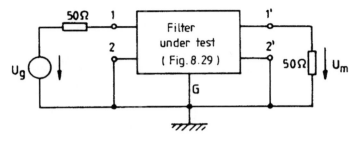

a. Basic Circuit

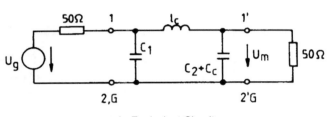

b. Equivalent Circuit

FIGURE 8.31 Insertion loss test method with grounded unused terminals

this insertion loss measuring, the equivalent circuit is as illustrated in Fig. 8.32b. The difference between the measurable insertion loss is already obvious, if only on the basis of equivalent circuits. The circuit $L - C_c$ in Fig. 8.30b provides considerably higher voltage attenuation than the $l_c - (C_2 + C_c)$ circuit in Fig. 8.31b, since inductance l_c is only the leakage between the windings of the common-mode choke coil. We would not expect the insertion loss as measured with open circuited versus connected unused terminals to vary a great deal.

For EMI filters with a common-mode choke coil, it is especially important to specify the noise suppression characteristics separately in terms of common-mode and differential-mode insertion loss. The common-mode and differential-mode insertion losses are measured as shown in Figs. 8.33 and 8.34, respectively. Because the commercial instruments have unbalanced (asymmetrical) inputs and outputs, the test method of Fig. 8.34 is rarely used, but the unbalanced instrumentation can be augmented for differential-mode insertion loss measurements with 180° power splitters. The splitter divides the input power equally between the two output ports while maintaining 180° phase difference between them. The splitter can be used as combiner, too, providing an output that is the difference between signals on the two input ports. The schematic diagram for insertion loss measure-

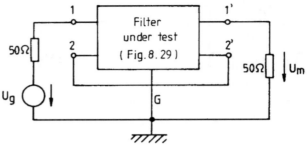

a. Basic Circuit

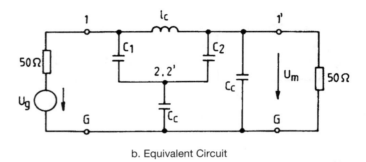

b. Equivalent Circuit

FIGURE 8.32   Insertion loss test method with grounded unused terminals

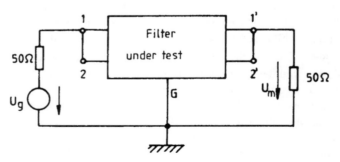

FIGURE 8.33   Common-mode insertion loss test method

ment with splitters is shown in Fig. 8.35 [286]. As of the time of this writing, splitters are commercially available only for 50 Ω measuring systems, so insertion loss measurements for EMI filters containing common-mode choke coils are usually made using 50 Ω systems.

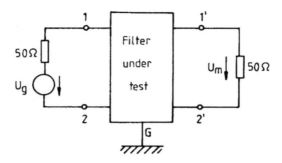

FIGURE 8.34 Differential-mode insertion loss test method

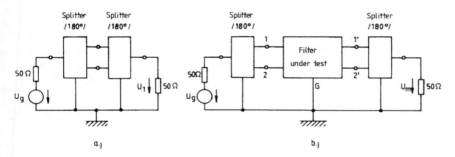

FIGURE 8.35 Schematic circuit for measuring differential-mode insertion loss with splitters

There are disadvantages in the use of splitter for such measurements, and one of the most annoying is that the worst-case insertion loss can be measured only with a very complicated test setup. During the latter part of the 1970s, new measurement techniques were suggested to simplify the insertion loss test method. One solution is the so-called *pseudo-differential measuring technique* [146]. The schematic diagram of pseudo-differential test method is shown in Fig. 8.36. Just as the mains impedance for conducted noise voltage measurements has been standardized, it is also proposed to terminate the common-core EMI filter under test with this standardized impedance (Z_L in Fig. 8.36). The source impedance, Z_g, would be calculated from the rated voltage and rated current of the EMI filter. In practice, the HF noises produced are essentially disturbances by nature and are coupled capacitively as common-mode noise. For such a case, the source impedance is suggested to be a capacitor. Of course, for worst-case insertion loss measurement, the source and load impedances are the required matching elements (see Section 8.3.3). It has been demonstrated that the pseudo-differential insertion loss and the ideal, true differential-mode insertion loss are in agreement, with reasonable approximation.

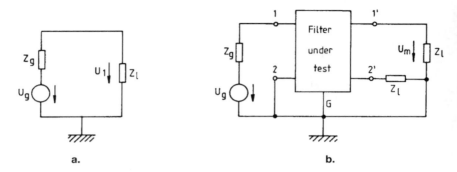

FIGURE 8.36 Schematic circuit for the pseudo-differential insertion loss test method

The other proposed test method for EMI filters with a common-mode choke coil is the so-called *normal-mode test* [286]. A schematic diagram of this method is shown in Fig. 8.37. The normal-mode test is actually a combination of common-mode and differential-mode insertion loss test methods. The normal-mode measurement sums the common-mode and differential-mode insertion loss. The normal-mode test is good for measuring the insertion loss of EMI filters designed to suppress of noises that are essentially coupled only to one line.

EMI filters manufacturers generally provide the common-mode and differential-mode insertion loss curves separately. For performance comparisons among different EMI filters, the applied test method should also be revealed, since the above discussion tends to prove that the rated insertion loss value depends considerably on the applied measurement technique. However, the lack of worst-case insertion loss curves is a much greater deficiency. The worst-case insertion loss curve for EMI filters with common-mode choke coils is more often neglected than for commercial EMI filters.

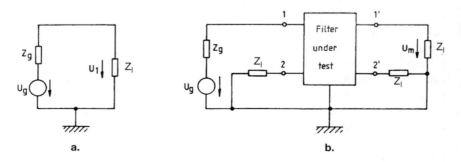

FIGURE 8.37 Schematic circuit for the normal-mode insertion loss test method

8.3.6 Insertion Loss Test System Layout

Because insertion loss measurement instrumentation measures very small signals, it is sensitive to disturbances from external electromagnetic fields. For this reason, such measurements are usually made in electromagnetically shielded enclosures.

Assuming proper shielding against external disturbances, there remain two basic hazards to avoid when designing or using insertion loss test systems: (1) coupling between input and output circuits and (2) grounding impedance.

Stray coupling between the signal source and the test instruments, especially for high insertion loss measurements, is very critical. This can be avoided by separating the input and output leads (which is more important when long wires are employed). The degree of parasitic coupling may be checked by performing an insertion loss measurement [286]. Parasitic capacitive coupling may be detected by leaving the filter output open circuited. Parasitic inductive coupling may be detected by shorting the output wires to ground. The measured isolation level should be at least 20 dB away from the intended measurement level.

It is also important to avoid common ground return impedance between input and output circuits. A bad solution for grounding is shown schematically in Fig. 8.38a, where a common grounding impedance exists for the current in the input circuit, I_i, and the output circuit, I_o. Here, the measurable voltage is:

$$U_m = U_o + (I_i - I_o) \times Z_g \qquad (8.34)$$

The filter performance is described only by the first term in the above relationship; the second term shows the effect of source current returning through the common grounding impedance. Even though this impedance is very low, the resultant voltage drop ($I_i \times Z_G$) can be on the order of the magnitude of the voltage appearing on the output port of the EMI filter under test (U_o).

The correct insertion loss measurement layout is shown in Fig. 8.38b, where the input and output ground connections are separated. With this layout, the voltage indicated by the measurement instrument is as follows:

$$U_m = U_o - I_o \times Z_{Go} \qquad (8.35)$$

The inaccuracy in insertion loss measurement will be much lower, since the current in the measuring circuit (I_o) is much lower than in the source circuit. The isolation between the input and output ground return may be evaluated by inserting the EMI filter and connecting it to the signal generator. If we ground the EMI filter output to the filter case, any signal so coupled should be well below the desired insertion loss measurement level.

The EMI filter under test must be grounded using the shortest possible wire with a suitable cross-section. Otherwise, the parasitic inductance of the grounding wire can form a resonant circuit with the capacitors of the EMI filter. The best, of

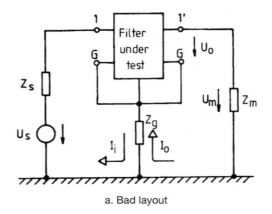

a. Bad layout

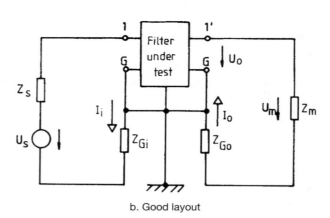

b. Good layout

FIGURE 8.38 Layout of an insertion loss test setup [286]

course, is to make Z_{Gi} and Z_{Go} as small as possible by using wide braided straps and a common ground plane (copper, aluminum, or galvanized steel plate) for the generator, filter, and measuring receiver.

9

EMI Filter Design

EMI filters are usually designed to provide a given insertion loss. Before determining the required insertion loss for a particular application, we first minimize the requirement by following appropriate EMI-preventive measures as discussed in Chapter 7. The next step is to decide whether the remaining emissions can be suppressed by an off-the-shelf filter chosen from a catalog, or if it is necessary to design and produce a custom filter. Finally, because many semiconductor devices may be operating at the same location, we must determine whether it is more economical to provide common EMI filtering or to place individual filters on each piece of electrical equipment. This chapter provides necessary guidelines for EMI filter design, for both common and individual filtering requirements.

The first step in designing an EMI filter is to determine the most appropriate filter configuration (see Chapter 8). Many design equations for lowpass filters are well known in telecommunication technology, but they frequently are not applicable to EMI filters. Primarily, this is because telecom filter equations were developed for matched source and load impedances, and this condition does not exist in power electronic equipment. In addition, these equations cannot handle the limitations for serial and parallel impedances that are common for powerline filters. EMI filter design is not simply a matter of calculating the element values with some applicable equation; many other points of view must be considered. For instance, a one-stage Butterworth filter with a cutoff frequency of 10 kHz calls for a capacitance of 0.25 μF and an inductance of 1.13 mH. Such an inductance can be designed for a low current rating but it is too great for high-power filters. Therefore, a different set of element values must be used.

In general, designing a filter to provide a given amount of insertion loss gives more satisfactory results than using methods developed for voltage attenuation.

To simplify calculations, relationships and charts have been developed with the assumption of resistive source and load impedances. Since this condition does not always exist in practice (because of mismatched impedance conditions), a broad margin of safety is usually added to the calculated insertion loss; that is, the filters are overspecified. To avoid overspecification, we will study some design methods geared specifically to impedance mismatch conditions. These methods are directed at analyzing the effect that the source and load mismatch has on insertion loss and designing the filter element values for worst-case conditions.

The frequency response of an EMI filter may be considerably distorted by parasitic resonances and nonideal characteristics of the noise filter elements. In extreme cases, parasitic resonance can even cause a gain in the stopband. Therefore, some HF losses should be incorporated in the filter configuration to prevent resonance development. The effects of a filter element's nonideal HF parameters are usually considered only after the design process by developing the complete circuit model and then analyzing it.

The EMI filter design is completed by determining its internal structure; i.e., the grounding-bounding schemes, component layout, cable routing and separation, and shielding. A bad layout may detract considerably from the calculated and expected insertion loss.

Although the attenuation of conducted emissions is the primary function of an EMI filter, a well designed filter also reduces the system's susceptibility to malfunction due to incoming conducted disturbances.

9.1 EMI FILTER DESIGN FOR INSERTION LOSS

EMI filters are generally characterized by their insertion loss rather than by voltage attenuation. Although design methods developed for voltage attenuation are well known, they are not applicable here, as voltage attenuation and insertion loss are not equivalent (see Subchapter 8.1).

9.1.1 Design of LC- and π-Filters Terminated by Resistive Sources and Loads

In EMI filters, the LC- and π-configurations are commonly used. Because industry practice is to define the insertion loss in a matched 50 Ω system, a graphical method has been developed for determining filter element values when the value of the needed insertion loss is known [289].

The insertion loss of the LC configuration terminated by equal and resistive source and load impedances (see Fig. 9.1) is as follows:

$$\text{IL} = 10 \ \log\left(1 + F^2 \times \frac{D^2}{2} + F^4\right) \qquad (9.1)$$

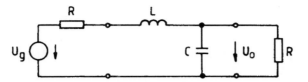

FIGURE 9.1 LC-configuration with resistive source and load impedances

In the relationship of Eq. (9.1), F is the normalized frequency in reference to the cutoff frequency of the filter circuit:

$$F = \frac{f}{f_o} = \frac{1}{2\pi} \times \sqrt{\frac{1}{L \times C}} \tag{9.2}$$

The quantity D in Eq. (9.1) is a function of damping ratio. This parameter relates the magnitudes of the filter element to the magnitudes of the source and load resistance:

$$d = \frac{L}{C \times R^2} \tag{9.3}$$

For ideal damping (i.e., for d = 1), the filter becomes a Butterworth lowpass filter. The quantity D is as follows:

$$D = \frac{1 - d}{\sqrt{d}} \tag{9.4}$$

The characteristic insertion curve is shown in Fig. 9.2. It is worth noting that no resonance occurs in the LC-configuration when the source and load impedances are equal and resistive. For d < 1, the low load resistance, and for d > 1, the high source resistance, also represents an effective damping. Values of d less than 1 result in insertion loss curves identical to those obtained when d is higher than 1; i.e.:

$$IL\,(d) = IL\,(\frac{1}{d}) \quad \text{for} \quad d \neq 1 \tag{9.5}$$

The chart in Fig. 9.3 presents the insertion loss as a function of the frequency with the parameter of the cutoff frequency (f_o) for an ideally damped (d = 1) one-stage LC-filter. The parameter f_o can be read on the scale across the chart. The insertion loss curve of a nonideally damped one-stage LC-configuration in the tran-

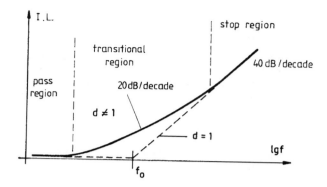

FIGURE 9.2 Characterization of insertion loss from Fig. 9.1

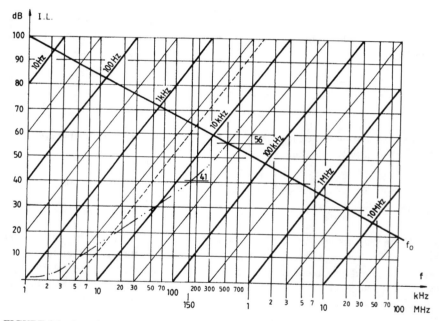

FIGURE 9.3 Insertion loss chart for ideally damped LC-configuration with resistive source and load impedances (Butterworth response) [289]

sition region varies by 20 dB/decade slope. The insertion loss chart in the transition region is shown in Fig. 9.4. The horizontal axis of the figure is the normalized frequency given by Eq. (9.2). Figure 9.4 is used with Fig. 9.3 to determine the insertion loss curves for a nonideally damped filter.

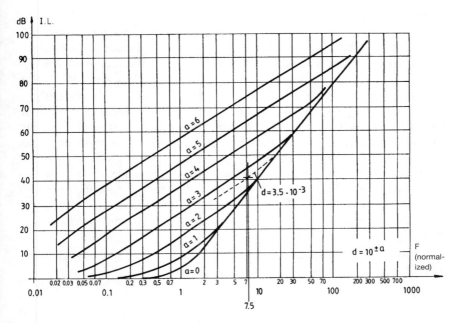

FIGURE 9.4 Insertion loss chart for nonideally damped LC-configuration with resistive source and load impedances [289]

The charts shown in Figs. 9.3 and 9.4 are not only useful for determining the insertion loss of one-stage LC-filter with known element values but for designing EMI filters as well. In general, the strictest insertion loss requirement is on the lowest limit of the frequency range to be filtered. The minimum insertion loss at this frequency determines a point in Fig. 9.3. By placing a straight line parallel to the graph lines and extending it through this point, the desired cutoff frequency, f_o, can be found on the scale. After plotting out the value of one element, taking into account practical considerations and impedance limitations, the other element value can be calculated using Eq. (9.2). After determining all element values, the damping ratio, d, is calculated. We then check to make sure the cutoff frequency is in the transition region and the actual insertion loss is not smaller than the desired value.

For a demonstration of how to use the charts of Figs. 9.3 and 9.4, let us examine two problems. First, assume we want to design a one-stage LC-configuration that provides an insertion loss of 60 dB at 150 kHz. The required cutoff frequency can be read from the horizontal axis by placing a straight line with 40 dB/decade slope through the 60 dB-150 kHz crossing point in Fig. 9.3. This line (shown as a dashed line in the figure) crosses the f_o line at $f_o = 5$ kHz. But before establishing the circuit element values, we first calculate the damping ratios for which the frequency

150 kHz is not in the transition region. Since F = 150/5 = 30, the chart in Fig. 9.4 shows that the required insertion loss can be provided when the condition $10^{-3} <$ d $< 10^3$ is met. After determining the cutoff frequency and damping ratio values, the filter elements can be designed within the practical limits for one element.

The second problem is to find the insertion loss of a one-stage LC-filter with a cutoff frequency of 20 kHz at f_1 = 150 kHz and f_2 = 500 kHz. Since F = 150/20 = 7.5, Fig. 9.4 shows that the frequency f_1 is in the transition region. Plotting with extrapolation the actual d value line, the insertion loss at 150 kHz is about 41 dB. The 500 kHz frequency is out of the transition region, so the insertion loss at this frequency can be determined from Fig. 9.3. Plotting a straight line with a 40 dB/ decade slope from the 20 kHz point on the horizontal axis, the desired insertion loss is about 56 dB. The insertion loss curve of this filter configuration is drawn with a dot-dash line in Fig. 9.3.

The charts of Figs. 9.3 and 9.4 are also useful for finding the insertion loss of CL-configurations [289].

The insertion loss of a π-configuration terminated by equal and resistive source and load impedances, shown in Fig. 9.5, is as follows:

$$IL = 10 \; \log \, [\, 1 + (F \times D)^2 - 2 \times F^4 \times D + F^6] \qquad (9.6)$$

For calculating the normalized frequency, F, first determine the cutoff frequency of the π-filter circuit. This is done as follows:

$$f_o = \frac{1}{2\pi} \times \sqrt[3]{\frac{2}{R \times L \times C^2}} \quad \text{for} \quad d \neq 1$$

$$f_o = \frac{1}{2\pi} \times \sqrt{\frac{2}{L \times C}} \quad \text{for} \quad d = 1 \qquad (9.7)$$

The damping ratio for calculating the constant d is:

$$d = \frac{L}{2 \times C \times R^2} \qquad (9.8)$$

The quantity D in Eq. (9.6) yields:

$$D = \frac{1 - d}{\sqrt[3]{d}} \qquad (9.9)$$

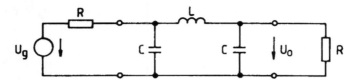

FIGURE 9.5 π-Configuration with resistive source and load impedances

The characteristic insertion loss curve of a π-configuration is shown for different damping ratios in Fig. 9.6. Contrary to a two-element filter, the character of insertion loss curve depends on the magnitude of the damping ratio. When d is equal to 1, the response is as the Butterworth response curve. When the damping ratio is higher than 1, the overdamped response arises. When the damping ratio is less than 1, the underdamped response has a hump with maximum value of:

$$IL_m = 10 \log \left(1 + \frac{4 \times D^3}{27}\right) \tag{9.10}$$

at the normalized frequency where

$$F_{max} = \sqrt{\frac{D}{3}} \tag{9.11}$$

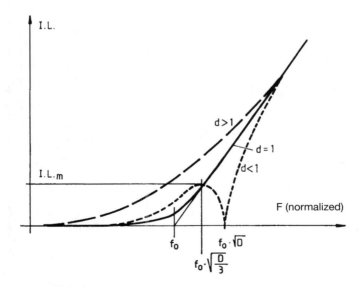

FIGURE 9.6 Insertion loss characterization of π-configuration shown in Fig. 9.5

The minimum insertion loss point of the hump falls at the normalized frequency where

$$F_{min} = \sqrt{D} \tag{9.12}$$

The value of the minimum point is zero for ideal components, but for a real case the depth of the dip is determined by the Q-factor of the circuit.

Figure 9.7 presents an insertion loss chart for an ideally damped π-configuration. Figures 9.8 and 9.9 give the same for an overdamped and underdamped π-configuration, respectively. These charts are used in the same manner as was discussed above for two-element filters.

Figures 9.7 through 9.9 are useful tools for designing element values of π-filters, although the knowledge of the cutoff frequency and the inductance or capacitance is not sufficient for designing the entire circuit; we must also determine the resistance value or damping ratio. The reason for this is that the cutoff frequency also depends on the source and load resistance.

The charts in Figs. 9.7 through 9.9 are also valid for T-configurations. The insertion loss can be calculated by Eq. (9.6), but in this case the cutoff frequency is calculated as:

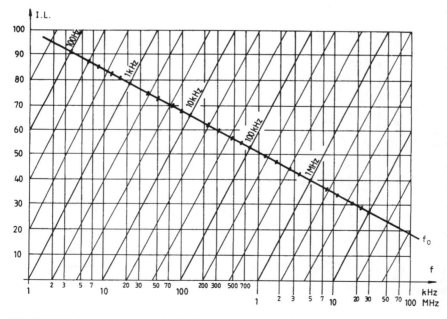

FIGURE 9.7 Insertion loss chart for ideally damped π-configuration with resistive source and load impedances (Butterworth Response) [289]

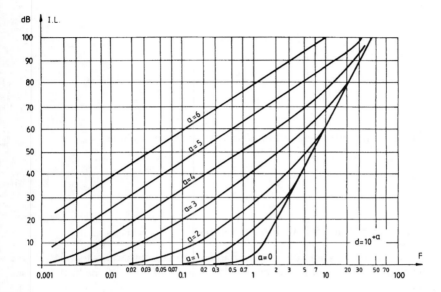

FIGURE 9.8 Insertion loss chart for overdamped π-configuration with resistive source and load impedances [289]

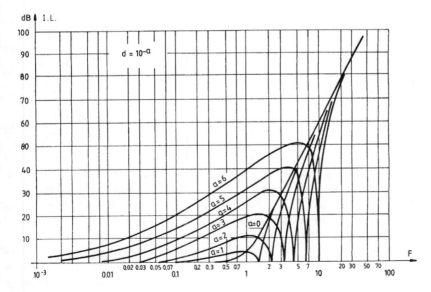

FIGURE 9.9 Insertion loss chart for underdamped π-configuration with resistive source and load impedances [289]

$$f_o = \frac{1}{2\pi} \times \sqrt[3]{\frac{2 \times R}{L^2 \times C}} \quad \text{for} \quad d \neq 1$$

$$f_o = \frac{1}{2\pi} \times \sqrt{\frac{2}{L \times C}} \quad \text{for} \quad d = 1 \tag{9.13}$$

and the damping ratio used in Eq. (9.9) is calculated as follows:

$$d = \frac{R^2 \times C}{2L} \tag{9.14}$$

9.1.2 Design Method for Multistage Filters

Powerline filters commonly are of multistage configuration when, despite the associated filter element limitations, a high insertion loss has to be provided (see Section 8.2.3). The multistage filter configuration is also designed to eliminate the effects of impedance mismatch.

The voltage attenuation of multistage LC-configuration consisting of identical inductances and capacitance is given by Eq. (8.27). This equation provides a conversion relationship for multistage filter design. Introducing the normalized frequency, F:

$$F = 2\pi \times f \times \sqrt{L_m C_m} \tag{9.15}$$

Eq. (8.27) then yields:

$$K_n(F) = 1 - \alpha_{n1}\left(\frac{F^2}{n^2}\right) + \alpha_{n2}\left(\frac{F_2}{n^2}\right)^2 - \alpha_{n3}\left(\frac{F_2}{n^2}\right)^3 + \ldots \pm \alpha_{nn}\left(\frac{F_2}{n^2}\right)^n \tag{9.16}$$

The voltage attenuation described by the above formula is plotted in Fig. 9.10, where the parameter n is the stage number. It should be noted that, although this chart gives the voltage attenuation and not the insertion loss, it can be used for designing the element values because these two characteristics may be considered equal for multistage filter configurations.

The optimal stage number can be determined as a function of the reactances and the lower limit of the frequency range to be filtered. The optimal calculation was based on obtaining minimal reactances [9]. This condition serves to minimize size, weight, and cost. The optimal number of filter stages can be found by reference to Fig. 9.11. For calculating the value of the normalized frequency, F, the

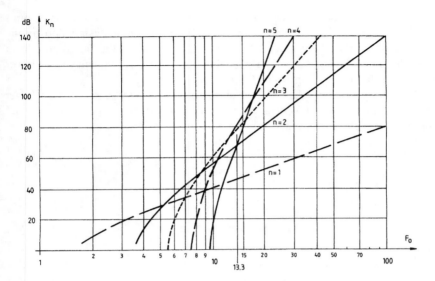

FIGURE 9.10 Voltage attenuation chart for multistage LC-filter configuration

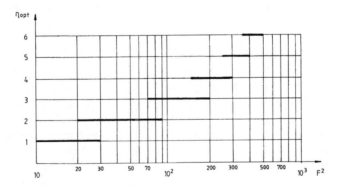

FIGURE 9.11 Optimal number of powerline filter stages as a function of lower frequency range limit and reactance limitations

lower frequency limit in hertz, the inductance in henries, and the capacitance in farads should be substituted into Eq. (9.15).

Sometimes there is no strict limitation on LC-product, but we know the lower limit of the frequency range to be filtered and the desired insertion loss value. In such a case, for economic reasons, the LC-product has to be designed so that the normalized frequency at the lower frequency limit will match as closely as possible the envelope shown in Fig. 9.10.

Let us see an example how to design a multistage LC-filter when a strict limitation exists on the value of the LC-product. It is desirable to design a powerline filter for which the LC-product is lower than 200 µHµF, and it should provide at least 60 dB insertion loss at 150 kHz. At the same time, this frequency is the lower limit of the frequency range to be filtered.

First, the optimal number of filter stages has to be determined. For the lower frequency limit of f_L = 150 kHz, the normalized frequency can be calculated according to Eq. (9.15), with good approximation, as follows:

$$F_L^2 (150\text{kHz}) = 4\pi^2 \times f_L^2 \times L_m C_m \approx L_{m'} C_{m'} \qquad (9.17)$$

where

$$L_{m'} = \text{sum of the inductances in } µH$$

$$C_{m'} = \text{sum of the capacitances in } µF$$

By this approximation formula and Fig. 9.11, for $F^2 = 200$, the optimal stage number is 4. The maximum amount of insertion loss at normalized frequency of

$$F = 2\pi \times 150 \times 10^3 \times \sqrt{200 \times 10^{-12}} = 13.3 \qquad (9.18)$$

is about 80 dB. That means that the desired insertion loss can be provided with the limitation of LC < 200 µHµF. The chart of Fig. 9.10 shows that a two-stage or three-stage filter configuration also can be used. In the example, a three-stage filter design can be offered because the 78 dB insertion loss represents an efficient safeguard, and for the three-stage filter configuration, the impedance mismatch hardly affects the insertion loss curve. After calculating the cutoff frequency of the LC-circuit and setting the value of the inductance or capacitance, the other filter element can be easily calculated.

9.2 CALCULATION OF WORST-CASE INSERTION LOSS

EMI filter insertion loss is highly dependent on the character and magnitude of the source and load impedances. Generally, it is difficult to secure design information about generator and load impedances. Although the HF impedance could be analyzed at a given mains point, many EMI filters are not designed to work at a specific location. Moreover, equipment designers are not concerned with impedance behavior outside the operating frequencies of their equipment. If impedance mismatch effects could be reduced through the use of proper filter configurations, it would be more difficult to justify insertion loss overspecification to compensate for the mismatched conditions. This situation indicates a need to calculate the worst-case insertion loss of a filter configuration with known elements at the design stage rather than simply measuring it after the filter is manufactured.

As shown in Chapter 8, the insertion loss of EMI filters operating under mismatched impedance conditions can be calculated using Eqs. (8.11) and (8.12). In practice, we do not perform a complicated functional analysis to determine the worst-case insertion loss; rather, we refer to relationships that are obtained with some simplification [31, 322, 323]. Although these relationships are valid only for certain source and load impedance combinations, they give quite a realistic description of EMI filter performance under impedance mismatch conditions.

To simplify the formulas, we will here introduce some new notation for the source and load impedances, the input impedances of the filter circuit seen from the input (subscript 1), and those seen from the output (subscript 2). Thus, let:

$$\overline{Z}_g = R_g + j \times X_g$$

$$\overline{Z}_L = R_L + j \times X_L$$

$$\overline{Z}_{1i} = R_{1i} + j \times X_{1i}$$

$$\overline{Z}_{2i} = R_{2i} + j \times X_{2i} \tag{9.19}$$

The input impedance at the primary terminals, with the secondary terminals open, can be given as:

$$\overline{Z}_{1o} = Z_{1o} \times \exp(-j\theta_{1o}) = R_{1o} + j \times X_{1o} \tag{9.20}$$

The input impedance at the secondary terminals, with primary terminals open, is:

$$\overline{Z}_{2o} = Z_{2o} \times \exp(-j\theta_{2o}) = R_{2o} + j \times X_{2o} \tag{9.21}$$

The above impedances can be expressed by the impedance and chain characteristics as follows [31]:

$$\overline{Z}_{1o} = Z_{11} = \frac{A_{11}}{A_{21}} \tag{9.22}$$

$$\overline{Z_{2o}} = Z_{22} = \frac{A_{22}}{A_{21}} \tag{9.23}$$

According to the above notations, let the input impedances for short-circuited, opposite-side terminations be:

$$\overline{Z_{1s}} = Z_{1s} \times \exp(-j\theta_{1s}) = R_{1s} + j \times X_{1s} = Z_{11} - \frac{Z_{12} \times Z_{21}}{Z_{22}} = \frac{A_{12}}{A_{22}} \tag{9.24}$$

$$\overline{Z_{2s}} = Z_{2s} \times \exp(-j\theta_{2s}) = R_{2s} + j \times X_{2s} = Z_{22} - \frac{Z_{12} \times Z_{21}}{Z_{11}} = \frac{A_{12}}{A_{11}} \tag{9.25}$$

Formulas serving for calculation the worst-case insertion loss have been elaborated for many impedance conditions and suggestions [31, 322, 323]. The relationships that are the most important for analyzing the effect of impedance mismatches on an EMI filter design are summarized below.

1. Low Source Impedance

For a very low source impedance ($Z_g \approx 0$), the worst-case insertion loss is achieved when

$$R_L = 0 \quad \text{and}$$

$$X_L = \frac{R_{2i}^2 + X_{2i}^2}{X_{2i}} \tag{9.26}$$

For a load impedance as given by Eq. (9.26), the worst-case insertion loss can be calculated as:

$$IL_{min} = |A_{11}| \times \left| \frac{R_{2s}}{X_{2s}} \right| = |A_{11}| \times \cos\theta_{2s} \tag{9.27}$$

2. High Load Impedance

For a nearly open-circuited secondary termination ($Z_L \approx \infty$), the worst-case insertion loss is achieved when

$$R_g = 0$$

$$X_g = -X_{1o} \tag{9.28}$$

For a source impedance as given by Eq. (9.28), the worst-case insertion loss is:

$$IL_{min} = |A_{11}| \times \left| \frac{R_{1o}}{X_{1o}} \right| = |A_{11}| \times \cos\theta_{1o} \qquad (9.29)$$

3. Known Source Impedance

In engineering practice, the magnitude and character of the source impedance may be approximated, and an EMI filter should be designed for load impedances that vary in a wide range. This may be the case when a special noise suppression design is required for power electronic equipments with high power ratings. In this case, the HF impedance of the EMI source can be approximated by measurement or calculation. Moreover, in many applications, the only real evaluation of the filter effectiveness into the equipment will be made by a conducted EMI test using a well defined LISN impedance (see "artificial networks" in section 4.3). So at least the filter can be optimized to "pass the test."

The worst-case insertion loss will be achieved when the load impedance is only reactive:

$$R_L = 0 \quad \text{and}$$

$$X_L = \frac{X_{2i}^2 - X_g^2 + R_{2i}^2 - R_g^2 + \sqrt{z}}{2(X_g - X_{2i})} = \frac{Z_{2i}^2 - Z_g^2 + \sqrt{z}}{2(X_g - X_{2i})} \qquad (9.30)$$

where

$$z = (X_g - X_{2i})^4 + (R_g^2 - R_{2i}^2)^2 + 2 \times (X_g - X_{2i})^2 \times (R_g + R_{2i})^2$$

The minimum envelope of the insertion loss for a load impedance as per Eq. (9.30) can be calculated by the relationship below, but only when the resistive part of the source impedance is not zero or very low. The worst-case insertion loss is:

$$IL_{min} = |A_{11}| \times \left| \frac{Z_g + Z_{1o}}{Z_{1o}} \right| \times \left[\frac{\sqrt{z_1}}{2 \times R_g} - \frac{\sqrt{z_2}}{2 \times R_g} \right] \qquad (9.31)$$

where

$$z_1 = (R_{2i} + R_g)^2 + (X_{2i} + X_g)^2$$

$$z_2 = (R_{2i} - R_g)^2 + (X_{2i} - X_g)^2$$

It should be noted that for a filter configuration with a high Q-factor ($R_{2i} = 0$), the worst-case load impedance is just the one used for worst-case measurements (see Section 8.3.3); i.e., the complex conjugate of the input impedance.

For a reactive source impedance (i.e., for $R_g = 0$), the worst-case performance of the EMI filter is when:

$$R_L = 0 \quad \text{and}$$

$$X_L = \frac{R_{2i}^2}{X_g - X_{2i}} - X_{2i}$$

(9.32)

The worst-case insertion loss for reactive source impedance is given by:

$$IL_{min} = |A_{11}| \times \left| \frac{Z_g + Z_{1o}}{Z_{1o}} \right| \times \sqrt{\frac{R_{2i}^2}{(X_g - X_{2i})^2 + R_{2i}^2}}$$

(9.33)

4. Known Load Impedance

In some cases, the load impedance can be approximated, and it might be suggested that the EMI source impedance varies in a wide range. This could be the case when we intend to design common noise filtering for multiple electrical equipments, and the impedance of the supply network at the connection point can be determined. In this case, the EMI source impedance for the filter may vary in a wide range, depending on the number of equipments switched on and their operational mode. The worst-case insertion loss for a known load and unknown source impedance can be determined via Eqs. (9.30) through (9.33), with the following substitutions:

$$\begin{matrix} Z_g \rightarrow Z_1 & A_{11} \rightarrow A_{22} \\ Z_{1o} \rightarrow Z_{2o} & Z_{2i} \rightarrow Z_{1i} \end{matrix}$$

(9.34)

9.3 DESIGN METHOD FOR MISMATCHED IMPEDANCE CONDITION

The standard EMI filter design procedure is first to determine the filter configuration, then calculate its element values, and finally to analyze the impedance mismatch effects. This is actually an iteration method, and a major disadvantage is that it does not consider all desired performance characteristics at once. If a particular derived characteristic needs improvement, the calculations must be repeated with another set of initial data. This iteration method is not ideal for engineering practice; therefore, a new method has been developed to account for impedance mismatch effects. This method, which its creator calls the *worst-case EMI filter design method* [9], makes it possible to design an EMI filter configuration for which the insertion loss does not drop below a given minimum value, even for a worst-case source and load impedance mismatch.

EMI filter performance (particularly that of powerline filters) can be formulated best as shown in Fig. 9.12. Around the operational (mains) frequencies, the EMI filter should provide neither loss nor gain higher than a given value. Between the operational and the lower limit of the frequency range to be filtered, it must not develop resonances. This can be formulated by saying that the gain in this frequency range should not exceed a given value, $-IL_1$. Because many national specifications do not include any EMI filter requirements in the transitional frequency range, a very high gain is regarded as acceptable. In the stopband, the insertion curve or its characteristic points can be regarded as known.

For the worst-case EMI filter design method, the performance in the stopband is formulated so that the EMI filter must provide a minimum insertion loss (IL_s) above the lower frequency limit (f_L) independent of the source and load impedances. This requirement is based on the suggestion that the insertion loss requirements in the stopband are the most stringent at the lower frequency limit of the range to be filtered. To simplify the calculations, the attenuation factor will be studied instead of the insertion loss (see Fig. 9.13). The attenuation factor creates

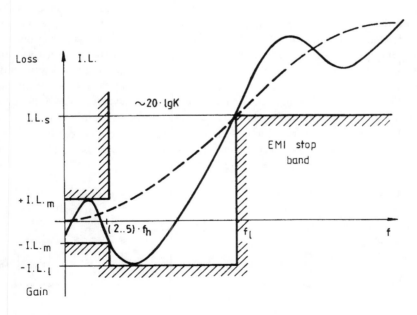

FIGURE 9.12 Insertion loss characteristics of EMI filters

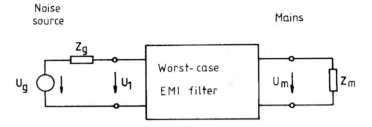

FIGURE 9.13 Attenuation factor interpretation for worst-case EMI filter design

a relationship between the noise source and the HF disturbance on the mains, and is defined as:

$$K = \frac{U_g}{U_m} \tag{9.35}$$

The attenuation factor is not only closely related to the insertion loss but also properly describes the real noise suppression situation, since engineering goal is to reduce the noise voltage to an acceptable value.

The worst-case insertion loss design method is based on the assumption that any injurious effects caused by mismatched source and load impedances can be reduced to an acceptable level by connecting an appropriate matching network to the input and output of the EMI filter. Given that the insertion loss reduction (and possibly signal gain) results from resonances, the simplest matching network consists of resistors only. Figure 9.14 is a schematic circuit where noise suppression by an EMI filter is completed with lossy matching circuits. With proper values of resistors R_p and R_s, the attenuation factor (and so the insertion loss) will not drop below a given value under any source and load impedance condition.

Although these lossy matching circuits could be designed quite easily, and they could provide proper impedance matching, this does not represent a practical engi-

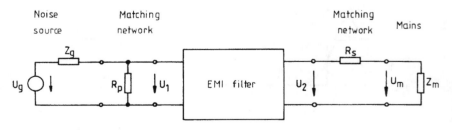

FIGURE 9.14 EMI filter with lossy matching elements

neering solution, since the operational frequency losses would be unacceptably high. For reducing these losses, the matching networks and the noise suppression components have to be built as shown in Fig. 9.15. The value of capacitor C_p must be selected so that its impedance will be much higher than R_p at operational (mains) frequency, and lower than R_p above the frequency limit F_L. The operational frequency losses of the resistor R_s are reduced by the parallel choke coil L_s which creates a low-impedance path at low frequencies. The inductance of L_s should be selected such that $L\omega$ is higher than R_s at frequency limit f_L. The the matching networks in Figs. 9.14 and 9.15 suggest a filter configuration with high primary input and low secondary input impedances. Such input impedances are typical of LC-circuits, which are the most commonly used EMI filters configuration.

Marking the attenuation factor (i.e., the insertion loss at zero source and infinitely high load impedances) with K_o, the attenuation factor of the EMI filter completed with matching networks as shown in Fig. 9.15 can be given as the product of three factors:

$$K = \left| \frac{U_g}{U_m} \right| = \left| \frac{U_g}{U_1} \right| \times K_o \times \left| \frac{U_g}{U_m} \right| \tag{9.36}$$

The minimum of the attenuation factor K (i.e., the worst-case insertion loss) can be calculated by determining the minimum values of the first and the third factors. The minimum of the first factor is to be analyzed as a function of the resulting impedance of Z_g and the "P" matching network, and that of the third factor as a function of the resulting impedance of Z_m and the "S" matching network.

The matching effect of the "P" and the "S" networks can be best characterized by the ratio of the worst-case attenuation factor after inserting the matching network to the ideal attenuation factor at the lower limit of the frequency range to be filtered (f_L). This ratio is always higher than 1 because, as a result of mismatched impedance condition, the voltage attenuation is always less than ideal. For a better matching effect, choose a lower ratio. By designating the matching effect of the "P" network as μ_p and that of "S" network as μ_s and then setting up the values of

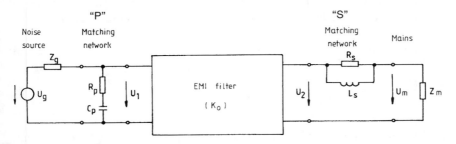

FIGURE 9.15 EMI filters with "P" and "S" matching networks

μ_p and μ_s, we can determine limitations for the required resistance of the serial and the parallel matching networks as a function of input impedances. These limitations for the resistors of the matching networks are as follows [9]:

$$R_p < X_{1o}(\omega_L) \times \sqrt{\mu_p^2 - 1}$$

$$R_s > \frac{X_{2s}(\omega_L)}{\sqrt{\mu_s^2 - 1}} \tag{9.37}$$

where

$X_{1o}(\omega_L) =$ the open-circuit impedance at the lower limit of the frequency range to be filtered, seen from the input

$X_{2s}(\omega_L) =$ the short-circuit impedance at the lower limit of the frequency range to be filtered, seen from the output

The above impedances are merely reactances, because the EMI filters consist only of choke coils and capacitors.

Connecting a "P" matching network to the input of the EMI filter, the source impedance (Z_{pr}) becomes the parallel resultant of "P" and Z_g, which is in magnitude and character of any kind. The load impedance of the EMI filter (Z_{sec}) is the resultant impedance of "S" and Z_m, which also can be any kind of magnitude and nature. We also may set up other limits for the resistors of the matching networks. The "P" matching network has to eliminate the mismatched effect of Z_g by reducing the input impedance seen from the noise source under the minimum of the primary input impedance, Z_{1i}. The "S" matching network reduces the mismatch effect of Z_m by increasing the input impedance seen from the mains to be filtered above the minimum of the secondary input impedance, Z_{2i}. The matching effect on the input impedances can be characterized by ε, which represents the ratio of the input impedances with and without the matching networks. Having introduced the ratio ε, the above limitations can be mathematically expressed as [9]:

$$|Z_{pr}| = |Z_g \times Z_p| < 2 \times \varepsilon_p \times Z_{1i_{min}} \tag{9.38}$$

$$|Z_{sec}| = |Z_m + Z_s| > \frac{Z_{2i_{min}}}{2 \times \varepsilon_s} \tag{9.39}$$

By setting a low ε_s and ε_p, the LC-filter configuration will operate in close to an ideally matched impedance condition, since the variation in Z_g and Z_m will not have any effect on the input impedances of the EMI filter with matching circuits. It should be noted that the design of a worst-case EMI filter will not become more difficult, even for ε ratios of less than 0.1. The voltage attenuation factor of the filter circuit and the matching networks is higher than or equal to the insertion loss. The LC-filter's desired ideal attenuation factor, K_o, can be calculated by the formula that gives the attenuation factor of the worst-case EMI filter:

$$IL_s \geq K = K_o \times \frac{(1 - \varepsilon_p) \times (1 - \varepsilon_s)}{\mu_m \times \mu_s} \tag{9.40}$$

To simplify the design method, a further assumption can be made. For a low ε_p, independent of Z_m, the minimum value of the primary input impedance is equal to the input impedance with open-circuited secondary terminals at the lower limit of the frequency range to be filtered:

$$Z_{1i_{min}} = Z_{1o}(\omega_L) = X_{1m} \tag{9.41}$$

For low ε_s, independent of Z_g, the minimum value of the secondary input impedance is equal to the input impedance with short-circuited primary terminals at the frequency f_L:

$$Z_{2i_{min}} = Z_{1o}(\omega_L) = X_{1m} \tag{9.42}$$

The element values of the matching network elements can be calculated by Eqs. (9.38) through (9.40) and by setting the values of μ_s, μ_p, ε_s, and ε_p. The element value limitations, using the notation of Eqs. (9.41) and (9.42), are as follows [9]:

$$C_p \geq \frac{1}{\varepsilon_p \times \omega_1 \times X_{1m}} \tag{9.43}$$

$$\varepsilon_p \times X_{1m} - \sqrt{(\varepsilon_p \times X_{1m})^2 - \frac{1}{(\omega_1 C_p)^2}} \geq R_p$$

$$\geq \varepsilon_p \times X_{1m} + \sqrt{(\varepsilon_p \times X_{1m})^2 - \frac{1}{(\omega_1 C_p)^2}} \tag{9.44}$$

$$R_p \geq \frac{1}{\omega_L C_p \times \sqrt{\mu_p^2 - 1}} \tag{9.45}$$

$$L_s \geq \frac{X_{2m}}{\varepsilon_s \times \omega_L} \tag{9.46}$$

$$\frac{1}{\dfrac{\varepsilon_s}{X_{2m}} + \sqrt{\left(\dfrac{\varepsilon_s}{X_{2m}}\right)^2 - \dfrac{1}{(\omega_L L_s)^2}}} \leq R_s$$

$$\leq \frac{1}{\dfrac{\varepsilon_s}{X_{2m}} - \sqrt{\left(\dfrac{\varepsilon_s}{X_{2m}}\right)^2 - \dfrac{1}{(\omega_L L_s)^2}}} \tag{9.47}$$

$$R_s \leq \omega_L \times L_s \times \sqrt{\mu_s^2 - 1} \tag{9.48}$$

The above design formulas are valid only for LC-configurations where the ratios of ε_p and ε_s are very small. Modifying the matching networks as shown in Fig. 9.16, the worst-case EMI filter design method also can be used for not-so-low ε ratios. Inductor L_p of the "P" matching network reduces the effect of Z_g variations on the input impedance seen from the secondary terminations; i.e., the value of the "resultant ε." Capacitor C_s of the "S" matching network reduces the interaction between the variation in Z_m and the input impedance seen from the primary

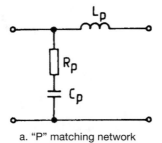

a. "P" matching network

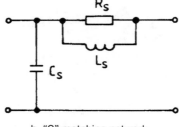

b. "S" matching network

FIGURE 9.16   Modified matching networks

terminations. The inductance of inductor L_p and the capacitance of capacitor C_s can be calculated as follows [9]:

$$L_p = \frac{1 + (\omega_L C_p \times R_p)^2}{2 \times R_p \times (\omega_L C_p)^2} \tag{9.49}$$

$$C_s = \frac{R_s^2 + (\omega_L L_s)^2}{2 \times R_s \times (\omega_L L_s)^2} \tag{9.50}$$

The voltage attenuation of the LC-filter configuration with modified matching networks as shown in Fig. 9.16 can be calculated as:

$$IL_s \geq K = K_o \times \frac{1}{\mu_p \times \mu_s} \tag{9.51}$$

Although the above relationships and train of thought refer to LC-filters, they are valid for other filter configurations, too, only the matching networks are to be otherwise selected. "P" matching networks providing a parallel loss are for filter terminations where the input impedance is high. "S" matching networks providing a serial loss are for filter terminations with low input impedance.

1. Low Input and High Output Impedance—CL-Configuration

The input of the CL-configuration is to be matched by the "S" network, and the output by the "P" network. The attenuation factor is:

$$IL_s \geq K = K_o \times \frac{1}{4 \times \varepsilon_p \times \varepsilon_s} \times \left(\frac{\omega}{\omega_L}\right)^2 \tag{9.52}$$

In the above formula, the last factor reflects the fact that an RC-attenuator on the input and an LR-attenuator on the output would be formed. The input attenuator consists of the serial resistor of the "S" network and the capacitor of the filter circuit. The output attenuator consists of the inductor of the filter and the parallel resistor of the "P" network.

Recall that the desired insertion loss provided by the EMI filter is geared to the lower limit of the frequency range to be filtered. Therefore, the voltage attenuation of the filter circuit can be calculated by the following relationship:

$$IL_s \geq K_o \times \frac{1}{4 \times \varepsilon_p \times \varepsilon_s} \tag{9.53}$$

2. Low Input and Output Filter Impedance—π-Configuration

The π-filter configuration is to be matched by "S" networks on both sides. The attenuation factor of the worst-case EMI filter can be calculated by:

$$IL_s \geq K = K_o \times \frac{1 - \varepsilon_{so}}{2 \times \varepsilon_{si} \times \mu_{so}} \times \frac{\omega}{\omega_L} \tag{9.54}$$

The second subscript in the above formula refers to the input or output (subscript i or o) of the filter circuit. The frequency-dependent factor is because an RC-attenuator would be formed by the serial resistor of the "S" network and the input capacitor of the filter circuit. For designing the filter circuit element values, the voltage attenuation is to be calculated from:

$$IL_s \geq K_o \times \frac{1 - \varepsilon_{so}}{2 \times \varepsilon_{si} \times \mu_{so}} \tag{9.55}$$

3. High Input and Output Filter Impedance—T-Configuration

The T-filter configuration is to be completed with "P" matching networks on both sides to obtain a worst-case EMI filter. The attenuation factor of the worst-case EMI filter is as follows:

$$IL_s \geq K = K_o \times \frac{1 - \varepsilon_{po}}{2 \times \varepsilon_{po} \times \mu_{po}} \times \frac{\omega}{\omega_L} \tag{9.56}$$

We need to determine the voltage attenuation of the T-filter configuration to provide the needed insertion loss value at the frequency f_L; i.e., the value of K_o can be calculated from:

$$IL_s \geq K_o \times \frac{1 - \varepsilon_{po}}{2 \times \varepsilon_{po} \times \mu_{po}} \tag{9.57}$$

It is advisable to design the worst-case EMI filter in the following steps:

A. Determine the proper filter configuration. The decision is usually influenced by the amount of insertion loss needed and, if known, the order of magnitude of the source and load impedances. Limitations on reactances in powerline filters also have to be considered. For worst-case EMI filters, a one-stage configuration can be employed, since the matching networks will provide the impedance matching (rather than the first and last stages of a multistage filter configuration).

B. Set the values of the ratios μ and ε,

C. Design the elements of the matching networks and the filter circuit. The values of the matching network elements are given by inequalities; thus, their actual values can be freely selected in the calculated range. The values of C_p and L_s should be as low as possible for minimizing low-frequency losses. If possible, μ_s and μ_p should be set lower than $\sqrt{2}$.

D. If an unsuitable set of μ and ε values results in unreal values for the matching elements, the design procedure should be repeated using another set of μ and ε ratios.

Let us examine an example for the worst-case EMI filter design procedure. The task is to design a noise filter that provides at least 40 dB insertion loss in the frequency range of 0.15 to 30 MHz. Since the demand on insertion loss is not too high, let the filter configuration be a one-stage LC-circuit. Thus, we will design and use a "P" matching network on the input and an "S" matching network on the output to eliminate the detrimental effects of impedance mismatch. We will set:

$$\varepsilon_p = \varepsilon_s = 0.05$$

$$\mu_p = \mu_s = \sqrt{2} \tag{9.58}$$

The needed voltage attenuation can be calculated using Eq. (9.40). This yields a minimum of 222.

The next step is to determine the value of the filter circuit elements. For acceptable voltage drop and rated current, let the inductance be $L = 1$ mH. While we design a worst-case filter, let the voltage attenuation be equal to the insertion loss. Furthermore, for safety reasons, let us calculate the value of the filter capacitor so that the voltage attenuation already will be 40 dB at 100 kHz, not 150 kHz. Since $\omega_1 = 2\pi \times 100 \times 10^3 = 6.28 \times 10^5$,

$$10^{-2} = \frac{1}{6.28^2 \times 10^{10} \times 10^{-3} \times C}$$

From the above equation, $C = 250$ nF.

The primary input impedance at open-circuited secondary terminations is:

$$Z_{1o} = j \times \omega L \times \frac{\omega^2 \times LC - 1}{\omega^2 \times LC} \tag{9.59}$$

and the secondary input impedance at short-circuited primary terminations is:

$$Z_{1s} = j \times \omega L \times \frac{1}{1 - \omega^2 \times LC} \qquad (9.60)$$

At the frequency f_L, the minimum of both input impedances is:

$$X_{1m} = 943 \ \Omega$$

$$X_{2m} = 4.25 \ \Omega \qquad (9.61)$$

First, the value of C_p has to be calculated by Eq. (9.43). Since this capacitance should be higher than 22.4 nF but as low as possible, let $C_p = 25$ nF.

Two limitations are given for the value of R_p by Eqs. (9.44) and (9.45):

$$26.3 \ \Omega < R_p < 68.3 \ \Omega, \text{ and}$$
$$42.3 \ \Omega < R_p$$

For reducing the operational frequency losses, let $R_p = 50 \ \Omega$.

The inductance of the "S" matching network should be higher than 90 µH, as can be calculated via Eq. (9.46). For minimizing low-frequency losses, physical size, and cost, inductor L_s should be as small as possible; thus $L_s = 100$ µH.

The limitations for the value of the resistor in the "S" matching network can be calculated using Eqs. (9.47) and (9.48). The following inequalities result:

$$59 \ \Omega < R_s < 152 \ \Omega, \text{ and}$$
$$R_s < 94.5 \ \Omega$$

Since the highest design value should be used to minimize low-frequency losses, let $R_s = 91 \ \Omega$ as the closest value in the standard value series.

Figure 9.17 shows the circuit of this worst-case EMI filter. It provides at least 40 dB insertion loss at 150 kHz, under any arbitrary source and load impedances. The worst-case EMI filter consists of a common one-stage LC-configuration completed with modified matching networks.

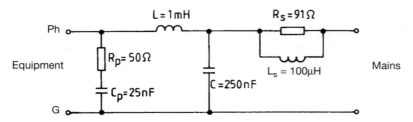

FIGURE 9.17 Worst-case EMI filter to provide at least 40 dB insertion loss at 150 kHz

9.4 DESIGN METHOD FOR EMI FILTERS WITH COMMON-MODE CHOKE COILS

High common-mode insertion loss usually can be provided only with common-mode choke coils if limitations exist for the serial or parallel reactances. The element values of EMI filters with common-mode choke coils cannot be designed using methods developed for filter configurations without mutual inductance. Although there are known relationships for calculating the insertion loss of a filter configuration with a common-mode choke coil, they are not useful for designing the filter element values [146] because they were derived for a particular filter configuration and impedance condition.

In general, the filter configuration should be designed first. Occasionally, we know the magnitude of the source and load impedances, and this offers a starting point. We do not use a multistage configuration for EMI filters with common-mode choke coils; because of the choke coil, the common-mode inductance can be practically as high as we like. Instead, we apply matching networks such as were introduced in Section 9.3 to eliminate the impedance mismatch effects. After deciding on the filter configuration, we transform the circuit into a four-terminal network as shown in Figs. 8.30 through 8.32. The value of the filter elements then can be calculated by some method on the basis of the common-mode equivalent circuit.

The accurate common-mode insertion loss of this filter configuration can be determined only via network analysis methodology [146]. Such calculations usually can be made only with computers since, even for a relatively simple filter configuration, at least four to six current loops have to be built and verified in many frequencies. The calculations grow more complex when the worst-case common-mode insertion loss also has to be determined.

9.5 DAMPED EMI FILTERS AND LOSSY FILTER ELEMENTS

Because they consist only of reactances with low loss, EMI filters can allow resonances to develop. These resonances can be caused by parasitic effects or impedance mismatch. Because parasitic resonances can highly distort filter response, they should be given due consideration. The most common method of avoiding spurious resonances is to employ some sort of dissipative mechanism. A combination of conventional and lossy filter elements can provide very good overall filter performance. We will discuss dissipative mechanisms separately for EMI filters with low and high current ratings.

9.5.1 Damped EMI Filters for Low Current Ratings

For EMI filters with low current ratings, use of the so-called *anti-interference wires and cables* has increased steadily from the early 1970s. Anti-interference wires offer a high impedance at high frequencies but a low impedance at low frequencies. The HF impedance is mainly resistive in nature, so they damp the HF parasitic resonant circuits. Three major techniques are known for developing HF losses: dielectric and magnetic absorption, pseudo-resonant loss, and artificial skin effect [62, 86, 200–205, 278, 281].

Ferromagnetic materials are composed of microscopic units separated by walls. These units, referred to as *domains,* act like tiny permanent magnets. Since these particles are oriented differently, their magnetic effects are summed, and macroscopic magnetism is zero. Exposing this material to an alternating magnetic force, these domains are subjected to alternating strengths in opposite directions, and they start to vibrate. For an applied field of a certain frequency, the vibration corresponds to the mechanical resonance of the domains. Analogous effects exist in some dielectric materials wherein dielectric absorption is caused by dipole rotation effects.

The structure of absorption-type anti-interference cable is shown in Fig. 9.18 [203]. Such anti-interference cables are commonly used for automobile ignition applications. The absorptive composite consists of fine magnetic ferrite powder in a plastic matrix. The performance of the anti-interference cable can be influenced by the selected ferrites and their granularity and distribution. It should be noted that the conductor is coiled not to increase the inductance but to lengthen it, thus increasing the absorptive effect. The cable's impedance is essentially resistive, which ensures that it will not resonate. The Q-factor of an absorptive-type anti-interference cable is usually near or less than 1.

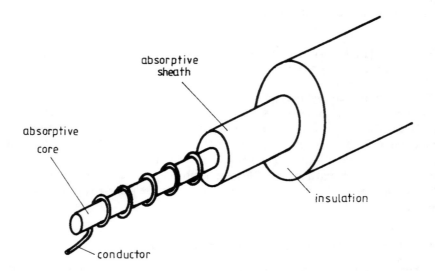

FIGURE 9.18   Structure of absorptive-type anti-interference cable

The other commonly used anti-interference cable structure is shown in Fig. 9.19 [205]. The two-layer structure offers the ability to decrease the cutoff frequency and provide an insertion loss-frequency curve with a higher slope. The one disadvantage of this structure is its limited absorption at higher frequencies.

High insertion loss can be achieved at high frequencies by building resonant circuits along the cable length. To form resonant circuits, small capacitors are introduced. These capacitors are implemented as a rings deposited on a thin insulator around the principal conductive coiling. The resonant circuits are damped by the absorptive effects. The schematic structure of pseudo-resonant cable can be seen in Fig. 9.20a [203]. The equivalent circuit shown in Fig. 9.20b illustrates the parallel circuits that absorb the frequencies near their resonant frequencies. Typical attenuation as a function in frequency is shown in Fig. 9.21. The low-frequency part of the attenuation curve is determined by the parallel resonant circuits, consisting of the conductor with absorptive core and the lumped capacitor. The high-frequency part is determined by the absorptive effects.

Investigations into improving the energy exchange condition between the conductor and the surrounding dissipative element led to the development of anti-interference cable using the so-called *artificial skin effect.* Artificial skin effect is based on the fact that slow propagation structures exchange energy to the medium with a higher wave propagation speed. This condition exists in the area around metallic conductors; that is why the metal radiates energy easily. The conductor of an artificial skin effect anti-interference cable is covered with successive dissipative layers. With proper structure, high performance can be obtained. These

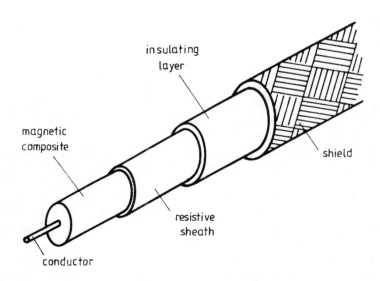

FIGURE 9.19 Structure of two-layer absorptive-type anti-interference cable

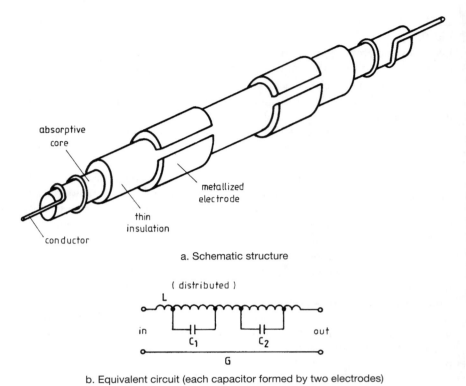

a. Schematic structure

b. Equivalent circuit (each capacitor formed by two electrodes)

FIGURE 9.20 Structure of pseudoresonant anti-interference cable

anti-interference cables can be used for suppressing HF disturbances above a few megahertz frequency.

Combining anti-interference cable and a lumped capacitor, high-performance noise filters can be realized. These noise filters simultaneously combine the advantages of the simple RC-filter and LC-filter configurations. They do not resonate and, in contrast to real RC-circuits, they have negligible resistance at low frequencies. They also provide an insertion loss curve with a slope of 40 dB/decade. The inner structure of such a noise filter is shown in Fig. 9.22 [203]. For improving HF performances, a feedthrough capacitor was used. The insertion loss curves in Fig. 9.23 were as measured in a system with low source impedance and high load impedance. The figure clearly shows that there is nearly zero resonance at the cutoff frequency.

Anti-interference cables with large HF losses are useful for noise filter choke coils. Such losses can be provided with no resonance developed. Figure 9.24 shows the Q-factor of two 10-turn, air-cored choke coils of identical dimensions,

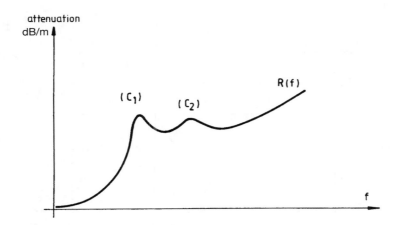

FIGURE 9.21   Typical attenuation vs. frequency of pseudoresonant anti-interference cable [203]

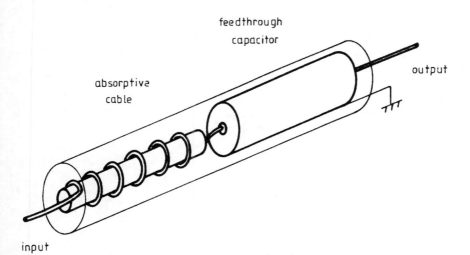

FIGURE 9.22   Structure of noise filter with anti-interference cable and feedthrough capacitor

with one made with copper wire and the other anti-interference wire [204]. An additional advantage of the anti-interference wire is that the insertion loss of the lossy inductor is often higher than one might expect from the number of turns and geometry. The reason for this is that the impedance of the choke coils is increased by the HF resistance of the anti-interference cable.

Excellent HF performance can be provided by dissipative distributed filters. Because of HF losses, they produce no resonances in the stopband. Distributed

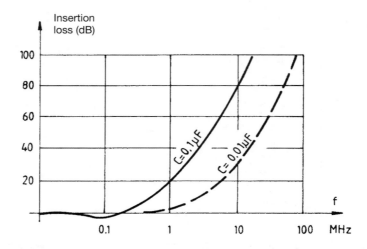

FIGURE 9.23 Typical insertion loss vs. frequency of lossy noise filter

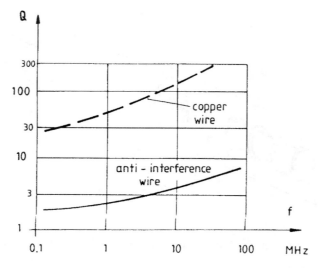

FIGURE 9.24 Q-factor vs. frequency for inductors wound with conventional and anti-interference wire [204]

noise filters usually provide high insertion loss (up to approximately 10 GHz) since parasitic effects do not decrease or even reverse the slope of the insertion loss curve. The basis of the distributed filtering is one or a combination of the above described absorptive effects [278, 281].

9.5.2 Damped EMI Filters for High Current Ratings

In some practical cases, the HF losses of EMI filters have to be analyzed not only to improve their HF characteristics but also to provide the necessary operational conditions for the power electronic equipment to be filtered. This brings us to the next problem, where EMI produced by an ac phase-controlled switching unit must be suppressed. For EMI filtering, we use a simple LC-circuit as shown in Fig. 8.5a (see Section 8.2.2). The layout of the problem is shown in Fig. 9.25.

The value of C is usually high and can reach a few microfarads. For analyzing the current of the SCR, we can neglect current loops to ground. The SCR control–load–EMI filter circuit can be regarded as an RLC-circuit where capacitor C is charged up and a switch is closed at the SCR firing moment. The current shape depends on the Q-factor given by:

$$Q = R\sqrt{\frac{C}{L}} \tag{9.62}$$

The current tries to reverse when $Q > 2.5$; thus, the SCR becomes nonconductive. This causes no problem for a continuously firing signal, but if we apply only a single, short signal (as is more common in practice), the phase control will not work. That is why SCR switch-off malfunction occurs more often in low-load situations. As Eq. (9.62) shows, the Q-factor can be decreased by increasing the inductance of the noise filter choke coil. However, to increase the serial inductance only for the purpose of decreasing the circuit's Q-factor is not a proper solution, both for economic reasons and because the mains frequency drop across the EMI filter usually would be too high. The SCR switch-off can be prevented by properly damping the EMI filter. Figure 9.26 illustrates two known ways to increase circuit damping. In practice, the configuration of Fig. 9.26a is more common because the decoupling capacitor, C_d, does not reduce filter insertion loss.

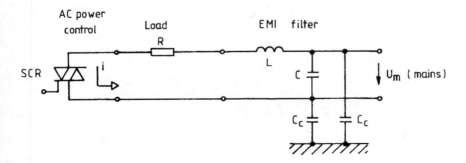

FIGURE 9.25   EMI suppression for ac power controller

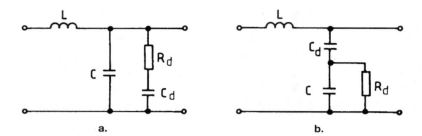

FIGURE 9.26 LC-filter configurations with HF losses

The damping element value can be designed with appropriate knowledge of filter circuit elements and the minimum load using the charts of Figs. 9.27 and 9.28 [37]. All points that lie under the curve for a given parameter of C_d/C represent parameter combinations that correspond to continuous current flow through the SCR. To use these charts, first calculate the Q-factor of the circuit without any damping. Then, drawing a horizontal line at this value, we intersect the minimum value of the ratio C_d/C. In Fig. 9.27, we can select a ratio of C_d/C of 3 or greater, but not 2 or less. (When the ratio is great, we have a greater range in which the circuit works well.) Next, we can read the limits for R_d/R. It is preferable to take an actual R_d/R ratio that falls at the halfway point between the range limits. That way, element tolerances do not push the set parameters outside the desired region.

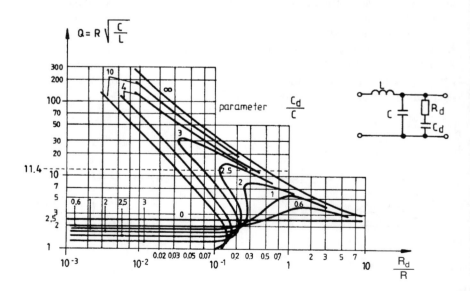

FIGURE 9.27 Design chart for lossy LC-filter of Fig. 9.26a [37]

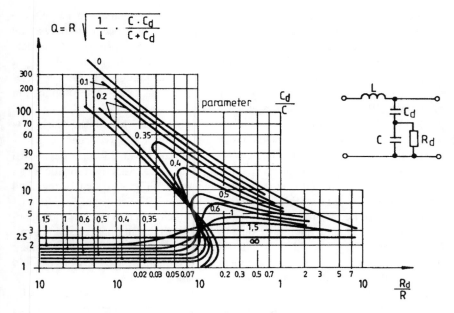

FIGURE 9.28 Design chart for lossy LC-filter of Fig. 9.26b [37]

For demonstrating the use of these charts, let the task be to design an EMI filter for a lamp dimming device. The SCR is fired by single signal, and the dimmer has to work well with load of 20 W on the 127 V mains. For the EMI filter, we use a one-stage LC configuration where the inductance of the noise filter choke coil is 1.6 mH and the capacitance of the X-capacitor is 0.22 µF. As was mentioned, we can neglect the Y-capacitors for analyzing the operational conditions of SCRs.

Calculating with the positive tolerance of the mains voltage, the resistance of the minimum load is 970 Ω. The Q-factor, calculated by Eq. (9.62), is 32, which considerably exceeds the acceptable value of 2.5. It is worth mentioning that if we would like to provide the same amount of insertion loss while reducing the Q-factor to less than 2.5, the inductance of the noise filter must be designed to 7.3 mH and the capacitance to 0.05 µF. Because of the high inductance, this element value set is rather uneconomical.

We will try to improve the transient behavior of the filter by employing the damping circuit of Fig. 9.26a. The horizontal line in Fig. 9.27 shows that $C_d/C =$ 2.5, or $C_d = 0.55$ µF, will ensure the required result. After designing $C_d = 0.6$ µF from the standard value series, the range of R_d/R goes from about 0.1 to 0.25. Therefore, by choosing an intermediate value of 0.15 for the actual ratio, $R_d =$ 150 Ω is suitable. The mains frequency loss across resistor R_d does not reach 0.1 W, which is negligible.

The spurious resonances in the EMI filter must be eliminated; otherwise, they can greatly limit EMI suppression performance. Resonances occur not only at the cutoff frequency of the filter configuration but also as a result of parasitic effects on higher frequencies. For reducing such resonances in EMI filters of high current ratings, the simplest and usually the least expensive technique is to use lossy elements. Two primary techniques are known in engineering practice: the application of damping RC-networks and lossy inductors.

Damping RC-networks are rarely used in high-power EMI filters. These networks consist of serially connected resistor and capacitor. Applying damping RC-networks such as the "P" matching networks of worst-case EMI filters or as shown in Fig. 9.26a, resonances at cutoff frequencies and other parasitic resonances can be well damped. In this sense, the filter configuration of Fig. 9.26a can be regarded as a one-stage LC-filter with a damping RC-network. The resistance in the RC-network is designed to provide sufficient loss for the resonant circuits. The capacitor is only for reducing the low-frequency losses of the damping resistor. Therefore, its impedance should be small relative to R at the supposed resonant frequencies, but high at the lower frequency. Such damping RC-networks should be designed for each stage in multistage filter configurations.

In EMI filters with high current ratings, the noise filter choke coils are usually made with HF losses. The Q-factor of EMI filter inductors is rather high but, as they often are built with an iron core, the Q-factor can be reduced, since HF losses usually arise in the iron core. HF damping of EMI filters with air-core choke coils can be provided best with the RC-networks mentioned above.

The losses of iron core transformer materials decrease the Q-factor of the inductor above the eddy current frequency limit (see Section 6.3.4) so that the inductor cannot resonate. Since the eddy current frequency limit is low relative to the frequency range to be filtered, a solenoid choke coil with iron core can be regarded as an effective lossy EMI filter element. The loss in iron transformer materials is usually enough to eliminate spurious resonances in EMI filter circuits.

The iron core of a common-mode choke coil does not have HF losses as high as those of iron core materials for normal transformers. This is because high-performance iron cores are needed for common-mode choke coils. These iron cores offer some damping only above the gyromagnetic frequency limit (see Section 6.3.4); i.e., above a few megahertz.

Special iron core materials have been developed that combine the advantages of the standard transformer iron core and ferrite materials. These composite iron core materials can be used advantageously for noise filter choke coils [38]. The qualities that are similar to those of transformer iron materials provide high saturation induction and HF losses for eliminating resonances. At the same time, the qualities characteristic of ferrite materials result in high permeability up to a few megahertz.

As of this writing, economic considerations have discouraged the use of anti-interference cables as described in Section 9.5.1 in the manufacture of lossy noise filter choke coils of high current ratings. However, successful experiments have been made using a special multilayer strip conductor structure [124].

It is well known that the current density in a conductor at high frequencies is not uniform but, rather, is concentrated toward the surface. This phenomenon is called *skin effect*. Maxwell's equations describe the current density and the magnetic field inside a current-carrying conductor. The nonuniform current density can be well characterized by the skin depth. The distance required for the wave to be attenuated to 1/e of its original value is defined as skin depth, δ, which can be calculated as:

$$\delta\,(\text{meter}) = \frac{1}{\sqrt{\pi \times f \times \mu \times \sigma}}\quad m \qquad (9.63)$$

where

σ = specific conductivity in Ω^{-1}/m

μ = absolute permeability = $\mu_0 \mu_r$

f = frequency in hertz

As a result of nonuniform current density, the resistance of a conductor increases at high frequency relative to its dc resistance. The increase of resistance can be expressed by the skin depth and the cross-section of the conductor. The HF resistance of a single strip conductor of thickness t can be approximated as follows [124]:

$$R = R_o \times \frac{t}{2\delta} = \frac{t}{2}\sqrt{\pi \times f \times \mu \times \sigma} \qquad (9.64)$$

Since the above formula gives the HF resistance, a corner frequency can be defined from which the increase in resistance with frequency exists. The value of this corner frequency is defined where the dc (R_o) and the HF resistances, calculated by Eq. (9.64) are equal, thus:

$$f_t = \frac{4}{\pi \times \mu \times \sigma \times t^2} \qquad (9.65)$$

When more than one current-carrying conductor is present, the current distribution in the conductors is interdependent. This electromagnetic induction phenomenon is known as *proximity effect*. Stacking strip conductors together, as in spiral coil arrangement, the current diverges along both surfaces according to the infinite strip formula, and the magnetic field becomes parallel to the surface. For a coil structure wound with strip conductor of width w, the resistance of a single layer under these conditions can be described as follows [124]:

$$R = R_o \times \frac{w}{2\delta} = \frac{R_o}{2} \times \sqrt{\pi \times f \times \mu \times \sigma \times w^2} \qquad (9.66)$$

The corner frequency of the infinite-stack model can be defined as f_t, thus:

$$f_t = \frac{4}{\pi \times \mu \times \sigma \times w^2} \qquad (9.67)$$

Although the infinite-stack model cannot be realized, measurements have shown that the model is quite good for a spiral coil arrangement as long as (1) the stack height is much larger than the individual thickness, and (2) the individual thickness is much larger than the insulating space between the layers.

Better HF damping can be provided if we lower the value of corner frequencies. As Eqs. (9.65) and (9.67) show, these can be reduced by using strip conductors with high permeability and good conductivity. To obtain high permeability, Permalloy materials were used. The relatively high dc resistance was decreased by combining the Permalloy layer with a conductor of good conductivity. Two-strip conductor structures are illustrated in Fig. 9.29 [124].

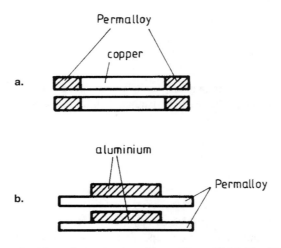

FIGURE 9.29 Composite strip conductor structures for providing artificial skin effect [124]

Interlayer parasitic capacitance also limits the HF performance of the spiral coil arrangement. To provide sufficient HF loss, the insulating space must be small compared to the thickness of the strip conductor, which aggravates the interlayer capacitance. The effect of parasitic capacitance can be decreased by a using a grounded interlayer or by connecting two smaller noise filter choke coils in series. The resonant frequency of experimental lossy inductors for very high current ratings has been successfully increased up to 12 to 15 MHz [124].

9.6 HF CHARACTERISTICS OF NOISE FILTER CIRCUIT ELEMENTS

The design methods discussed in Sections 9.1 through 9.5 presumed that the circuit elements would be ideal but the HF characteristics of real noise filter components would be distorted by different parasitic effects in the targeted insertion loss frequency range of 0.15 to 30 MHz (see Chapter 6). But because of nonideal element characteristics, the actual insertion loss curve can differ considerably from the calculated or expected one. To illustrate this, we can study the character of the voltage attenuation with frequency of an LC-configuration consisting of real noise filter elements. In the equivalent circuit shown in Fig. 9.30, capacitor C_p represents the parasitic turn capacitance of the inductor, and the inductor L_p represents the parasitic inductance of the noise filter capacitor. Capacitor C_g is the stray capacitance between the circuit elements and ground. As the figure clearly shows, in the practice, instead of an LC-circuit, we must consider at least a filter configuration with four to five elements. If, for example, the resonant frequency of the noise filter capacitor is lower than that of the circuit L–C_p, then the result can be voltage gain instead of voltage attenuation (insertion loss).

Full consideration of the actual HF characteristics of noise filter elements would require design methods too complicated for practical engineering, so the nonideal HF characteristic effects are usually studied only after the design is com-

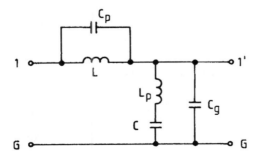

FIGURE 9.30 Equivalent circuit of an LC-filter configuration consisting of nonideal elements

plete. This is accomplished by analyzing the actual voltage attenuation (insertion loss) curve. Calculations of this type can be made by computer-aided analysis programs. Lacking an applicable network analysis/design/synthesis program, a simple programmable relationship was devised for calculating the voltage attenuation or insertion loss of a two-stage π-configuration EMI filter consisting of real elements (Fig. 9.31). The filter elements are inductors L_1 and L_2, and the capacitors C_3 through C_5. Resistors R_1 and R_2 represents the losses of the noise filter choke coils, and capacitors C_1 and C_2 represent the parasitic turn capacitances. Inductors L_3 through L_5 depict the parasitic inductances of the filter capacitors. The source impedance is composed of the elements of R_g and L_g, and the load impedance is made up of the circuit elements R_1, L_1, and C_1. The transfer function of the circuit is as follows:

$$\frac{U_1}{U_2} = \frac{A(\omega) \times J(\omega) \times (1 - \omega^2 \times L_4 C_4) \times (1 - \omega^2 \times L_1 C_1 + j \times \omega R_1 C_1)}{Z_g \times F(\omega) \times G(\omega) + j \times \omega C_3 \times Z_g \times E(\omega) \times H(\omega) + J(\omega) \times I(\omega)}$$

$$(9.68)$$

where

$$A(\omega) = (R_1 + j \times \omega L_1) \times (1 - \omega^2 \times L_5 C_5) \times (1 - \omega^2 \times L_2 C_2 + j \times \omega R_2 C_2)$$

$$B(\omega) = (R_2 + j \times \omega L_2) \times (1 - \omega^2 \times L_5 C_5) \times Z_L$$

$$C(\omega) = (1 - \omega^2 \times L_4 C_4) \times (1 - \omega^2 \times L_5 C_5)$$
$$\times (1 - \omega^2 \times L_2 C_2 + j \times \omega R_2 C2) \times Z_L$$

$$D(\omega) = j \times \omega C_5 \times (R_1 + j \times \omega L_1) \times (1 - \omega^2 \times L_4 C_4)$$
$$\times (1 - \omega^2 \times L_2 C_2 + j \times \omega R_2 C_2)$$

$$E(\omega) = A(\omega) + B(\omega) + j \times \omega C_5 \times (R_1 + j \times \omega L_1) \times (R_2 + j \times \omega L_2)$$

$$F(\omega) = C(\omega) + D(\omega) + j \times \omega C_4 \times E(\omega)$$

$$G(\omega) = (1 - \omega^2 \times L_1 C_1 + j \times \omega R_1 C_1) + \times [J(\omega) + j \times \omega C_3]$$
$$\times (R_1 + j \times \omega L_1)$$

$$H(\omega) = (1 - \omega^2 \times L_1 C_1 + j \times \omega R_1 C_1) \times (1 - \omega^2 \times L_4 C_4)$$

$$I(\omega) = E(\omega) \times (R_1 + j \times \omega L_1) + F(\omega) \times G(\omega)$$

$$I(\omega) = (1 - \omega^2 \times L_3 C_3)$$

$$Z_g = R_g + j \times \omega L_g$$

$$Z_L = 1 - \omega^2 \times L_1 C_1 + j \times \omega R_1 C_1$$

The voltage attenuation of the circuit shown in Fig. 9.31 is the absolute value of the complex transfer function. The insertion loss of the noise filter configuration can be determined by calculating the voltage attenuation with and without the filter circuit. The voltage attenuation without the noise filter is given by the source and load impedances. The above relationship can be applied for filter configurations other than a two-stage π-filter, only zero should be substituted for the value of missing circuit elements. Using Eq. (9.68), insertion loss charts can be plotted with a known parameter such as, for instance, one special parasitic element. Using these charts, one can analyze the effects of parasitic phenomena and impedance mismatch.

For determining the complete equivalent circuit, the HF parameters of the noise filter elements must be known. Three major modeling techniques for passive components including their parasitic effects have been developed: (1) direct calculation, (2) engineering approximation, and (3) analytical approximation [341]. The applicability of a particular modeling technique may depend on the type of component data available in manufacturers' catalogs and data sheets.

The direct calculation method requires the minimum amount of parametric data and is the simplest approach for generating a complete model. This modeling method is based on equivalent circuits of noise filter elements as discussed in Chapter 6. For calculating the lumped element value representing the different parasitic phenomena, knowledge of the resonant frequency is usually enough. If

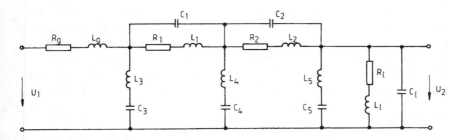

FIGURE 9.31 Equivalent circuit of a two-stage π-filter configuration consisting of nonideal elements

TABLE 9.1 Typical element values of realistic filter capacitors in HF equivalent circuit [341]

Capacitance of Noise Filter Capacitor	Parasitic Inductance*
< 10 nF	10 to 20 nH
10 nF–1 μF	40 nH
> 1 μF	30 to 100 nH

*This inductance is an order of magnitude lower (1/10) if the filter capacitor is a feedthrough type.

TABLE 9.2 Typical element values of realistic noise filter choke coils in HF equivalent circuit [341]

Choke Coil Inductance	Serial Resistance	Parasitic Capacitance
< 50 μH	1.5 mΩ	2 pF
50–200 μH	10 mΩ	5 pF
> 200 μH	0.5 Ω	10 to 30 pF

information is available with regard to the noise filter elements' HF characteristics, and if we require knowledge about the actual or expected insertion loss curve, analysis can be made using approximate values. The approximate value of a lumped inductor representing the parasitic inductance of the noise filter capacitors is summarized in Table 9.1. The information in this table does not include the parasitic inductance of the wiring to the noise filter capacitors, so this should be added. Table 9.2 gives the approximate value of a lumped capacitor representing the turn capacitance of the noise filter choke coils and of the serial resistance [341]. It should be noted that it is preferable to calculate the serial resistance of the noise filter choke coils for high current rating rather than to use this approach.

The engineering approximation modeling technique is of moderate complexity but more accurate. Therefore, it is often used by EMC experts. This method presumes that the engineer will select ideal components and interconnections that correlate with the behavior of real components and interconnections as a function of frequency (e.g., impedance magnitude). However, it requires that the characteristic curve of the noise filter element be given as a function of frequency. The complexity of the model can vary from a simple three-element version to a complex multi-element model, depending on the type of desired response curve and the detail in which it is analyzed. For generating the complex model, first approximate the impedance curve with ideal component asymptotes; i.e., with horizontal lines, and lines with +20 dB/decade or −20 dB/decade slope. The next step is to determine the configuration in which to connect the ideal components. The model selection criteria are summarized in Table 9.3 [341]. The engineering approximation method can be used not only with the limited data available from manufacturers but also with measured data.

TABLE 9.3   Directives for determining the substitute network from the frequency function [341]

Frequency Function	Substitute Circuit
0 dB/decade	Resistor
−20 dB/decade	Capacitor
+20 dB/decade	Choke coil
0 and −20 dB/decade	Parallel RC-circuit
0 and +20 dB/decade	Serial RC-circuit
−20 and 0 dB/decade	Serial RC-circuit
+20 and 0 dB/decade	Parallel RL-circuit
−20 and +20 dB/decade	Serial LC-circuit
+20 and −20 dB/decade	Parallel LC-circuit

The analytical approximation technique is somewhat sophisticated and rigorous. Therefore, it is rarely used in engineering practice.

9.7   EMI FILTER LAYOUT

The design of an EMI filter is completed by determining its layout. To avoid ruining the filter's HF performance, some important layout aspects must be considered.

EMI filters should be placed as near as possible to the terminals where noise is to be suppressed. EMI filter units are designed for connection to the mains or supply networks without fuses. This operational mode is best for EMI suppression, but the consumer must understand and accept it. The filter capacitors must be designed to enhance safety, and the filter cover should be marked with a warning that it remains connected to the power source even when the electrical equipment is switched off. The cover should be designed to preclude the accidental contact with hazardous elements. The cover should be constructed of a noncorrodible material. For low current rated EMI filters, it is particularly important to provide effective electromagnetic shielding.

EMI filters must be designed and constructed so that they are not damaged by transients coming from the power mains or caused by the electrical equipment (e.g., inrush current). It is usually necessary to specify two operational limits separately: conditions under which the EMI filter will continue to function properly, and conditions under which it will merely survive. One should pay some attention to specifications for bleeder resistors that are used to discharge the filter capacitors after the EMI filter is switched off.

Losses for an EMI filter of high current rating can be quite high. Most losses are generated by noise filter choke coils. For predicting the loss of iron-cored solenoid choke coils, the iron losses should be added to the copper losses. Since excess heat generation may damage neighboring components or the filter itself, special attention should be paid to cooling effectiveness. The maximum allowable temperature rise is limited mainly by the insulation material used for the choke coil winding. When designing the cover and determining the mounting position, the airflow characteristics should also be considered. It may be necessary to mount the filter capacitors so that they are not warmed up by the flow of air over the noise choke coils.

Power electronic equipments often produce a nonsinusoidal mains current. The lower harmonics are relatively powerful, which may generate considerable acoustic noise. Although we don't often think of EMI filters as likely to generate a significant level of acoustic noise, special care should be taken for powerline filters that may operate in areas where spoken communication is important or where intense concentration is required. The acoustic noise emanating from powerline filters can be reduced by impregnating the noise filter choke coils, gluing the iron laminations together, and applying flexible fasteners.

For determining the internal arrangement of components in an EMI filter, three there are three major concerns: (1) stray magnetic fields from solenoid choke coils, (2) filter capacitor lead routing, and (3) grounding layout.

Solenoid-type filter choke coils generate intense stray magnetic fields. These fields are concentrated at the end of the inductor; therefore, the filter capacitor should be located at maximum distance, as should all electrical leads. Operational frequency magnetic fields can be enclosed with structural iron elements. However, the losses in these assembly elements increase the losses of the EMI filter and, what is worse, can constitute a local heat source. Undesired heat development can be reduced by using nonmagnetic fastening materials.

In EMI filters for a three-phase mains, solenoid choke coils are generally placed close together. Therefore, it is reasonable to connect them in the circuit so that a "current compensation effect" (see Section 6.3.4) will develop. In this manner, the mains frequency leakage magnetic field can be reduced, and the insertion loss can be somewhat increased.

Capacitor performance in an EMI filter depends a great deal on the length of the connection leads (see Section 6.2). Capacitor layout, especially in the case of Y-capacitors, should aim for the shortest possible leads. Keep in mind that a 3 to 5 cm increase in lead length can significantly degrade the insertion loss at frequencies above a megahertz.

I/O circuits should be separated as much as possible. Maximizing the distance between these circuits and components is important for two reasons: radiated coupling between the noisy and filtered circuits is decreased, and common I/O ground paths can be avoided. Input and output circuits separation is especially important when laying out EMI filters for low-current applications.

Lead routing and grounding schemes are closely related. I/O circuit currents should not flow through a common grounding impedance, even if the value of this impedance is very low (see Section 8.3.6). The same point of view is applicable to the grounding design. In some installations, the mains grounding and the chassis ground are not directly connected in the electrical equipment. When the EMI filter is connected only to the mains grounding, a relatively high ground impedance develops at high frequencies. Since the EMI measurement is generally made relative to the chassis ground, a high EMI level can exist across the grounding path of the EMI filter. This EMI can be reduced by connecting a capacitor between the grounding of the EMI filter and the actual grounding to which the EMI level will be referenced. This capacitor shunts the impedance of the grounding path at high frequencies. However, the best is to *always* guarantee a low-impedance bonding of the filter case to the equipment chassis.

Power line filter capacitors have to be shunted by some type of bleeder resistor. These resistors are used to discharge the capacitors after switching off the EMI filter. The bleeder resistance is usually about 0.1 to 1 MΩ, if dielectric strength specifications do not requires a higher value. Such values ensure a drop-off of the capacitor charge within a few seconds of switch-off.

Dielectric strength tests of electrical equipment with EMI filters should be made with special care, since some capacitors are connected between the output terminals and the ground and may be stressed by the full line voltage even when the electrical equipment is switched off. The dielectric strength test should be made with dc voltage to avoid overloading the Y-capacitors and prevent a false indication of faulty insulation. As an alternative, the dielectric strength test can be made with an ac voltage but with the Y-capacitors disconnected. In this case, the test voltage of the Y-capacitors must be verified from the manufacturer's catalog or data sheet.

10

Testing for Susceptibility to Power Line Disturbances

As mentioned earlier, EMC can be divided into the two major categories of emissions and susceptibility. While EMI refers to potentially disruptive emissions generated by electrical apparatus, electromagnetic susceptibility (EMS) is concerned with analyzing and improving a particular device's ability to withstand unwanted emissions. A device may fail to meet EMC specifications not only by producing excessive EMI levels but also by displaying unacceptable susceptibility to its intended environment. The issue of EMS is usually subdivided into conducted and radiated disturbances (see Fig. 1.1).

Susceptibility test programs have been in existence about a decade for civil applications, and at least 20 years for the military. In the past, most consumer electronic devices connected to the power mains were not terribly sensitive to HF disturbances, and it was generally sufficient to protect them against atmospheric and switching transients only. For this reason, the first test methods based EMS measurements on a specified surge voltage.

EMS specifications are necessarily based on compromise. We cannot require an electrical equipment to be immune to absolutely any level of disturbance that might be encountered. But at the same time, it must be capable of operating in the real world, where electromagnetic noise is increasingly common. In practice, electrical equipments are qualified as acceptable if they survive disturbances of specified parameters without malfunction or breakdown. Since HF disturbances usually travel along the public mains, EMS test signals have been developed primarily to address that type of conducted emission.

10.1 SURGE VOLTAGES IN AC POWER MAINS

Surge voltages on electrical mains have been studied systematically for about 50 years. The subject of transient disturbances has received increased attention since the late 1970s because of the proliferation of electronic components in household and industrial electrical equipment. Most malfunctions and failures in such equipment can be traced back to voltage surges (spikes and transients).

Transient disturbances on public ac power mains can be classified per various parameters. At first, transients were qualified in terms of maximum amplitude, and the only distinction made was whether they originated from load switching or lightning effects [39, 99, 100, 126, 362].

Using data collected over many years and from many sources, we can plot surge occurrence rates by voltage level; however, reported rates vary widely, depending on the power system examined and many other factors. Such a plot is shown in Fig. 10.1 [362]. Comparing the three plot lines, we can see the frequency of occurrences at voltage levels relative to particular classifications of power distribution systems (*high, low,* and *extreme* exposure). The low-exposure systems are in geographical areas known for minimal thunderstorm and load switching activities. They also correspond to power distribution systems using mostly buried instead of overhead cables. High-exposure systems are geographical areas known for prolific lightning activity and frequent and severe switching transients. The ex-

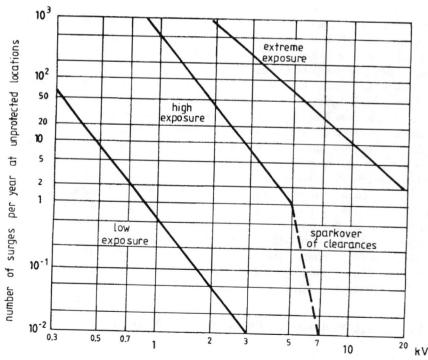

FIGURE 10.1 Surge occurrence rates by voltage level

treme-exposure systems, usually supplied by long transmission cables, are rare, and their installation characteristics result in high sparkover levels [362]. Figure 10.1 generally supports the standard engineering assumption that transient occurrences at 3 kV amplitude can be expected to range from 0.01 to 10 per year, depending on whether they are counted at a high- or low-exposure location.

In recent years, we have observed that electronic unit susceptibility cannot be fully explained on the basis of these transient occurrence curves. Measurements and investigations have proven that transient disturbances have changed—not only in number but also in character. Noise on ac power mains has become a much more significant problem with the rapid spread of surge-sensitive electronic systems.

The most comprehensive investigations of electronic unit disruption and failure indicate that, for purposes of analysis, it is not sufficient to determine only the maximum amplitude of the surge voltages; the energy and the frequency content also must be considered. Because information could not be derived from early studies, a new survey was made in Europe on 230–240 V mains to obtain the required parameters of surge voltages occurring on public ac mains [257]. On the basis of the stored data, these measurements determined the maximum amplitude, rise time, oscillatory frequency, energy, and the amplitude density function. Although the primary purpose of the survey was to describe the environment, a collateral result was to give base data for elaboration of standard EMS test methods.

Most of the observed transients were aperiodic. Periodic transients were generated mostly by switching processes. Because of contact shower effects, the switching transients were repeated with a frequency of approximately 100 Hz for a period of several milliseconds. It was stated that common-mode and differential-mode disturbances do not differ significantly. The results of these measurements were summarized in charts of (1) maximum amplitude versus highest slope, (2) maximum amplitude versus energy, and (3) highest slope versus energy. The first two of these charts are recreated in Fig. 10.2 [115]. Interestingly, the measurements showed that power electronic equipments are not the worst culprits in transient production. In addition, they produced the unexpected conclusion that surge voltages are often higher in offices and computer facilities than in heavy industrial plants.

These measurements indicated that the highest surge voltage slope occurs when the highest amplitude is developed. The rise time of the observed transient disturbances were quite high, and the slope of a surge voltage of moderate amplitude can reach a value of about 10 V/ns. The highest slope of a transient rapidly decreases along the wiring. An easily measurable decrease was observed across a wiring path of only 10 m. The energy integral of the observed aperiodic surge voltages with amplitudes of 100 to 1,000 V was in the range of 100 to 1,000 V and could be characterized by a $V^2 \times$ duration product of 10^2 to 10^5 V^2-ms. An obvious correlation between rise time and energy could not be demonstrated, but many surge voltages were observed with relatively long rise times and high energy levels.

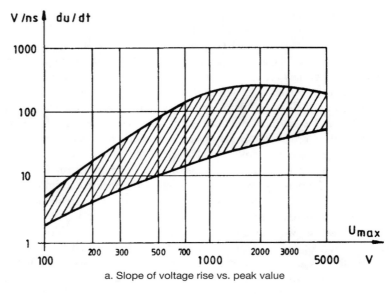

a. Slope of voltage rise vs. peak value

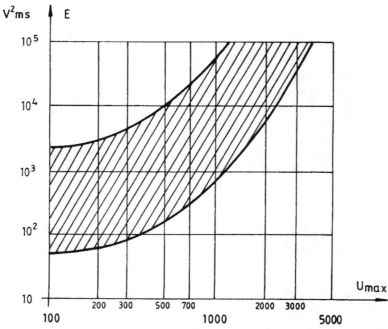

b. Energy content vs. peak value

FIGURE 10.2   Parameters of measured transients on ac mains

The amplitude density function (spectrum) of recorded common-mode transient disturbances was also determined. Some examples are shown in Fig. 10.3, measured at a point near a typical transient noise source [257]. In Fig. 10.3a, Curve 1 was measured in a heavy industrial plant, Curve 2 in telecommunications installation, and Curve 3 in a laboratory. In Fig. 10.3b, Curve 1 was derived in a power station, Curve 2 in a chemical plant, and Curve 3 in a private household.

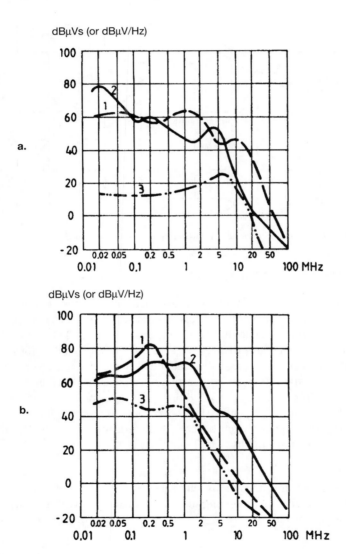

FIGURE 10.3 Spectra of common-mode transient disturbances on ac mains [257]

The given spectra are composed of the envelope of the detected and analyzed transients. Since the worst case must always be considered for EMS tests, the envelope of all detected spectra was also determined. Figure 10.4 shows the spectra of common-mode and differential-mode surge voltages occurring in ac power mains.

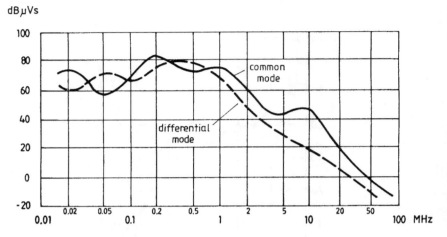

FIGURE 10.4 Envelope of transient disturbance spectra on ac mains [257]

10.2 EMS TESTS PER IEC SPECIFICATIONS

Because there are few limitations on transient emissions, it is very important to test the transient immunity of electrical equipment. EMS tests methods are usually based on various applicable IEC or IEEE specifications [352–356] for civilian applications, and MIL-Std 461/462 CS tests for military applications. Although these requirements were originally elaborated for relays and control units, they can be used for transient withstand testing of semiconductor equipments.

To cover different operating conditions, IEC specifications include three test voltage levels (Class I, II, and III) and two kinds of tests (impulse voltage withstand and high-frequency disturbance). These tests are recommended for various types of equipment as follows [353]:

Class I
Relays and control units in this class are exempted from transient voltage test since the equipment incorporating these relays and units will be tested in accordance with its own class.

Class II
Relays and control units in this test voltage class may be used where:

a. the auxiliary circuits are connected to a separate voltage supply, or the leads are short and, due to the absence of switching operations on other circuits connected to the supply, transient voltage levels on the supply leads will be low;
b. input circuits employ good shielding and earthing on the connecting leads;
c. the output circuits are connected to a load by short leads; or
d. a voltage test is not normally required but extra security is desired.

Class III

Relays and control units in this test voltage class should be used where:
a. the auxiliary circuits are connected to a central power supply and the leads are long, and transient voltages of high amplitude may arise from switching in other circuits to the same supply source;
b. input circuits where long leads are involved and effective screening and earthing are not employed;
c. output circuits are connected to a load by long leads so that longitudinal transients of high amplitude may appear;
d. a lower test voltage normally is sufficient, but extra security is required.

The test signal amplitudes are summarized for the three classes in Table 10.1.

TABLE 10.1   EMS test voltages

		Damped Oscillatory Wave	
Class	*Impulse Voltage (kV)*	*Differential-mode*	*Common-mode*
I	0	—	—
II	1	0.5	1.0
III	5	1.0	2.5

10.2.1 Impulse Voltage Withstand Test

The impulse voltage withstand test shall be regarded as a type test only. This test is recommended to determine whether the relay and its components will withstand without damage transient disturbances of high amplitude.

The shape of the recommended test signal, often referred to as a *1.2/50 μs impulse,* is shown in Fig. 10.5, and its parameters are given in Table 10.2 [353]. The standard test circuit for producing the test impulse is shown in Fig. 10.6. For testing the impulse withstand capability of an electrical equipment or its components, three positive and three negative impulses should be applied at intervals of not less than 5 s. The test leads should not be longer than 2 m.

The test impulses are to be applied to appropriate points of the equipment under test (EUT) that are accessible from outside the case; i.e., this test should be made

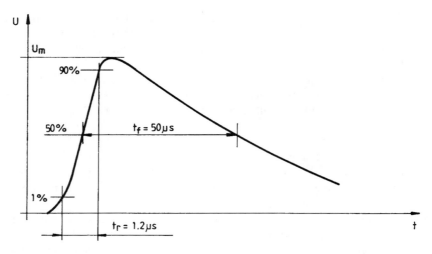

FIGURE 10.5   Voltage test impulse waveform

among all terminals and from terminals to the earthing point (i.e., chassis), as well as between independent circuits with terminals connected. The EUT should not be in operation, and all auxiliary energizing circuits should be disconnected from it. After the test, the EUT should still comply with all relevant specifications. It is to be noted that flashover does not necessarily constitute failure—especially if it occurs where it causes no damage. In this case, the manufacturer must decide whether to eliminate the cause of flashover. The repeated impulse test should be limited to a maximum of 60 percent of the class voltage, since the repeated stressing may reduce the performance of the item or its components.

Impulse withstand test recommendations frequently specify a unidirectional discharge current impulse, often called a *8/20 μs current impulse* [362]. This type of EMS test is usually applied to low-impedance devices. The shape of the impulse current test waveform is represented in Fig. 10.7, and the layout of the recommended impulse generator is shown in Fig. 10.8.

TABLE 10.2   IEC impulse test voltage parameters, no load (see Fig. 10.5)

Voltage rise time (t_r)	1.2 μs ± 30%
Voltage fall time (t_f)	50 μs ± 20%
Source impedance	500 Ω ± 10%
Source energy	0.5 Ws ± 10%
Tolerance of peak value of the test voltage before the test item is connected to the test circuit terminals	+ 0% to −10%

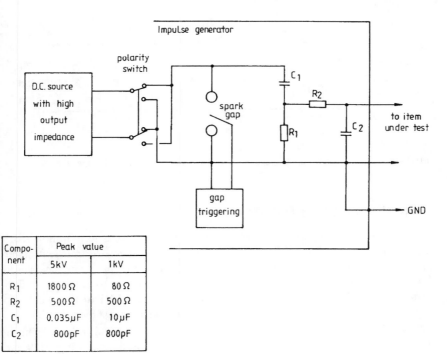

Compo-nent	Peak value	
	5kV	1kV
R_1	1800 Ω	80 Ω
R_2	500 Ω	500 Ω
C_1	0.035μF	10μF
C_2	800pF	800pF

FIGURE 10.6 Impulse generator for producing impulse test voltage waveform

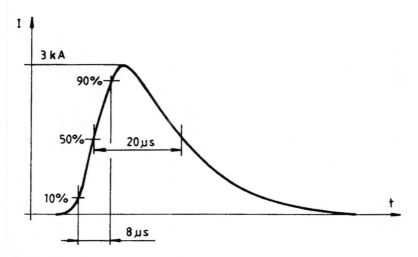

FIGURE 10.7 Test impulse current waveform

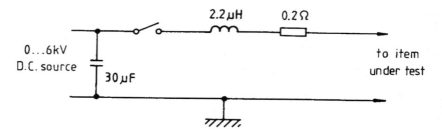

FIGURE 10.8 Layout of impulse generator for producing test impulse current waveform

10.2.2 High-Frequency Disturbance Test

The high-frequency disturbance test is recommended for determining whether the electrical equipment will operate properly when a specified transient disturbance is applied under reference conditions. Although only one frequency is used, it is considered to represented the real-world conditions and to give a basic indication of the ability of the EUT to withstand environmental HF disturbances. The high-frequency disturbance test is also be regarded as type test only.

The shape of the recommended HF disturbance test signal, often called the *1 MHz damped oscillatory wave* or *1 MHz ring wave*, is shown in Fig. 10.9, and its parameters are given in Table 10.3 [353]. The test circuit for producing the damped oscillatory impulse is represented in Fig. 10.10.

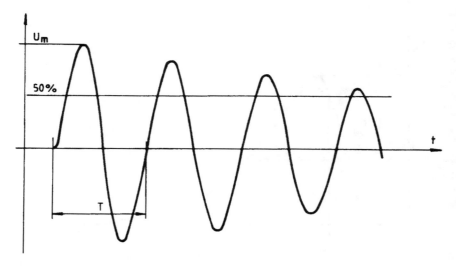

FIGURE 10.9 1 MHz damped oscillatory wave voltage

TABLE 10.3   IEC damped oscillatory wave test voltage parameters (see Fig. 10.9)

Frequency	1 MHz ±10%
Source impedance	200 Ω ± 10%
Number of oscillations for the envelope to decay to half the value of the initial amplitude	3 to 6
Repetition rate	400 Hz
Duration of the test	2 s ± 10%
Tolerance of the initial test voltage peak value before the EUT is connected to the test circuit terminals	+0% to −10%

The HF disturbance signal should be applied for 2 s at an appropriate point that is accessible from outside the EUT. The test leads should not be longer than 2 m. The HF disturbance test is carried out between each set of input and output terminals and earth (longitudinal), between all independent circuits (longitudinal), and between terminals of the same circuit (transverse). The HF longitudinal coupling mode for testing the terminals of the same circuit is shown in Fig. 10.11, and longitudinal coupling for EMS testing between the terminals and earth is shown in Fig. 10.12. The inductors, L, are for HF decoupling of the terminals under test and the auxiliary circuits. A newly proposed modification of the coupling network circuit scheme is shown in Fig. 10.13 (SN 31158, Teil 03–Siemens Standard). The

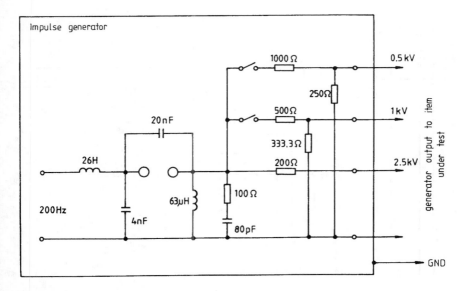

FIGURE 10.10   Layout of impulse generator for producing test damped oscillatory waveform

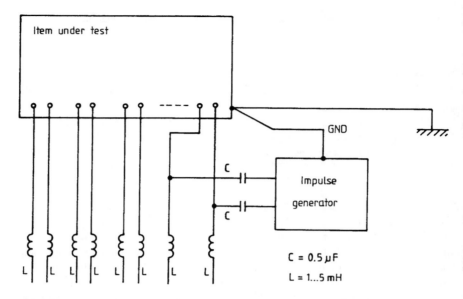

FIGURE 10.11   HF disturbance test signal coupling between terminals of the same circuit

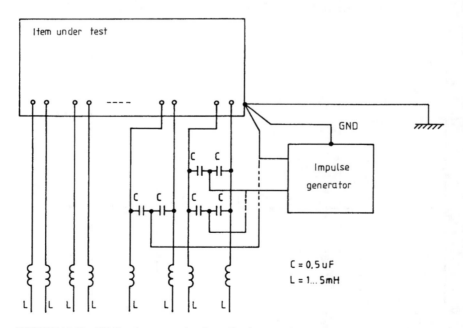

FIGURE 10.12   HF disturbance test signal coupling between the tested terminals and earth

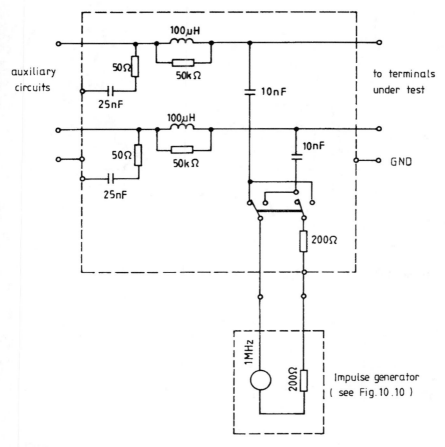

FIGURE 10.13   Proposed modified coupling network

purpose of the modification is to develop HF damping and thereby eliminate resonances during the disturbance test. This network connects the non-measured terminals to earth by a 200 Ω resistor.

Some specifications recommend other waveforms. A well known test signal
for HF disturbance testing is the 0.5 μs/100 kHz ring wave shown in Fig. 10.14
[362]. The duration of the first half-cycle is not defined. The time of the first half-
cycle has to be fitted to the parameters of the test item. The recommended circuit
scheme for generating this ring wave test signal is presented in Fig. 10.15. The introduced generator couples the HF disturbance test signal on the mains voltage
through the 5 μH inductor. This impulse generator is capable of producing both
voltage and current test signals: a voltage transient is delivered to high-impedance
test items, and a current transient to low-impedance test items.

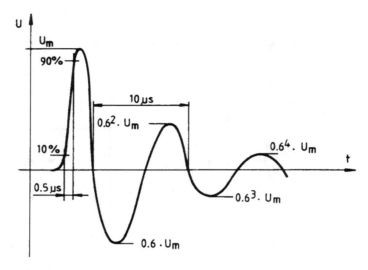

FIGURE 10.14   100 kHz ring wave test pulse

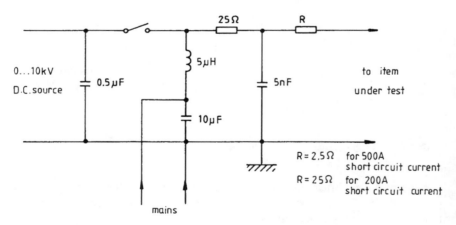

FIGURE 10.15   Layout of impulse generator for producing test ring wave

10.2.3 Electrostatic Discharge Test

New electronic systems containing more sensitive electronic units have come into service in recent years. Because many are susceptible to electrostatic discharge (ESD) events that occur when the operator touches the equipment, the ESD test of Ref. 357 (Part 2) was developed. The two types of ESD tests are (1) contact and (2) air discharge.

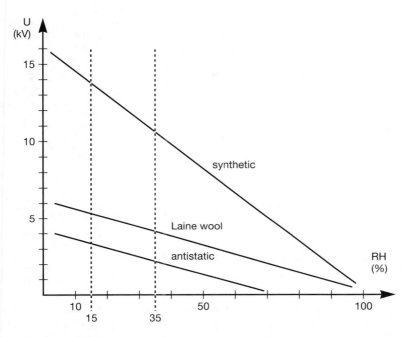

FIGURE 10.16 Maximum static voltage levels to which the human body may be charged while in contact with various insulating materials

Several severity levels are defined that relate to different environmental and test setup conditions. These are shown in Table 10.4. If voltages higher than those shown in the table are specified, special test equipment may be required.

Static levels to which the body may be charged, and therefore to which equipment may be subjected, depend on the quality of the insulation material and the relative humidity (RH) of the ambient air. Figure 10.16 shows how maximum charge levels vary according to RH.

A simplified circuit of an ESD test generator is given in Fig. 10.17. The energy storage capacitor charge, including the distributed capacitance between the

TABLE 10.4   Severity Levels for Contact and Air ESD Tests

Level	Test Voltage, Contact Discharge	Test Voltage, Air Discharge
1	2 kV	2 kV
2	4 kV	4 kV
3	6 kV	8 kV
4	8 kV	15 kV

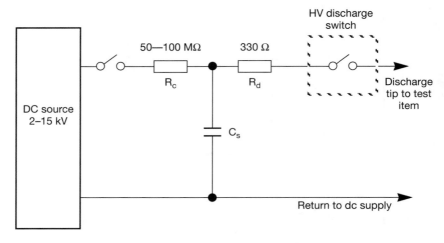

FIGURE 10.17 Simplified circuit of ESD test generator

generator and test item, ground, and coupling wiring, may be a maximum of 150 pF ± 10 percent. The discrete storage capacitor, the discharge resistor, and the discharge HV switch must be placed as close as possible to the discharge electrode.

The dimensions and shape of the discharge electrode for contact and air ESD tests are specified by the IEC standard. An adequately insulated test cable must be used, and its length generally cannot exceed 2 m, although in special cases it may be as long as 3 m. The recommended ESD test current output waveform is shown in Fig. 10.18, and the mains characteristics of the ESD test current are summarized in Table 10.5. The current values are to be measured with a 1,000 MHz bandwidth measuring instrument.

In the case of air discharge testing, the ambient temperature must be within the range of 15 to 35° C, the relative humidity between 30 and 60 percent, and the air pressure from 680 to 1,060 bar.

TABLE 10.5 ESD Test Current Waveform Characteristics

Level	First Peak (A, ± 10%)	Current at 30 ns (A, ±30%)	Current at 60 ns (A, ±30%)
2	7.5	4	2
2	15	8	4
3	22.5	12	6
4	30	16	8

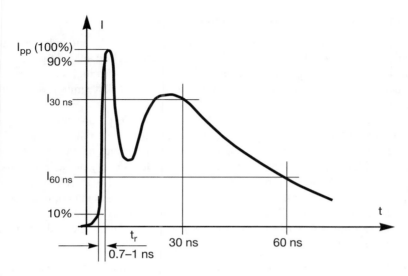

FIGURE 10.18   Typical output current waveform of ESD test generator

The ESD test is performed according to the test plan, which should specify the representative operating condition, the scheme of the test setup, the test method (contact or air discharge), the point at which the discharges are to be applied, the severity level, and the number of ESD test impulses to be applied. The specified test points and surfaces of the test item must be accessible to personnel during normal service. The application of a test pulse to any point that is accessible only for maintenance purposes is not allowed without prior consent.

The test voltage is increased from the minimum to the selected test severity level. The test is performed with single discharges on a preselected point, with at least 10 discharges of the most sensitive polarity applied.

For contact ESD testing, the time interval between single test signals is recommended to be 1 s. The point at which the discharges should be applied may be selected by means of an exploration carried out at a repetition rate of 20 test signals per second or more. The tip of the test electrode must touch the test item before the test discharge switch is operated. If the test surface is painted that is not declared to be an insulating coating, the pointed tip of the ESD test generator must penetrate and make contact with the conductive substrate.

For air-discharge ESD tests, the rounded tip of the test electrode is moved as quickly as possible toward the test item surface, short of causing mechanical damage. After applying the test signal, the electrode is removed.

The test results should be classified on the basis of the operating conditions of the test item. In the case of acceptance tests, the qualifying program and the interpretation of the test results are subject to agreement between the manufacturer and user.

10.2.4 Electrical Fast Transient and Burst Test

Investigations into electronic malfunctions in commercial and industrial applications has shown that many failures can be traced back to switching transients. Because switching operations generate fast transients with high repetition rates, it became necessary to develop an EMS test method that simulates switching transients.

The electrical fast transient/burst (EFT/B) test [357, part 4] couples bursts of multiple fast transients into the power supply, control electronics, and signal terminals of electronic equipment. Significant characteristics of the EFT/B test include short rise time, specific repetition rate, and low energy of the transients.

Table 10.6 shows the test severity levels. for the EFT/B test. If higher voltages than those shown are expected in the operational environment and, therefore, required for testing, special test equipment may be needed.

The recommended selection of EFT/B test severity levels is based on the electromagnetic environment as follows:

Level 1. Well protected environments where all contacts are protected by suitable circuits; the power supply and control, measurement, and signal lines are well separated; and the power supply cables are shielded and properly earthed. A typical example would be a computer room.

Level 2. Partial suppression in control circuits that are switched only by relays, separation of all circuits from other circuits that are associated with higher severity level environments, and physical separation of unshielded power supply and control cables from the signal and communication lines. A typical example is a control or terminal room of an industrial or power plant.

Level 3. No suppression in control circuits that are switched only by relays; poor separation of the industrial circuits from other circuits associated with higher severity level environments; dedicated cables for power supply, control, signal, and communication lines; and an earthing system represented by conductive pipes, ground conductors in the cable trays, and a ground mesh.

Level 4. No suppression in control circuits that are switched by relays and contactors; no separation of the industrial circuits from other circuits associated

TABLE 10.6   Severity Test Levels for EFT/B Test

	Open Circuit Test Voltage ±10%	
Level	Power Supply	I/O Signal, Data, and Control Lines
1	0.5 kV	0.25 kV`
2	1 kV	0.5 kV
3	2 kV	1 kV
4	4 kV	2 kV

with higher severity level environments; dedicated cables for power supply, control, signal, and communication lines; and earthing system represented by conductive pipes, ground conductors in cable trays, and a ground mesh.

A simplified circuit of the EFT/B test generator is shown in Figure 10.19. The dc supply should deliver 4 mJ of energy per pulse at 2 kV on a 50 Ω load, with both positive and negative polarity. To allow comparison of EFT/B test results from different test generators, and because the value of the circuit elements may depend on the particular realization, only the characteristics of the test pulse are given. For verifying rise time, impulse duration, and repetition rate of the impulses within a burst, the EFT/B test generator output should be measured with an oscilloscope of at least 400 MHz bandwidth, connected through a 50 Ω coaxial attenuator.

Figure 10.20 shows the recommended waveform of a single EFT/B test pulse. The repetition rate of the EFT/B pulses depends on the test voltage. The recommended rates are:

5 kHz ±20%	0.25 kV
5 kHz ±20%	0.5 kV
5 kHz ±20%	1.0 kV
2.5 kHz ±20%	2.0 kV

It should be mentioned that the actual phenomenon of a burst occurs with repetition rates of the single pulse changing from 10 kHz up to 1 MHz. Because this relatively high repetition rate is difficult to achieve with a single air gap, the relatively slow repetition rates cited above are specified for the EFT/B test. The burst duration is to be 15 ms, ±20 percent, and the burst period is 300 ms, ±20 percent.

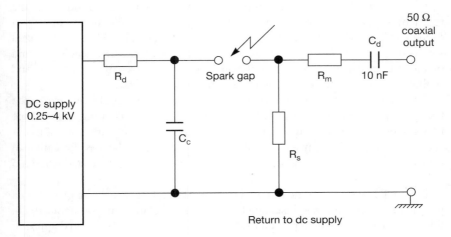

FIGURE 10.19   Simplified circuit of EFT/B test generator

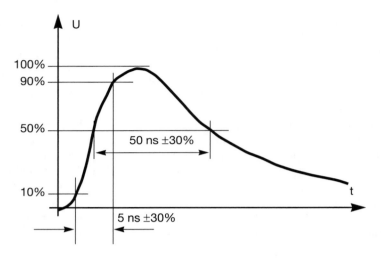

FIGURE 10.20   Typical output current waveform of EFT/B test generator onto a 50 Ω load

The EFT/B test signal must be coupled between earth (reference point) and the test points, including the equipment earthing. For laboratory tests, a coupling/decoupling network is recommended for ac and dc supplies. A three-phase solution is shown in Fig. 10.21.

The 33 nF coupling capacitors must have good HF characteristics up to 100 MHz and be tested with a 5 kV, 1.2/50 μs test pulse. The coupling/decoupling circuit shown has less than 2 dB coupling attenuation and more than 30 dB crosstalk attenuation from one line to the next.

Using a capacitive coupling clamp, the EFT/B test signal can be coupled to the test item without a galvanic connection to the terminals, shield, or any other part of the test item. The capacitive coupling clamp unit is composed of a galvanized steel, brass, copper, or aluminum housing placed above the tested cable and laid on the reference ground. The ground plane is to have an area of at least 1 m² and shall extend beyond the clamp by at least 10 cm on all sides.

The EFT/B generator is connected to the end of the clamp that is nearest the test item. The clamp itself is installed to provide maximum coupling capacitance between the cable and clamp. In the IEC standard, the given clamp dimensions ensure a typical coupling capacitance of 50–200 pF, and it can be used for cables of 4 to 40 mm in diameter. The capacitive coupling clamp is to be tested using a 5 kV, 1.2/50 μs test pulse.

For EFT/B tests performed in a laboratory, the test item must be placed on a ground reference plane and insulated from it by an insulating support about 10 cm thick. In the case of tabletop equipment, the test item should be located approximately 0.8 m above the reference ground.

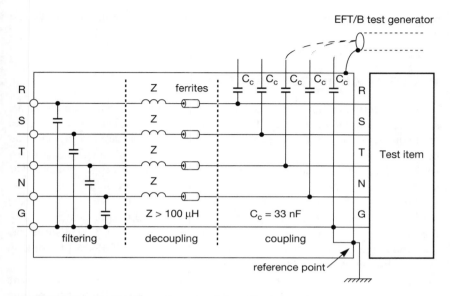

FIGURE 10.21   Coupling/decoupling network for ac/dc power mains

The EFT/B test can be performed in the final installed location without coupling/decoupling networks. This simulates the natural interference environmental conditions as closely as possible.

The power mains of the test item should be tested by applying the test signals between the reference ground plane and each of the power terminals, including the protective or functional earth (e.g., the cabinet). The EFT/B generator must be located on the ground plane such that the hot wire does not exceed 1 m. The connection is unshielded but well insulated. If blocking capacitors are needed, their capacitance is specified at 22 nF.

The EFT/B test on I/O circuits and communication lines will be performed, if possible, with a capacitive coupling clamp. If the clamp cannot be used (due to size, cable routing, or other restrictions), it may be replaced with a tape or conductive foil that envelops the tested lines. The capacitance of this coupling arrangement should be 50 to 200 pF. In other cases, the EFT/B test generator can be connected to the test item via a discrete 100 pF capacitor.

Lacking any other prearranged specifications, the climatic conditions are to be 15 to 35° C, 45 to 75 percent relative humidity, and air pressure of 680 to 1,060 bar. Before running the EFT/B test, the characteristics of the actual test generator should be verified.

The duration of the EFT/B test cannot be less than one second. The test plan should show the test voltage and type, polarity of the test voltage, test duration, lines tested, representative operating conditions, and the test setup scheme. Test results should be classified on the basis of the operating conditions of the test item.

10.3 OTHER EMS TEST METHODS

In engineering practice, EMS testing of electrical equipments with electronic subsystems is usually done per IEC, IEEE, or Mil-Std specifications as described in the previous section. This is primarily because such standard test instrumentation is available off the shelf. HF decoupling networks, usually available as accessories, are capable of superimposing the test pulse on the supply voltage. The repetition rate and phase shift of HF test pulses generally are adjustable.

National EMC regulations and standards require EMS testing of electrical equipments using various test signals as stipulated by IEC specifications [115]. For instance, in addition to the standard 1.2/50 μs surge voltage, 8/20 μs surge current, and 0.5 μs/100 kHz ring wave transient immunity tests, CENELEC (valid in Western Europe) suggests three additional pulses.

The so-called *100/1300 μs test pulse*, with an amplitude of 1.3 times that of the mains voltage peak value, is suggested to represent transient disturbances of long duration but relatively low amplitude. Such transients are generated mainly by the breaking of low-voltage fuses. Bursts of rapid transients, the IEC 801/4 *5/50 ns test signal,* represents a short rise time with high repetition rate. Such a transient is generated by switching off inductive loads. The *75 ns/0.1–1 MHz test transient* was developed for testing equipments installed in HV substations. However, it is known that many transients occurring on ac mains are oscillatory in nature, so application of this pulse has been extended to electrical apparatus operating on public power distribution lines. Originally, only a 1 MHz test frequency was applied to the EUT, it is now considered advisable at least to apply an additional 0.1 MHz test signal, and preferably several signals in the frequency range of 0.1 to 1 MHz.

Most transient disturbances occurring on ac power mains are damped oscillatory transients generated by switching processes. To obtain better information about the ability of electronic units to withstand transient disturbances caused by load switching, another test method is also used [365]. This is the Swedish recommendation SEN 361503 (1.4.1977), which can be regarded as a modified EFT/B test (IEC 801/4). The layout of the impulse generator is shown in Fig. 10.22, and the test signal waveforms are presented in Fig. 10.23a and 10.23b. The characteristics of the test pulse train depend mainly on the magnitude of the stray capacitance, C_S, and the quality of the relay contacts. The spectrum of the test signal may spread up to 10 to 100 MHz. The signal coupling and the test method can be identical to those described for the HF disturbance withstand test (see Section 10.2.2). This EMS test method was elaborated for testing electronic control systems for power plants, but it can be well applied to semiconductor circuits and equipment operating with a relay control system nearby.

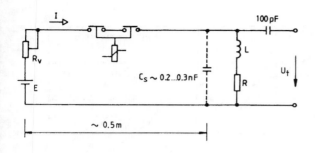

Circuit	U_{tm}	
elements	2-4 kV	4-8 kV
E (VAs)	110-220	220
T (mA)	5...10	50...100
L (H)	300	100
R (kΩ)	10...20	2...4

FIGURE 10.22   Layout of impulse generator for producing switching transient test signals

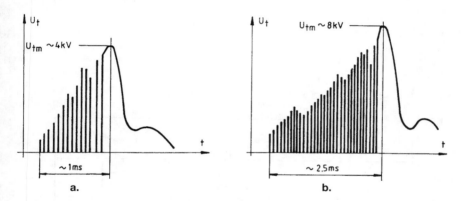

FIGURE 10.23   Switching transient test waveforms

For comparing the results of different test methods, we first consider the important parameters of the applied test signals [115]. The spectra of the most common test transients are summarized in Fig. 10.24. The spectrum of the probable common-mode transient (from Figs. 10.3 and 10.4) is also shown in the figure. The spectra clearly show that the IEC 1.2/50 μs surge voltage suitably covers the requirements up to about 200 kHz. However, HF EMS is best tested using fast transients.

Electromagnetic coupling is closely associated with the rise time of the transient disturbance. Figure 10.25 shows the voltage rise gradients of different test transients and of transients measured on public ac mains. This comparison shows that only a fast transient burst with a 5 ns rise time corresponds to the maximum mains condition, and the damped oscillatory test signal with a 75 ns rise time could be accepted for modest requirements.

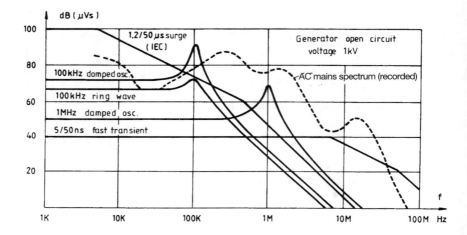

FIGURE 10.24 Test transient spectra [115]

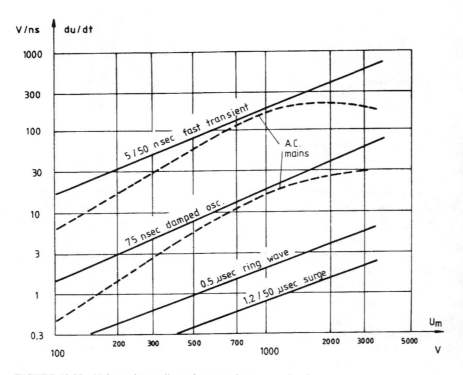

FIGURE 10.25 Voltage rise gradient of test transients vs. peak value

The energy of the discussed test signals as a function of the peak value is plotted in Fig. 10.26. Under test conditions, these energy values will be shared between the test generator impedance and the test item. Therefore, it is difficult to draw a realistic comparison. But, in general, the 1.2/50 μs IEC surge voltage can be viewed as the most suitable test signal. Other EMS test signals are of low energy, but the level can be increased through rapid repetition.

From Figs. 10.24 through 10.26, some general conclusions can be drawn about the different EMS test methods and test signals. First, the impulse withstand test method using a 1.2/50 μs surge appears to be relevant, but it should be augmented by a fast transient test to cover the requirements of rapid voltage rise. With regard to the HF disturbance test, perhaps the 0.1 MHz damped oscillatory transient seems adequate, but the repetition rate and time may be high with regard to the energy delivered to the test item. The alternative is the 1 MHz damped oscillatory transient originally designed for electronic systems installed in HV substations.

Electrical equipment malfunctions can be caused not only by transient disturbances but also by voltage distortions and short-duration breaks. So-called *network simulators* have been developed for testing the immunity of electronic units

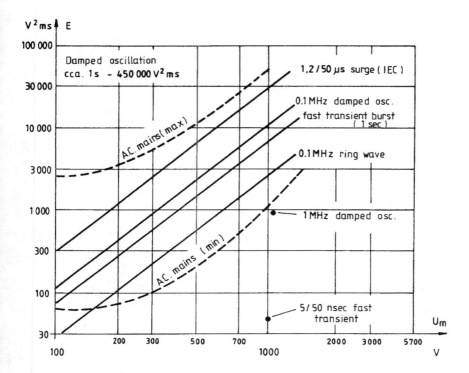

FIGURE 10.26 Energy content vs. peak value of test transients

against these disturbances. These EMS test instruments generate ac mains voltage distortions shown in Fig. 10.27. Although there are no exact requirements for semiconductor or power electronic equipments, these devices usually should operate without malfunction through a supply voltage break of at least one-half cycle.

FIGURE 10.27   Generated ac mains voltage distortions

11

Reduction Techniques for Internal EMI

The widespread use of semiconductor modules in power electronic equipment has required a more detailed EMS analysis. Because these adjacent high-power circuits and sensitive electronic components interact so profoundly, the design engineer must consider not only external environmental disturbances but the EMS of semiconductor circuits to internally generated noise.

EMI suppression techniques should be considered in early design stages rather than waiting until problems appear during testing or field operation. It is much more economical to solve EMS shortcomings early on. As the design process progresses, available EMS reduction techniques become fewer while possible source-victim paths become more numerous and more difficult to analyze. A system designed without regard to EMS will almost surely have some noise problems, some of which may be difficult and expensive to solve.

The first step in analyzing an existing EMS problem is to determine how the noise source (culprit) and the receiver (victim) are coupled. From an EMS perspective, noise generated by the culprit can no longer be suppressed, and the victim cannot be made less sensitive. Therefore, we must modify the transmission channel. Coupling occurs by conduction, radiation, or both. Conductive noise coupling can be decreased via the use of proper circuit connections, whereas radiated EMI is reduced through shielding techniques.

11.1 CONDUCTIVE NOISE COUPLING

An obvious situation in which noise can be coupled from one circuit to another is when two different circuit currents flow through a common impedance. In engineering practice, common impedance coupling typically occurs when separate circuits are connected as shown in Fig. 11.1. Here the impedances represent the resistance and inductance of the wiring. This conductive coupling can be well characterized by standard network analysis methods. In general, the higher the common impedance coupling, the higher the current flowing through the common impedance and the higher the common impedance value. For calculating the conductive coupling, we will neglect the impedances Z_{i1} and Z_{i2} (representing the interconnection impedances), which results in the equivalent circuit shown in Fig. 11.2. The parallel combination of Z_{1N}, Z_{2N} is now labeled Z_c. The impedances

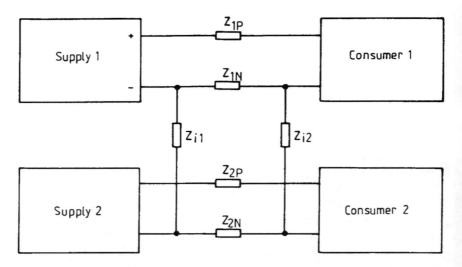

FIGURE 11.1 Conductive noise coupling between two circuits

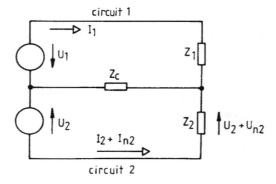

FIGURE 11.2 Equivalent circuit for calculating conductive noise coupling

Z_{1P} and Z_{2P} are not considered since they do not influence the conductive coupling and, in comparison to Z_1 and Z_2 (which represent the impedances of the power consumers), are small enough to be neglected. It should be noted that a similar equivalent circuit can be set up when the conductive noise coupling occurs across a common grounding impedance.

Assuming that circuit 1 disturbs the circuit 2, the noise voltage appearing across Z2, by common impedance coupling, is:

$$U_{n2} = I_1 \times \frac{Z_c \times Z_2}{Z_c + Z_2} = \frac{U_1}{Z_1 + Z_c} \times \frac{Z_c Z_2}{Z_c + Z_2} \tag{11.1}$$

The noise current coupled into circuit 2 can be calculated as follows:

$$I_{n2} = \frac{U_1 Z_c}{Z_c Z_1 + Z_1 Z_2 + Z_c Z_2 + Z_c^2} \tag{11.2}$$

In most practical cases, the noise coupling impedances are the resistance and inductance of the wiring. At low frequencies, it is sufficient to consider only the dc resistance of the wiring since it is only above 10 to 100 kHz that impedance increases by the skin effect and the self-inductance of the wire.

It is well known that the resistance of a conductor increases at high frequencies due to skin effect (see Section 9.5.2). The skin depth, δ, can be calculated using Eq. (9.63). If this value is smaller than the radius of the wire, the HF resistance of a round conductor of diameter d is as follows:

$$R = R_o \times \frac{d}{4\delta} \tag{11.3}$$

For a solid, round copper conductor of diameter d (in millimeters), a dc resistance, $R_o(\Omega)$, and at a frequency of f, the ac resistance can be approximated by the following expression [5]:

$$R = R_o (0.25 + 0.00376 \times d \times \sqrt{f}) \tag{11.4}$$

$$\text{for } d \times \sqrt{f} > 300$$

The relative increase in resistance of the round copper conductor as a function of the frequency, described by the above relationship, is shown in Fig. 11.3.

The self-inductance of a conductor may considerably influence the noise and transient performance of electronic circuitry. This self-inductance consists of two parts: internal and external [6]. The internal inductance of a straight wire with circular cross-section is about 0.5 μH/m, independent of conductor size. This part is negligible when compared to the external inductance, except in the case of very

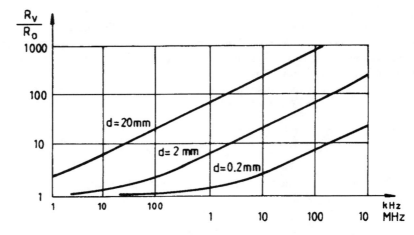

FIGURE 11.3 Increase in HF resistance vs. frequency of a round copper conductor

close wire spacing. The external inductance of a straight, round conductor above a ground plane, when the ratio of the conductor height and the conductor diameter is higher than 1.5 [6], can be calculated as follows:

$$L_e = \ell \times \frac{\mu_o}{2\pi} \times \log_n \left(\frac{4a}{d}\right) \tag{11.5}$$

where

a = center-to-plane spacing

d = conductor diameter

ℓ = conductor length in meters

μ_o = permeability of free space ($4\pi \times 10^{-7}$ Vs/m^2 for air)

The external inductance of two parallel conductors is as calculated below, with notation as used in Eq. (11.5), when the ratio of the conductors spacing "a" and the conductor diameter is higher than 3:

$$L_e = \ell \times \frac{\mu_o}{\pi} \times \log_n \left(\frac{2a}{d}\right) \tag{11.6}$$

Note that the inductance of one single wire in the pair will be half this value.

The self-inductance of the wire arrangement common in power electronic equipments can be accurately calculated if there are no ferromagnetic materials between the conductors. For practical reasons, the self-inductance is often related to unit length:

$$\frac{L_s}{\ell} = 0.4\,(\log_n \frac{2a}{d} + 0.25) \times 10^{-6} \qquad \text{H/m} \qquad (11.7)$$

Figure 11.4 gives the self-inductance of parallel conductor pairs as a function of the ratio $r = a/d$. For a rectangular cross-section, d in the ratio represents the circumference of the conductor.

Another obvious but often overlooked source of conductively coupled noise is the power supply network of the electronic module. The designer of semiconductor equipments with on-board electronics must pay special attention to the layout of the power supply leads. Such a problem is illustrated in Fig. 11.5. The electronic unit is supplied through a relatively long wire, and a display lamp is also oper-

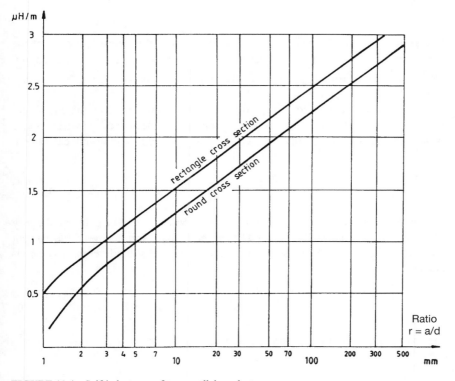

FIGURE 11.4   Self-inductance of two parallel conductors

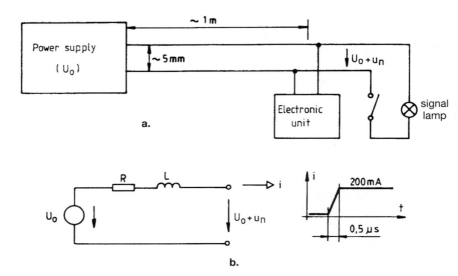

FIGURE 11.5 Noise coupling through power supply wires

ated from these supply leads by a transistor switch. Let the inrush current of the sign lamp be 200 mA and the switching time of the transistor 0.5 μs. The supply leads are 0.5 mm^2 round copper wires. The dimensions of the wiring are shown in Fig. 11.5a.

The conductively coupled noise voltage can be calculated using the equivalent circuit shown in Fig. 11.5b, where R represents the resistance of the wiring and L is the inductance of the supply circuit. The noise voltage can be simply calculated:

$$u_n = R \times i + L \times \frac{di}{dt} \tag{11.8}$$

The dc resistance of the 0.8 mm dia. supply wires, as can be calculated easily from the given data, is $R_0 = 0.07 \ \Omega$. The equivalent frequency of the current change is $0.35/t_r = 0.7$ MHz, so we must check to see if skin effect has to be considered. Given that

$$d \times \sqrt{f} = 0.8 \times \sqrt{0.7 \times 10^6} = 670 > 300 \tag{11.9}$$

the ac resistance should be approximated by Eq. (11.4):

$$R = 0.07 \times (0.25 + 0.00376 \times 0.8 \times \sqrt{0.7 \, (2) \times 10^6}) = 0.2 \ \Omega \tag{11.10}$$

The self-inductance can be read from Fig. 11.4. For r = 5/0.8 = 6.25, the self-inductance is 1.1 μH/m. Substituting the ac resistance value and the self-inductance into Eq. (11.8), the noise voltage generated by the switch-on of a little display lamp is:

$$u_n = 0.2 \times 0.2 + 1.1 \times 10^{-6} \times \frac{0.2}{0.5 \times 10^{-6}} = 0.04 + 0.44 \cong 0.5 \quad V \quad (11.11)$$

Although this value is not very high, it can exceed the noise threshold level of some electronic units. It is worth mentioning that the noise voltage is mainly inductive in nature, in spite of the high ac resistance.

The noise voltage developed across a common impedance can be reduced by the layout shown in Fig. 11.6. The improvement is obtained by decreasing (or even eliminating) the magnitude of the common line impedance. Further improvement can be achieved by slowing down the switching process.

Conductive noise coupling can be reduced best by minimizing the common impedances between the source and the receiver. The most important recommended methods for connecting separate circuits are summarized in Fig. 11.7 [6–8]. Figure 11.7a shows a further solution for decreasing the conductive noise coupling, based on an analysis of Fig. 11.5. The common impedance coupling (R_w in the figure) is shunted out by capacitor C. The preferred circuit connection method is shown in Fig. 11.7b. Using only one connection wire, the common impedance shown in Fig. 11.1 can be eliminated. Two conditions are necessary for these "star" or single-point ground practices to work: (1) in 11.7a, circuits 1 and 3 must have *no other interconnection* than the power feed; otherwise, this would recreate a ground loop, and (2) in 11.7b, power supplies 1 and 2 must have completely isolated outputs, including their 0V.

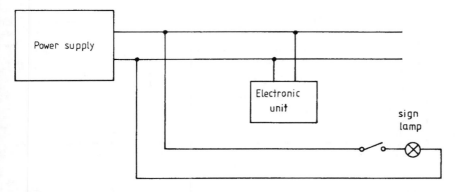

FIGURE 11.6 Layout to minimize conductive noise coupling through power supply lines

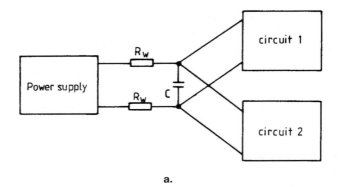

a.

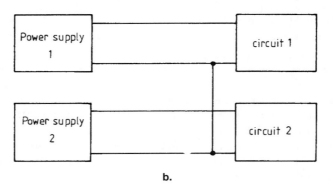

b.

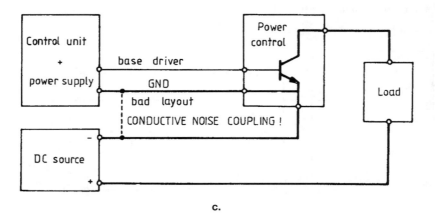

c.

FIGURE 11.7 Circuit connection layouts for conductive noise coupling reduction

This solution is recommended for equipment with power semiconductor control units as shown in Fig. 11.7c. The dc power source and the base driver unit of the power switching transistor should be connected quite near the power transistor pins.

The other conductive noise coupling mode is the ground loop, and the aspects shown in Fig. 11.7 are also useful for designing a proper grounding scheme [6]. A ground loop is formed when all circuits are separately grounded. Grounding problem may arise when low-level signals must be transmitted a relatively long distance. The loop can best be broken by an isolation transformer, but this solution is not practical when dc or low-frequency continuity is also required. Under this condition, a common-mode choke coil (often called *longitudinal choke*) or a neutralizing transformer or balun should be used. This is shown in Fig. 11.8.

The common-mode choke coil presents a low impedance to differential signal current I_s and allows dc coupling. However, to any common-mode (longitudinal) noise current (I_{g1} and I_{g2} in Fig. 11.8a), it is a high impedance (see Section 6.3.4). For designing common-mode choke coils, the current division between resistor R_2 and ground impedance Z_g should be considered (see Fig. 11.8b). Resistor R_2 now represents the sum of the lead resistance and the resistance of one winding of the common-mode choke coil. Although R_2 is usually small, for designing a common-mode choke coil, this current division should be considered because the imped-

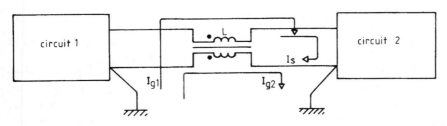

a. Layout of the neutralizing choke coil

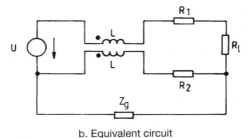

b. Equivalent circuit

FIGURE 11.8 Neutralizing choke coil for reducing conductive noise coupling [6]

ance of the ground path is also very small. If a current division develops, the condition of current compensation will not be met (see Section 6.3.4) and the choke coil will not act as a common-mode choke coil.

We should design the common-mode inductance of the choke coil to be sufficiently high for noise currents but not high enough to disturb a signal transmission. This condition is met [6] if:

$$L \geq 5 \times \frac{R_2}{\omega_L} \qquad (11.12)$$

where

R_2 = resistance of the return path on ground potential

ω_L = the lowest signal frequency

Under the above condition, the ground current is nearly zero. This means that virtually all the signal current will be returned to the source through the second conductor and not via the ground; i.e., no conductive noise coupling is expected.

11.2 ELECTROMAGNETIC COUPLING

Noise coupling may occur between circuits or wires that are electrically independent (isolated) and therefore have no common impedance. All circuit elements, including conductors, radiate electromagnetic fields whenever current moves through them. When the radiated fields encounter other circuit elements, including conductors, a current will be generated in them. This phenomenon is called *electromagnetic coupling*. Electromagnetic coupling is discussed in two primary ways, depending on whether the source and the receiver are in close or distant proximity.

At a point relatively distant from the source (far field), the properties of the electromagnetic field are determined primarily by the medium through which the field travels. But close to the source (near field), these properties depend mainly on the source characteristics. The far field is said to begin at a distance of wavelength divided by 2π.

The far and near fields also can be defined by the wave impedance, which is the ratio of the electric field to the magnetic field strength. In the far field, this ratio is a constant 377 Ω for air or free space. In the near field, the wave impedance is determined not only by the characteristics of the source but also by the distance. If the source has high current and low voltage, the near field is dominantly magnetic, and its wave impedance is smaller than that of the far field. In this case, the magnetic field changes at a rate of the cube of the distance, and the electric field changes at a rate of the square of the distance. As a result, wave impedance is proportional to distance. Conversely, if the source has low current and high voltage, the near field is electric in nature, and the wave impedance is higher than that of the far field. The wave impedance is also proportional to distance, only the electric field changes at a rate of the cube of the distance and the magnetic field at a rate of the square.

Within power electronic equipments (and other electric apparatus) most electromagnetic coupling is expected in the near field, given that the transition region between the near and far fields is about 50 m at frequencies of approximately 1 MHz. In the near field, electric and magnetic fields must be considered separately. Therefore, two separate coupling modes must be studied: capacitive (electric field) and inductive (magnetic field) coupling.

11.2.1 Capacitive Coupling

Figure 11.9 shows a simplified model illustrating the principal means of undesirable capacitive (electric field) coupling. Capacitor C_{12} is the stray capacitance between the conductors. The capacitors with a subscript g represent the capacitance between each conductor and its return, which can be a ground plane or earth. Circuit 2 (the receiver) is loaded toward ground by resistor R. The capacitive noise coupling can be best analyzed by the equivalent circuit shown in Fig. 11.9b. This equivalent circuit aptly illustrates that any capacitance connected between source and earth has no effect on noise coupling, as long as it does not affect the amplitude and waveform of U_G (recognizing that for very high frequencies, C_{1g} may have a filtering effect on U_G).

Considering the voltage U_g as the interference source, the noise voltage coupled onto the conductor 2 can be expressed as follows [6]:

$$U_n = U_G \times \frac{j\omega \times \dfrac{C_{12}}{C_{12} + C_{2g}}}{j\omega + \dfrac{1}{R\,(C_{12} + C_{2g})}} \qquad (11.13)$$

The above expression can be simplified, allowing us to study how the pickup voltage depends on the various parameters when:

$$R \ll \frac{1}{j\omega\,(C_{12} + C_{2g})} \qquad (11.14)$$

This condition is met in most practical cases, since the stray capacitances are usually very small. This yields:

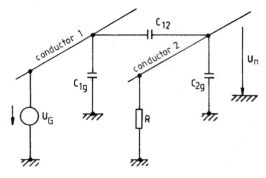

a. Physical representation

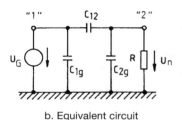

b. Equivalent circuit

FIGURE 11.9 Capacitive noise coupling between two conductors

$$U_n \approx U_G \times j\omega \times RC_{12} \qquad (11.15)$$

The simplified expression clearly shows that the capacitively coupled voltage is directly proportional to the frequency of the noise source, the resistance of the affected circuit to ground, the capacitance between the source and receiver conductors, and the magnitude of the source voltage.

At high frequencies, or when the load resistance of receiver circuit toward the earth is so high that the condition Eq. (11.14) is not met, the coupled noise voltage is determined only by the source voltage and the stray capacitances:

$$U_n(\omega \approx \infty) = U_G \times \frac{C_{12}}{C_{12} + C_{2g}} \qquad (11.16)$$

Under this condition, the capacitively coupled noise voltage is independent of frequency and is of a larger magnitude than Eq. (11.15) describes. The frequency dependence of the noise voltage is shown in Fig. 11.10. The cutoff frequency can be calculated from the condition of Eq. (11.14), but an "equals" sign should be considered (see Fig. 11.10).

The capacitance between the conductors and between conductors and earth can be calculated with good approximation. The stray capacitance of a conductor above a ground plane is [6]:

$$C_{2g} = \frac{2\pi \times \ell \times \varepsilon_r \times \varepsilon_o}{\text{arch} \dfrac{2a}{d}}$$

(11.17)

where

a = distance from the earth

d = wire diameter in the same units as a

ℓ = conductor length in meters

$\varepsilon_o = 8.85 \times 10^{-12}$ F/m for air

ε_r = relative permittivity

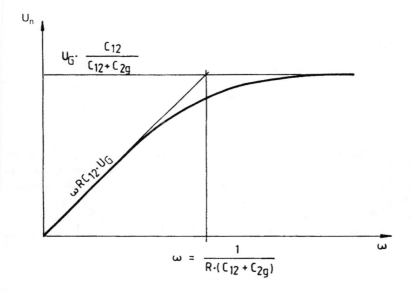

FIGURE 11.10 Frequency response of capacitively coupled noise voltage

In practice, if $a > 3 \times d$:

$$C_{2g} \approx \frac{55.6 \times \varepsilon_r}{\log_n \dfrac{4a}{d}} \left(\frac{pF}{m}\right)$$

ε_r is factored in only if the entire space between conductor and ground is embedded in insulating material. Otherwise, $\varepsilon_r = 1$.

The other determinant parameter for capacitive noise coupling is capacitance between the wires. The stray capacitance of two parallel conductors embedded in a medium ε_r can be calculated as [1]:

$$C_{12} = \frac{\ell \times \varepsilon_r \times \pi \times \varepsilon_o}{\operatorname{arch} \dfrac{a}{d}} \tag{11.18}$$

The stray capacitance between wires is often related to length. The above expression can be simplified if the ratio of the wire distance and the conductor diameter is higher than 3 and if of C_{12} occurs in air. Thus, the stray capacitance related to length yields:

$$\frac{C_{12}}{\ell} = \frac{27.8 \times 10^{-12}}{\log_n \dfrac{2a}{d}} \quad F/m \tag{11.19}$$

where

$\ell =$ wire length in meters

$a =$ distance between the wires

$d =$ wire diameter in the same units as a

The effect of conductor spacing on capacitive coupling for parallel wires is shown in Fig. 11.11. This chart also can be used for determining the earth stray capacitance, only the value read from the chart should be multiplied by two.

Capacitive noise coupling is often studied in terms of crosstalk, which is the ratio of the coupled noise voltage to the source voltage. In this case, instead of stray capacitance, crosstalk is plotted [1]. Crosstalk is expressed mathematically as:

$$k = \frac{U_n}{U_G} \tag{11.20}$$

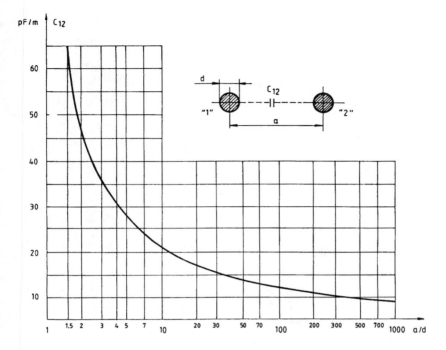

FIGURE 11.11 Chart for determining the stray capacitance of two parallel conductors

11.2.2 Inductive Noise Coupling

Inductive noise coupling occurs when a current flow in a closed circuit (source) excites a magnetic flux, and the magnetic flux change induces voltage in another closed circuit (receiver). This type of noise coupling is often called *magnetic coupling* or, with misleading terminology, *electromagnetic coupling.*

Figure 11.12 shows the inductive noise coupling between two circuits. The voltage induced in the receiver circuit (right) is due to a magnetic flux excited by a current flow in the source circuit (left). The induced voltage can be given as [6]:

$$U_n = \frac{d}{dt} \int_A \overline{B} \times \overline{A}\ dA = M \frac{di_1}{dt} \tag{11.21}$$

where

A = area of the disturbed circuit (vector)

B = flux density in area A (vector)

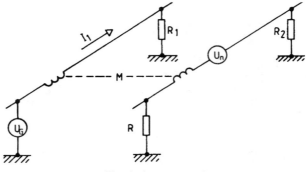

a. Physical representation

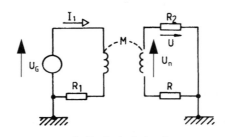

b. Equivalent circuit

FIGURE 11.12 Magnetic coupling between two circuits

The mutual inductance, M, characterizing the interaction between the circuits, depends on the geometry of the circuits and magnetic properties of the medium between the circuits. The inductive noise coupling is usually studied using the equivalent circuit shown in Fig. 11.12b. If the closed loop of the receiver circuit is stationary, and the flux density varies sinusoidally in time but is uniform over the area of the receiver loop, Eq. (11.21) reduces to:

$$U_n = j\omega \times B \times A \times \cos\theta = j\omega \times M \times I_1 \qquad (11.22)$$

where

θ = the angle between the magnetic field and the loop area

As the above relationship clearly shows, the magnitude of the inductively coupled voltage is determined by the geometry of the circuits, which are characterized by mutual inductance, the amplitude of the current in source circuit, and the frequency. There are two important differences with capacitive coupling. First, the coupled noise voltage is influenced by the impedances in the source circuit, since these impedances dictate the source current, I_1. Second, the coupled noise voltage does not depend on the impedances of the receiver circuit. However, as Fig. 11.12b shows, if resistor R represents the impedance of the sensitive electronics, the disturbing noise voltage can be influenced by the ratio of R and R_2 acting as a divider.

The mutual inductance between the circuits can be determined in the knowledge of their geometry. In engineering practice, however, these calculations for U_n are seldom performed because they are rather complex. The mutual inductance of several circuit arrangements can be found in various handbooks and other materials [1, 5, 260]. Some circuit arrangements that are typical in power electronic equipments are shown in Fig. 11.13. Figures 11.13a and b show the mutual inductance of single wires. Figure 11.13c illustrates the mutual inductance of coils.

The inductive noise coupling situation is often characterized not by the mutual inductance but by the transfer impedance. For inductive noise coupling, transfer impedance is defined by the ratio:

$$Z_T = \frac{U_n}{I_1} \qquad (11.23)$$

In some handbooks [1], instead of mutual inductance, transfer impedance is plotted in diagrams for different wire arrangements.

11.3 ELECTROMAGNETIC COUPLING REDUCTION METHODS

The three primary ways to minimize EMI coupling are to (1) increase the distance between the source and receiver, (2) apply shielding, and (3) design a symmetrical circuit arrangement. Although the simplest technique is to separate the circuits, this is not always a practical solution. Many electronic systems are subject to size restrictions, and the reduction effect decreases with distance. EMI coupling usually can be reduced to acceptable levels by properly applied shielding. To design shielding for conductors, circuits, or sensitive components, we first must identify the coupling mechanism. This is because capacitive and inductive noise coupling require quite different EMI reduction techniques. This chapter is primarily concerned with the shielding of conductors, although the same basic principles apply to any shielding application.

11.3.1 Conductor Shielding

First, let us study a situation where the receiver conductor is totally shielded and the shielding is connected to ground. The capacitive coupling can be calculat-

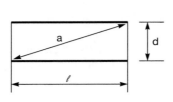

$$M = \frac{\mu_o}{4\pi} \times \log_n \left(\ell \times \log_n \frac{\ell + a}{d} - a + d \right)$$

if $\ell \gg d$

$$M \sim \ell \times \frac{\mu_o}{4\pi} \times \left(\log_n \frac{2\ell}{d} - 1 \right)$$

a.

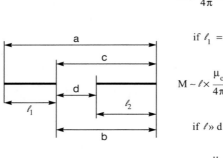

$$M = \frac{\mu_o}{4\pi} \times \log_n \frac{a^a \times d^d}{b^b \times c^c}$$

if $\ell_1 = \ell_2 = \ell$

$$M \sim \ell \times \frac{\mu_o}{4\pi} \times \left[2 \times \log_n \frac{2\ell + d}{\ell + d} - \frac{d}{1} \times \log_n \frac{(2\ell + d) \times d}{(\ell + d)^2} \right]$$

if $\ell \gg d$

$$M \sim \ell \times \frac{\mu_o}{2\pi} \times \log_n 2$$

b.

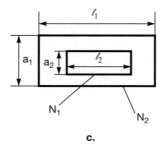

if $(a_1 > a_2)$ and $(\ell_1 > \ell_2)$

$$M \sim \mu_o \times \pi \times N_1 N_2 \times \frac{a_2^2}{4\ell_1}$$

N_1, N_2 = number of turns

c.

FIGURE 11.13 Mutual inductance calculations

ed by the equivalent circuit set up as the basis of Fig. 11.9b and shown in Fig. 11.14. The capacitance C_{12} now consists of two parts: (1) the capacitance between the source conductor and shield and (2) the capacitance between the shield and the receiver conductor. Z_s represents the grounding impedance of the shield. Ideally, when $Z_s = 0$, the noise voltage on the receiver conductor is reduced to zero.

In practice, the receiver conductor is not totally buried within the shielding since the conductors' ends extend beyond the shielding. Even if the shielding is grounded by a very small impedance, there is a noise coupling path between the source and receiver conductor. This path is represented by the stray capacitance (C'_{12} in Fig. 11.15) that exists between the portions of the source wire and receiver conductor that extend beyond the shield. For this situation, the capacitively coupled noise voltage can be calculated using Eqs. (11.13), (11.15), and (11.16), assuming $Z_s \ll R$. However, the substitutions offered by Eq. (11.24) should be considered:

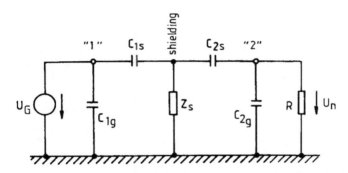

FIGURE 11.14 Equivalent circuit for calculating capacitively coupled noise voltage into a totally shielded conductor

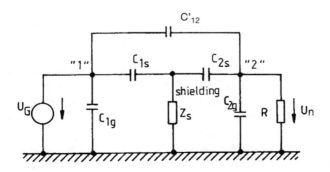

FIGURE 11.15 Equivalent circuit for calculating capacitively coupled noise voltage into an less than totally shielded conductor

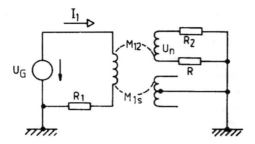

FIGURE 11.16 Equivalent circuit of shielded conductor grounded at one point

$$C_{12} = C'_{12}$$

$$C_{2g} \rightarrow C_{2g} + C_{2s} \tag{11.24}$$

To provide a shield that effectively reduces capacitive noise coupling, the designer must minimize the lead length of any conductor that extends beyond its own shielding. In addition, the shield must be properly grounded. For power electronic equipments, a single-point ground usually is sufficient; multiple grounding is recommended only if the cable is longer than one-twentieth of the wavelength.

However, such an electric-field shield provides no protection against magnetic-field noise coupling. For a shield grounded at one point only, the magnetic coupling can be studied by an equivalent circuit as shown in Fig. 11.16, where M_{1s} is the mutual inductance between the source wire and the shield. Since the shield does not change the magnetic field around the receiver conductor, and the geometry between the source and receiver conductors has not changed, the induced voltage will be the same as without shielding.

To analyze the requirements for a good magnetic shield, we will first examine magnetic coupling between a hollow conductive tube (the shield) and a conductor placed inside the tube. The most important observation is that the mutual inductance between the shield and the center conductor is almost equal to the self-inductance of the shield [6]. The validity of this statement depends only on the assumption that no magnetic field exists inside the tube due to shield current. This is true when the tube is cylindrical and the current density is uniform around the circumference of the tube; i.e., the inside conductor does not have to be coaxial.

Now, the voltage induced magnetically into the center conductor can be calculated on the basis of the equivalent circuit shown in Fig. 11.17. First, a noise voltage is induced into the shielding by the noise source. This voltage, U_s, can be calculated using Eqs. (11.21) and (11.22). This voltage generates a current (I_s) across the resistance (R_s) and inductance (L_s) of the shield. Thus, the current in the shield can be given as [6]:

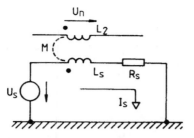

FIGURE 11.17   Equivalent circuit of magnetic shielded conductor

$$I_s = \frac{U_s}{L_s} \times \frac{1}{j\omega + \dfrac{R_s}{L_s}} \tag{11.25}$$

The noise voltage induced into the receiver conductor can be calculated using Eq. (11.22) and by $L_s \cong M$:

$$U_n = U_s \times \frac{j\omega}{j\omega + \dfrac{R_s}{L_s}} \tag{11.26}$$

The noise voltage as a function of the shielding current is shown in Fig. 11.18. As the figure also shows, the voltage induced into the receiver conductor is nearly equal to the shield voltage above the cutoff frequencies. Since these two voltages are almost equal (within 99 percent or better) and opposite, the noise reduction factor is large. It should be noted that this frequency range is at approximately the high end of the audio frequency band for most shielded cables.

The magnetic noise coupling also can be prevented by shielding the source conductor. If this shield is grounded only at one point, the shielding will reduce the electric-field noise coupling but will have very little effect on the magnetic-field noise coupling. Grounding the shield at both ends, as shown in Fig. 11.19, a current division will develop between the shield and the ground path. For providing a good shielding effect, the return current must flow in the shield and not into the ground path. The current division can be analyzed by the equivalent circuit of Fig. 11.19b. After substituting $M = L_s$, the shield current is as follows [6]:

$$I_s = I_1 \times \frac{j\omega}{j\omega + \dfrac{R_s}{L_s}} \tag{11.27}$$

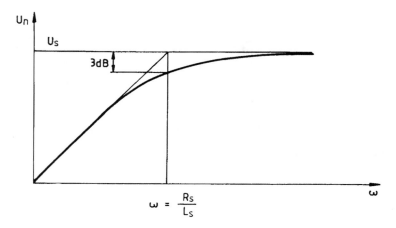

FIGURE 11.18 Frequency response of a center conductor inside a shield as a function of shield current

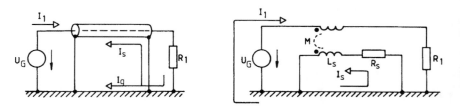

FIGURE 11.19 Current division between shield and ground path [6]

As this expression indicates, the shield current approaches the center conductor current (i.e., to the noise source current) only above the shield cutoff frequencies. As the frequency decreases below five times the cutoff frequency, the shield will be increasingly less effective at eliminating the radiated field from the source conductor.

It should be noted that if we remove the ground from the one end of the circuit (e.g., from the resistor R_1 side) and connect the low side of R_1 directly to the shield, all of the return current must flow in the shield. This shielding method is especially good for low frequencies.

11.3.2 Circuit Balancing

In addition to the use of shielding, we can minimize electromagnetic coupling by designing proper circuit geometry. If the potential source and receiver conductors are not placed in geometric parallel, both capacitive and magnetic noise coupling can be reduced effectively.

Figure 11.20 shows the layout of a circuit that has been balanced to reduce capacitive coupling. The capacitive coupling will be zero between circuits 1 and 2 and between circuits 3 and 4; i.e., the circuits can be considered balanced if the following condition is met for stray capacitances [7]:

$$C_{13} : C_{23} = C_{14} : C_{24} \tag{11.28}$$

The conditions for eliminating magnetic noise coupling can be derived from Eq. (11.22). First, the circuit area should be minimized, and the distance between them should be maximized. But there is another opportunity to reduce magnetic noise coupling. Placing the source and receiver circuit perpendicular to each other, the angle between the excited magnetic flux and the pickup circuit (θ) will be 90°; i.e., the term $\cos\theta$ will be zero. Figure 11.21 shows two balanced circuit arrangement examples. These circuits are balanced for magnetic noise coupling reduction if, for the distances, the following condition is met [7]:

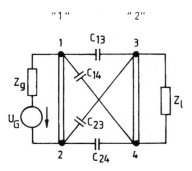

FIGURE 11.20 Balanced circuit arrangement for reducing capacitive noise coupling [7]

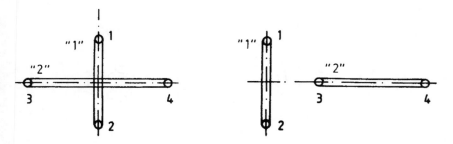

FIGURE 11.21 Balanced circuit arrangements for reducing inductive noise coupling [7]

$$h_{14} \times h_{23} = h_{13} \times h_{24} \qquad (11.29)$$

There is still another method for reducing magnetic noise coupling between circuits. Since the magnitude of the magnetic coupled noise voltage is proportional to the area of the receiver circuit, the pickup area can be effectively decreased by twisting the wires. The decrease in noise voltage for a twisted conductor can be studied in Fig. 11.22. For identical distances and parallel circuits, noise voltages induced into the two circuit parts are identical in magnitude but opposite; therefore, the resultant noise voltage is zero. This ideal case is rarely met in practice because the conductors usually are not perfectly in parallel, and the areas of the circuit parts are not absolutely equal.

11.4 WIRING LAYOUT METHODS TO REDUCE EMI COUPLING

In power electronic equipments, electromagnetic noise coupling is usually a problem because the high-current cabling (source) cannot be laid far enough from the control wiring (receiver) to reduce the noise coupling to an acceptable level. For providing the required EMC performance, one should focus special attention on wiring layout for noise coupling reduction.

To illustrate various noise reduction techniques, the effects of EMI on different circuit configurations and shield arrangements have been measured and compared [1, 6]. A test setup for inductive coupling was as shown in Fig. 11.23. For all of the wires tested, a measurement frequency of 50 kHz was selected because it is greater than five times the shield cutoff frequency. The test data is helpful in determining the best design for coupling reduction.

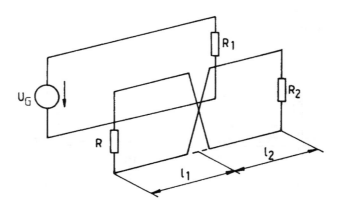

FIGURE 11.22   Magnetic noise coupling into twisted conductor [7]

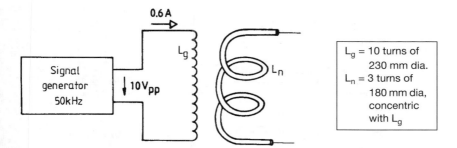

FIGURE 11.23 Test setup for inductive coupling experiment [6]

The relative EMI coupling susceptibility of various circuit and shield configurations is summarized in Fig. 11.24 [1, 6, 191]. The shielded test wire pair was terminated on one end by a resistor of 100 Ω, and the noise voltage was measured on the other end, which was terminated by a resistor of 1 MΩ. Configuration 1 provides essentially no shielding, so the measured pickup of 0.8 V of this configuration was used as a reference (0 dB) for comparison with other configurations.

As the test with configuration 2 clearly shows, grounding the shield at one end only has no effect on noise coupling because it does not reduce inductive coupling [1, 6]. Interestingly, the test showed that very little coupling reduction is expected when twisted pair wire is used in the shield, grounded at one end only (configuration 4). This is because the inductive noise coupling in the pair-to-ground loop remains unchanged since both source and load return wire ends are also grounded, and here is no current in the shield; i.e. the same configuration is developed as in circuit 2. Some noise pickup reduction is provided by twisting the wire pair, but its effect is masked by the ground loop, as configuration 3 shows. Adding a shield (with one end grounded) to the twisted wire pair has no effect (see configuration 4). Better EMI reduction can be provided by grounding the shield at both ends as shown in configuration 5, but the performance will be improved only slightly by using a shielded twisted wire-pair as in configuration 6. In general, configurations 1 through 5 do not provide really good inductive noise coupling protection because of the existence of ground loops. If the circuit must be grounded at both ends, configuration 5 or 6 should be used.

Improved performance can be provided in configurations grounded at one end only. The not so good performance of a twisted wire pair alone (configuration 7) can be explained by the fact that some capacitive noise coupling is beginning to show up. This assumption is supported by configuration 9, where an additional shield, optimized for electrical coupling reduction, improves the noise pickup protection considerably. The lesser performance of configuration 8 is due to the high shield current in the ground loop because the magnetic cancelling of the shield has not reached its best yet (this will occur above several hundred kilohertz). A really

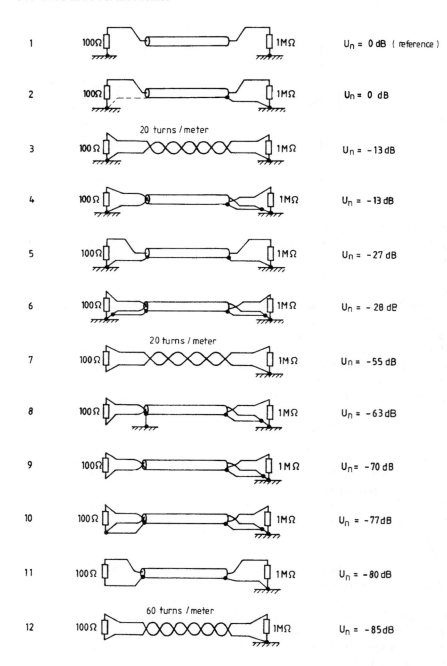

FIGURE 11.24 Results of inductive coupling experiment [1, 6]

good shielding effect can be achieved by configurations 10 through 12. Configuration 10 combines the features of twisted wire-pair with those of coaxial cables. In general, it almost always better to connect the shield and the signal wire together at just one point. This point should be designed so that the shield current does not have to flow across the signal conductor to reach ground.

The twisted pair or shielded twisted pair is very useful at frequencies below 100 kHz. In some specialized applications, twisted pairs are useful up to 10 MHz [6]. But, in general, coaxial cables should be used for frequencies in the megahertz range and above, especially for high-impedance circuits.

For very strict EMS requirements, we may need to calculate potential interference levels of the designed wiring layout or to check the effectiveness of different hardening methods [220]. But for general reference, simple graphic representations have been worked out that permit quick and sufficiently accurate determination of induced noise voltages. These convenient expressions are derived for various combinations of shielded and unshielded lines as well as for sinusoidal (ac) and transient noise signals. Coupling levels were determined on the assumption that magnetic coupling was predominant, and experimental data has confirmed their accuracy in power electronics applications. Measurements have indicated that the analysis is accurate from dc to about 200 kHz for sinusoidal interference. It is also useful for predicting transient peak values—even for the fastest transients.

The wire and shield configurations for which the noise coupling analysis has been made are shown in the left-hand column Fig. 11.25. The center column drawings represent the equivalent circuit of the noise coupling, and the right-hand column gives the expression for calculating the coupled noise voltage. As the source, a sinusoidal voltage generator with an internal voltage of

$$e_g = U_g \times \sqrt{2} \times \sin \omega t \qquad (11.30)$$

was assumed For predicting the transient noise coupling, a voltage source with time function of

$$e_g = U_g \times 1\,(t) \qquad (11.31)$$

was supposed. For a transient source voltage, the typical time function of the induced noise voltage is shown in Fig. 11.26, where the parameter is the ratio of the time constants of the source and receiver circuits. In this graph, the vertical axis represents U_n/U_{max}, where U_{max} is the noise voltage for $R_1 = 0$ as the maximum induced disturbance voltage.

For calculating the RMS value of the sinusoidal noise voltage (as well as the peak value of the induced transient voltage), the values of the circuit inductance

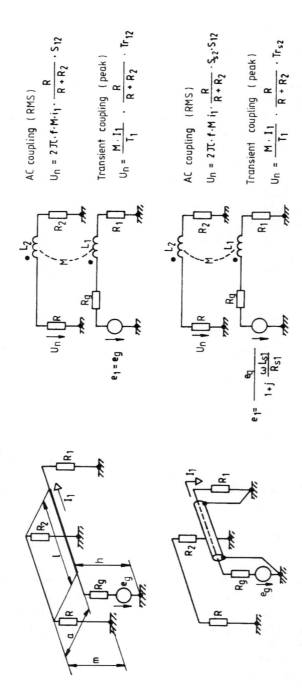

AC coupling (RMS)

$$U_n = 2\pi \cdot f \cdot M \cdot i_1 \cdot \frac{R}{R + R_2} \cdot S_{12}$$

Transient coupling (peak)

$$U_n = \frac{M \cdot I_1}{T_1} \cdot \frac{R}{R + R_2} \cdot Tr_{12}$$

AC coupling (RMS)

$$U_n = 2\pi \cdot f \cdot M \cdot i_1 \cdot \frac{R}{R + R_2} \cdot S_{s2} \cdot S_{12}$$

Transient coupling (peak)

$$U_n = \frac{M \cdot I_1}{T_1} \cdot \frac{R}{R + R_2} \cdot Tr_{s2}$$

$$e_1 = \frac{e_g}{1 + j \frac{\omega L_{s1}}{R_{s1}}}$$

$e_1 = e_g$

FIGURE 11.25 Induced EMI, shielded and unshielded wire [220]

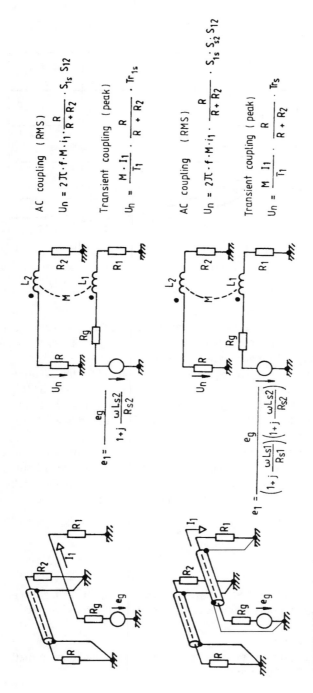

AC coupling (RMS)

$$U_n = 2\pi \cdot f \cdot M \cdot i_1 \frac{R}{R + R_2} \cdot S_{1s} \, S_{12}$$

Transient coupling (peak)

$$U_n = \frac{M \cdot I_1}{T_1} \cdot \frac{R}{R + R_2} \cdot Tr_{1s}$$

AC coupling (RMS)

$$U_n = 2\pi \cdot f \cdot M \cdot i_1 \cdot \frac{R}{R + R_2} \cdot S_{1s} \cdot S_{s2} \cdot S_{12}$$

Transient coupling (peak)

$$U_n = \frac{M \, I_1}{T_1} \cdot \frac{R}{R + R_2} \cdot Tr_s$$

$$e_1 = \frac{e_g}{1 + j\,\dfrac{\omega L_{s2}}{R_{s2}}}$$

$$e_1 = \frac{e_g}{\left(1 + j\,\dfrac{\omega L_{s1}}{R_{s1}}\right)\left(1 + j\,\dfrac{\omega L_{s2}}{R_{s2}}\right)}$$

FIGURE 11.25   continued

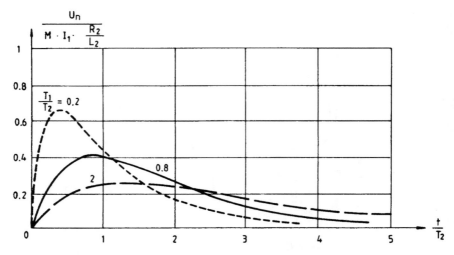

FIGURE 11.26 Typical time function of transient interference of open-to-open wire coupling [220]

and self-inductance can be read of from Fig. 11.27. Figure 11.27a shows the inductance of a shielded wire over a ground plane. The figure serves for determining both the shield and the inner conductor inductance. Figure 11.27b gives the mutual inductance of two wires over a ground plane.

For a sinusoidal source voltage, the factors required for the equations shown in the right-hand column of Fig. 11.25 are given in Fig. 11.28, where f is the frequency of the source voltage. The continuous line, as shown by the arrows, refers to the lower horizontal and the left vertical axes. The dashed line refers to the upper horizontal and right vertical axes.

For a transient voltage disturbance, the factors required for the equations shown in Fig. 11.25 can be found in Fig. 11.29. The factors refer both to shielded and unshielded conductors, since the noise reduction effect of the shielding is represented by the factor Tr. The time constant, T_1, refers to the source circuit:

$$T_1 = \frac{L_1}{R_g + R_1} \tag{11.32}$$

For a shielded source and victim wire configuration, the Tr factor can be taken from Fig. 11.29, but only when both shields have the same construction.

Illustrative Example

Let us use an example to show how this coupling calculation method works. Let the layout of the examined circuit be as follows:

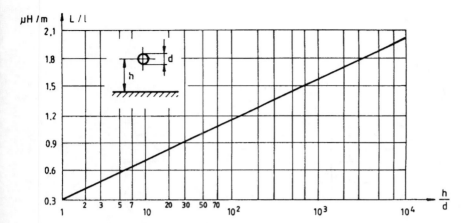

a. Inductance of a wire over a ground plane

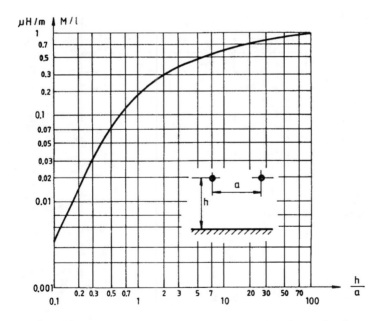

b. Mutual inductance of two wires over a ground plane for a ≥ 3 × dia.

FIGURE 11.27 Charts for calculating self- and mutual inductance

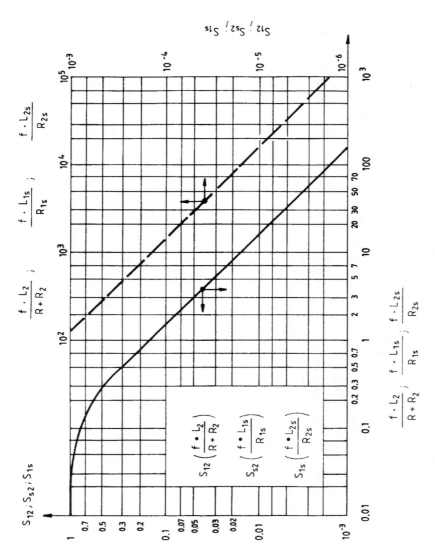

FIGURE 11.28 Inductive and shield attenuation factors for a sinusoidal disturbance [220]

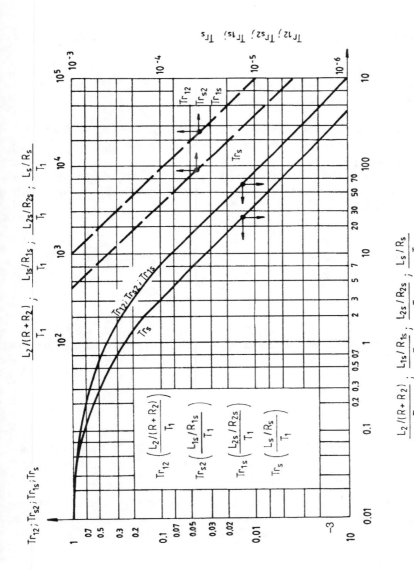

FIGURE 11.29 Inductive and shield attenuation factors for a transient disturbance [220]

- coupling length, $\ell = 3$ m
- distance between wires = 2 cm
- height over the ground plane, h = 10 cm
- resistors in the victim circuit are $R_2 = 100\ \Omega$, $R = 1\ k\Omega$

The victim wire is shielded, and its characteristics (taken from the vendor's data sheet) are as follows:

- diameter, $d_s = 1$ mm
- shielding resistance, $R_s = 10$ mΩ/m

The disturbing circuit is not shielded. A circuit analysis showed that a transient disturbance peak value of $I_1 = 20$ A can be expected, and the time constant, T_1, of the circuit is 0.5 µs.

The peak value of the noise voltage can be calculated for the transient disturbance from Fig. 11.25. The mutual inductance can be determined from Fig. 11.27b for h/a = 5 as follows:

$$L_2 = 0.9\ \mu H/m \times 3\ m = 2.7\ \mu H$$

For reading off the coefficient Tr_{1s}, we first must calculate the ratio ($R_s = 30$ mΩ):

$$\frac{L_s/R_s}{T_1} = \frac{2.7 \times 10^{-6}/30 \times 10^{-3}}{0.5 \times 10^{-6}} = 180$$

Knowing this ratio, the coefficient Tr_{1s}, from Fig. 11.29, is about 0.005. We now have all the data needed to calculate the peak value of the noise voltage on the victim resistor, R:

$$U_n = \frac{1.8 \times 10^{-6} \times 20}{0.5 \times 10^{-6}} \times \frac{1000}{100 + 1000} \times 0.005\ V = 0.327\ V$$

Supposing that T_1 is much smaller than T_2 (see Fig. 11.26), this peak value certainly will cause no breakdown in solid state elements, and we would not expect it to cause a failure in most digital logic.

In practice, industrial process control systems are made up of several modular subsystems. The electronic unit is usually a separate component connected to the others by cables. A detailed analysis of malfunctions and breakdowns in such systems has shown that reliability depends greatly on the grounding scheme as well as the intermodule cable layout. To determine the best grounding and cabling layout for EMS reduction, a test setup was assembled and several measurements were made in a room-size Faraday cage [222]. The shielded room was employed to eliminate the disturbing effects of external electromagnetic fields.

To draw the most general conclusions about the relationship between wiring layout and the EMS performance, 12 different grounding and wiring combinations were set up. These included two power supply configurations [star (radial) and grid] and three grounding layouts. In addition, tests were performed with (a) power supply and signal wiring in a common canal and (b) the signal wiring placed 50 cm above the power supply cables. The various combinations are summarized in Fig. 11.30 [222].

The EMS measurements were performed by coupling a number of test impulses into the different wiring points. EMS performance was rated on the number of malfunctions and other electronic errors, and derived performance qualification numbers are also shown in Fig. 11.30. According to these measurements, the best EMS performance may be expected if the power supply cables and the signal wiring are separated. Performance also improves if there is a separate power supply ground system that is isolated from the equipment enclosure grounds.

11.5 PCB DESIGN CONSIDERATIONS

For improving the EMC performance of electrical equipments with electronic modules, it is not enough to address only the electromagnetic noise coupling between the main circuits and the signal wiring; we must also consider EMI reduction techniques at the printed circuit board (PCB) level [6, 240, 330, 337, 339]. It is worthwhile to discuss these because many designers (particularly those who specialize in analog circuits) are not aware that different grounding, power distribution, and interconnection techniques are required for digital circuits. These techniques serve to minimize susceptibility to both internal and external noise sources.

The EMS performance of analog circuits tends to be poor because they contain high-gain amplifiers and operate at a very low signal level. Digital circuits are usually better, with the worst-case noise margin for TTL circuits typically being 400 to 600 mV. For CMOS circuits the margin is even greater—approximately 0.3 times the supply voltage.

In contrast to analog systems, internal noise sources are the major concern in digital circuits. This is because a current transient of very short rise time is generated at every logic gate switching operation. To reduce the noise voltage due to these (and other) current surges in the grounding system, the ground impedance must be minimized. This requires a ground configuration of low inductance because, given the frequency of current transients, the inductive part of the ground impedance usually is many times greater than the resistive part. The most practical way to produce a low-inductance ground system is to provide many parallel paths between all elements that are to be grounded [240]. Although a ground plane would provide optimum performance, it is not always realizable in practice. A more practical and almost equally effective ground layout is the grid-type ground

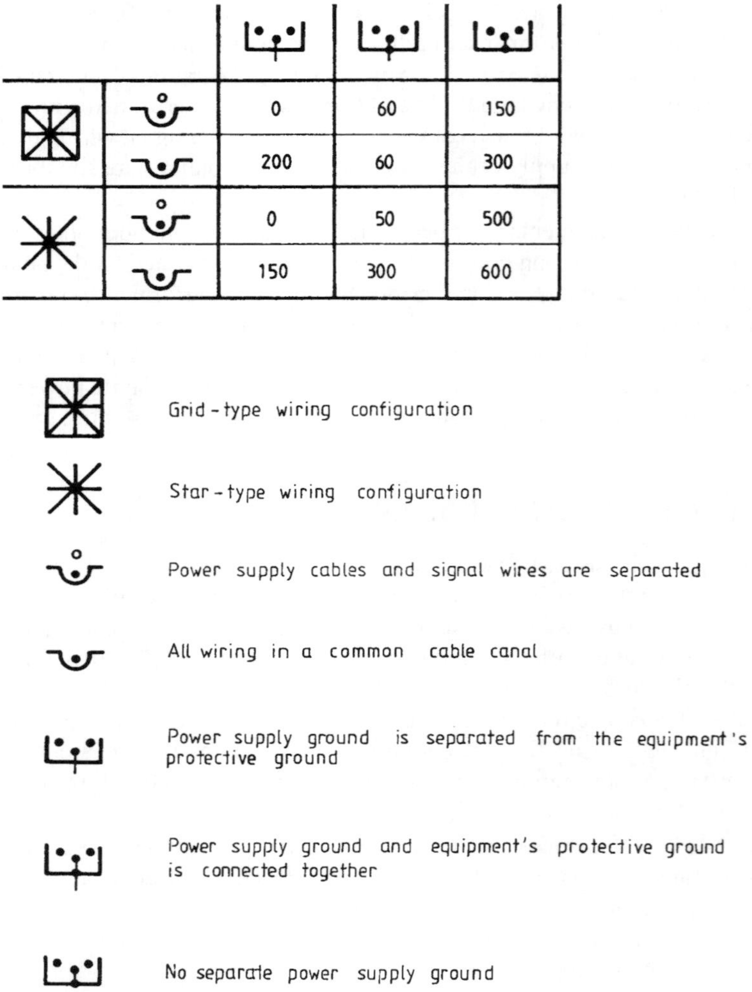

		⌊•⌐•⌋	⌊•⌐•⌋	⌊•⌌•⌋
⊠	~○⌣~	0	60	150
	~⌣~	200	60	300
✳	~○⌣~	0	50	500
	~⌣~	150	300	600

⊠ Grid-type wiring configuration

✳ Star-type wiring configuration

~○⌣~ Power supply cables and signal wires are separated

~⌣~ All wiring in a common cable canal

⌊•⌐•⌋ Power supply ground is separated from the equipment's protective ground

⌊•⌐•⌋ Power supply ground and equipment's protective ground is connected together

⌊•⌌•⌋ No separate power supply ground

Note: This figure is derived from Ref. 222. The measured failure ratio was defined as any failure in the function of the equipment. Power electronic equipment with digital control circuitry was analyzed in an electromagnetically "clear" environment. The figure was composed as follows:

Many kinds of disturbances were injected at many points, and many operating conditions (switching transients generated, etc.) were analyzed with the objective of finding the most sensitive condition. From these measurements, the authors derived a fixed operating condition that was the basis for further EMS measurements. They then built up the layouts shown in the figure, and the tests were performed under the standard operating condition. The EMS characteristics of the various layouts are characterized by the numbers given in the figure. Other pertinent information (e.g., power electronic equipment type, whether HF filtering was used, etc.) was not provided, but the figure offers some useful guidelines for cable installation in a complex system.

FIGURE 11.30 EMS dependence on cable layout between electronic equipments

system as shown in Fig. 11.31. The grid-type ground effectively produces multiple ground paths between all points to be grounded.

The other important performance grounding system consideration is to eliminate common-impedance noise coupling. This is especially important in analog circuits. In PCBs containing both analog and digital circuits, the grounding of both traces are separated [337] and connected only at the PCB terminal.

To prevent noise coupling due to logic gate switching current surges, it is important to minimize not only the grounding impedance but also the impedance of the power distribution circuit. A common (but poor) power supply layout scheme is shown in Fig. 11.32. The inductance of such a layout is fairly high because of the large loop area. A better power distribution system layout is shown in Fig. 11.33. Here, the power supply and return path are located close together to form a circuit that considerably reduces the self-inductance of the power supply system [337].

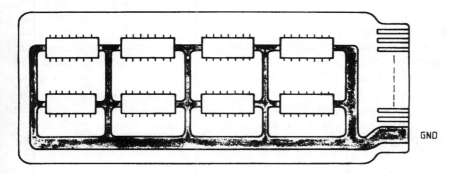

FIGURE 11.31   Grid-type ground system

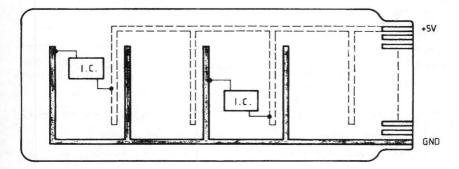

FIGURE 11.32   Bad power distribution system layout

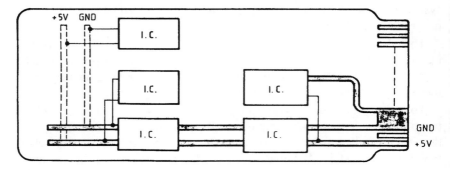

FIGURE 11.33   Recommended power distribution system layout

Because of the very short rise times, noise voltage induced by current surges in the self-inductance of the power distribution system may exceed acceptable levels. A complementary way to reduce conductive noise coupling in a power supply system is to apply decoupling capacitors. The noise reduction effect of decoupling capacitors connected to the supply wires is shown in Fig. 11.34. Without the decoupling capacitor, the surge current flows across the inductance of the power supply wire. For very short rise times, the induced voltage can be rather high, even when the inductance of power supply wire or trace (L_p) is very low. With a decoupling capacitor between the power supply wires, the surge current will not flow across the noise coupling inductance. Since the decoupling capacitor will be the source of the surge current for logic gates, it is necessary to minimize the impedance (i.e., the inductance) between the capacitor and the digital circuit as shown in Fig. 11.35 [240, 337].

Typically, 0.5 to 10 nF capacitance is satisfactory for digital logic ICs; the higher value is needed for TTL ICs, while the lower one is for CMOS [330]. The

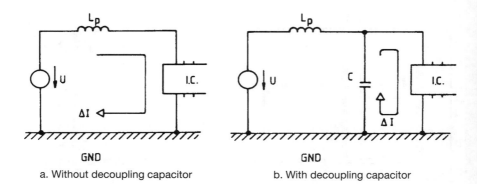

a. Without decoupling capacitor   b. With decoupling capacitor

FIGURE 11.34   Power supply distribution decoupling

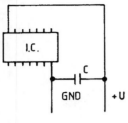

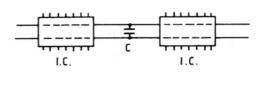

a. Poor placement   b. Recommended placement

FIGURE 11.35   Integrated circuit decoupling on a PCB

decoupling capacitors should exhibit good HF characteristics, so small disc ceramic or monolithic ceramic capacitors are good choices. The capacitance between the layers of multilayer PCBs can also help decoupling capacitance, since its value reaches 10 to 15 pF/cm². When a single-layer PCB is used and space is limited, an elevated power distribution technique may be recommended [330, 337]. This techniques applies two isolated copper strips connected to the power supply wires. This assembly can be placed under the IC package to save space.

In power electronic systems, the electronic circuits are usually assembled in a rack unit that secures the PCBs. The internal layout of interconnection wires can also compromise EMC performance. The primary design objective for the cable layout is to provide a return current path that encloses the least possible area. The best way to handle digital signals is to use coaxial cable, but twisted pairs may be an acceptable alternative. If a shielded wire is applied, the shielding should be grounded at both ends to provide effective protection against inductive noise coupling (see Section 11.3.1).

Magnetic noise coupling is often reduced by electromagnetic shielding, which limits magnetic flux leakage to other circuits and systems. The rack assembly itself is often used as electromagnetic shielding, but the transformer and choke coils inside the electronic unit often are shielded as well. The shielding performances of different materials and assemblies are described by shielding effectiveness (SE) and given in decibels. The SE for a shield without any seams, holes, or other apertures is composed of reflection loss, absorption loss, and secondary reflection loss.

We can determine the magnetic field reflection loss of a shield without apertures from the following relationship [260], provided that the shield thickness is greater than a skin depth:

$$R\,(dB) \;=\; 20\log\left(\frac{1.17}{r}\sqrt{\frac{\mu}{f\times G}} + 0.054\times r\times\sqrt{\frac{f\times G}{\mu_r}} + 0.354\right) \qquad (11.33)$$

where

r = distance from field source to shield in centimeters

μ_r = relative permeability

G = conductivity of shield material relative to copper

f = magnetic field frequency in hertz

The absorption loss can be calculated for the same shield as follows [260]:

$$A\,(dB) \;=\; 0.131 \times T \times \sqrt{\mu_r \times f \times G} \qquad\qquad (11.34)$$

where

T = thickness of the shield in millimeters

μ_r, f, G = as in Eq. (11.33)

The secondary reflection loss becomes significant only when the absorption loss is less than 6 dB and is not usually considered in shielding designs for power electronic equipment [260].

A common misconception is that material of relatively high permeability is needed to provide a good magnetic shield [260]. This is true at low frequencies, but at high frequencies most metallic conductors of adequate thickness may provide sufficient SE. In addition, high-permeability materials will saturate at relatively lower flux levels.

An SE comparison of different types of shields is provided in Fig. 11.36 [339]. Figure 11.36 also shows that the same shielding effectiveness may be achieved by increasing the thickness of a copper or iron shield.

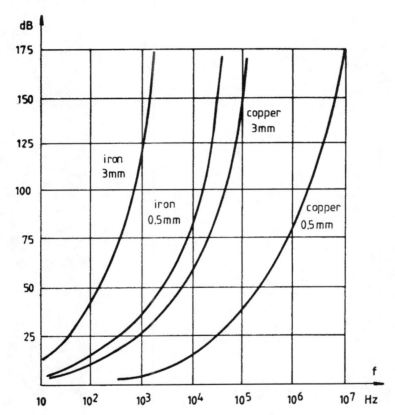

Shielding Effectiveness
against H-field

FIGURE 11.36   Shielding effectiveness of various materials [339]

12

Transient Susceptibility Analysis Method

The EMS performance of electrical apparatus can be determined by various disturbance withstand tests. By means of such EMS tests, we can gather accurate data about the equipment reliability under realistic operational conditions. However, to validate system performance in terms of applicable EMC specifications, we must consider the test methodology employed [239].

EMS specifications usually cannot be met purely by employing noise decoupling techniques as discussed in Chapter 11. Therefore, we apply noise filters to the input (and sometimes to the output) of the electronic units. In engineering practice, it is rather difficult to specify the appropriate filter parameters, given that we cannot predict the exact nature of the environmental disturbances to which the equipment will be exposed and, therefore, do not know exactly how much susceptibility reduction is needed. The rate of malfunction may certainly be reduced by improving filter performance. But because costs increase exponentially with performance, noise filter parameters should be determined with great care.

From the point of view of analyzing the EMC performance, an electrical apparatus or system can be regarded as a collection of interconnected disturbance sources and noise victims. The EMS of a system consisting only of one source and one receptor is relatively easy to predict—at worst, we simply need to compare emission values to susceptibility thresholds.

But EMS analysis becomes more complicated if there are more disturbance sources and noise victims. The recommended way to ascertain the required susceptibility performance is to set up an EMS matrix for both conducted and radiated phenomena [239]. Such a matrix summarizes the emission performances of single disturbance sources and the susceptibility characteristics of single noise victims. The matrix elements describe the relationship of each source and receptor.

The data for setting up the matrix elements can be based on EMC specification limits or actual test data. Overall system performance then can be predicted by summarizing applicable matrix elements. Both emission and susceptibility threshold levels can be well described in terms of frequency spectrum. For a system with many sources and receptors, the calculations become complex and time consuming. But by dividing the frequency spectrum into decades and entering the data for each decade, the calculations can be automated via programs such as Lotus 123 and other mathematical spreadsheets [239].

It is relatively simple to determine the EMS matrix elements for continuous broadband disturbances, given that it usually is sufficient to give amplitude limitations. If amplitude is the only characteristic to be limited, noise filter circuits can be designed by methods known in telecommunication technology or as discussed in Chapters 8 and 9. However, typical disturbances, whether originating from environmental phenomena or electromagnetically coupled into the electronics, are impulsive in nature. Therefore, this chapter focuses on noise filter design methods for reducing the transient susceptibility of electrical equipments with electronic components.

Transient disturbances cannot be adequately described by amplitude only; we must also consider the energy content, frequency content, and time duration. The nature of the required calculations suggests the use of the Laplace transform. Since the root of the Laplace transform usually has to be known, Section 12.2.1 describes a simple method for determining these roots, presented in power series form. For better characterization of transient noise signals, an approximate energy calculation method is given in Section 12.2.2. The energy of transient signals that pass through bandpass filters can be simply calculated by introducing the energy density function as described in Section 12.2.3. Section 12.3.1 offers simplified formulas for deriving the various impulse characteristics for noise filter design, and Section 12.3.2 summarizes a design method based on the introduced impulse characteristics. Finally, for a more precise analysis, Section 12.3.3 describes methods for plotting the approximation time function of transient noise signals, given their Laplace transforms or spectra.

12.1 NOISE FILTER DESIGN METHOD FOR VOLTAGE ATTENUATION

In many practical cases, when the victim circuit impedance is not very low (e.g., $\geq 50\ \Omega$), EMC performance is improved by connecting a filter capacitor in parallel to the input terminals of an electronic unit. The capacitance is usually designed for voltage attenuation. If there is no other specification, the standard EMS energetic impulse test (see Section 10.2.1) often will be regarded as the basis for noise filter design. The time function of the IEC test impulse may be approximated as follows:

$$u_g(t) = 1(t) \times U_g \times \exp(-\alpha t) \tag{12.1}$$

This approximation is valid for transient disturbances with very short rise times and exponentially damped voltages. For designing the capacitance of the filter capacitor, for cases when the load resistance is much greater than the resistance of one noise generator R_g, it can be assumed that the transient disturbance acts on a simple RC network, as shown in Fig. 12.1a. The value of α in the above expression can be calculated for a standard IEC test impulse from the condition that in 50 μs the voltage $u_g(t)$ decreases to half of the peak value. The peak value of the capacitor voltage generated by the input signal given by Eq. (12.1) can be calculated relatively simply[9]. This relationship is plotted in Fig. 12.1b, which shows the voltage attenuation of the RC network as a function of the filter capacitance. The parameter is the value of the serial resistance; i.e., the pulse generator internal resistance. This

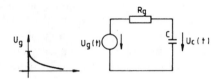

a. equivalent circuit

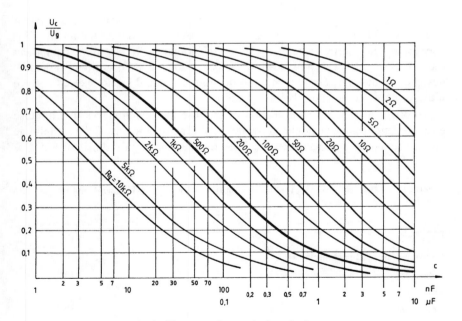

b. filter capacitance design chart

FIGURE 12.1 Noise filtering with capacitor against the standard 50 μs IEC pulse

chart is useful for designing the filter capacitor to reduce noise voltage of a given value for the standard IEC energetic impulse withstand test. The thick line (500 Ω parameter) gives the capacitance of a filter capacitor for the serial resistance represented by the standard test impulse generator. Note that the ratio U_c/U_g, as shown, assumes a quasi-perfect capacitor without parasitic inductance (i.e., no overshoot induced by the pulse rise time).

12.2 CALCULATING THE ENERGY CONTENT OF TRANSIENT DISTURBANCES

Detailed analyses of failures and breakdowns in electronics have revealed that they generally can be traced back, not to broadband, continuous noise, but rather to impulse-like disturbances. For semiconductor devices, the permissible amplitude of an impulse-like disturbance depends on its time duration. This time dependence of the peak value can be best characterized by the energy content of the transient-like disturbance. The explanation for this is that the breakdown of semiconductor devices is closely related to the energy flowing into the p-n junction [126, 326]. For some typical semiconductor devices, noise energy levels that may cause malfunction or breakdown are summarized in Table 12.1.

Because the energy content of impulse-like signals is such an important factor in susceptibility performance, research has been conducted with the intent of incorporating the energy content aspect into EMS analysis of electronic equipment [60, 179, 218, 219, 313, 314, 324]. It has been shown that noise filters should be designed so that the energy content of the output signal does not exceed a given limit for known or supposed transient noises.

To arrive at a better qualification of impulse-like disturbances, the impulse characteristics may be calculated by means of time function as well as amplitude density function (see Section 4.7.1). In practice, the impulse-like noise passes

TABLE 12.1 Disturbance energy withstand limits (in joules) for various semiconductor devices [126]

Semiconductor Device	Malfunction Limit	Destruction Limit
CMOS ICs	10^{-7}	10^{-6}
Small power switching transistors	10^{-6}	10^{-5}
Switching diodes, small power transistors	10^{-5}	10^{-4}
Regular diodes	10^{-4}	10^{-3}
Zener diodes	10^{-3}	10^{-2}
Regular transistors	10^{-2}	10^{-1}
Power transistors	1	10
Power rectifiers, power SCRs	10	100

through a transmission network onto the sensitive semiconductor. Calculations usually are made in the frequency range, but use of the amplitude density function does not impose any limitations, since the broken-line envelope of the amplitude density function for the signal appearing on the output of a transmission network can be easily plotted in log-log coordinate system: it is the sum of the spectrum of the input signal and the frequency function of the network. The spectra of impulsive signals can be taken from handbooks or determined by means of inequalities relating to the Fourier transform (see Section 5.3.1). The broken-line envelope of the frequency function of a transmission network can be determined by the plotting rules of a Bode diagram.

12.2.1 Laplace Transform Root Calculation Method

In the course of engineering calculations, the Laplace transform often arises in the form of rational fraction function and, for using the energy calculation method, the broken-line envelope of the amplitude density function has to be known. Therefore, the root of Laplace transform has to be determined.

The Laplace transform has two basic forms. One is the rational fraction function:

$$L(s) = \frac{A(s)}{B(s)}$$

where

$$A(s) = \Sigma a_i s^i$$

$$B(s) = \Sigma ab_i s^i$$

$$s = j\omega \text{ for most functions } (s \equiv \sigma + j\omega)$$

For engineering practice, the other and better form is:

$$L(s) = A \times \frac{(1 + sT_{n1})(1 + sT_{n2})(1 + sT_{n3}) \cdots}{(1 + sT_{d1})(1 + sT_{d2})(1 + sT_{d3}) \cdots}$$

There is no formula for conversion between the two forms.

For simplicity, the root calculation method will be introduced for a Laplace transform that contains power series only in the denominator. The method also can be used for calculating the root of the numerator. Let the Laplace transform be in the following form:

$$F(s) = \frac{A}{1 + a_1 s + a_2 s^2 + \dots + a_{n-1} s^{n-1} + a_n s^n} \tag{12.2}$$

where

$$A, a_i = \text{constants}$$

$$s = j\omega \text{ (or in some cases, } s = \sigma + j\omega)$$

For an s-quotient higher than three, explicit formulas are not known for calculating the roots. The approximate root calculation method is based on the root separation; i.e., on the assumption that the roots can be in sequence taken out of the power series.

Now, the task is to convert the denominator in the form of:

$$\prod_i (1 + sT_i) \times \prod_k (1 + s \times 2\zeta_k T_k + s^2 T_k^2) = 0 \qquad 0 \le \zeta_k \le 1 \tag{12.3}$$

First step is to determine the nature of the roots. To do so, the following determinants must be calculated:

$$D_i = a_i^2 - 4 \times a_{i-1} \times a_{i+1} \qquad (a_o = 1) \tag{12.4}$$

When all determinants are positive (i.e., all $D_i > 0$), the roots are real and can be determined according to Procedure 1 (which follows in subsequent paragraphs). When $D_i = 0$ represents a double root, with the number not successively negative, D_i points to complex conjugated roots which can be calculated by Procedure 2. Successively negative determinants indicate multiple roots which should be determined by Procedure 3.

Procedure 1: Real Roots

The time constant of Eq. (12.2) can approximated (omitting the negative sign) as follows:

$$T_i = \frac{a_i}{a_{i-1}} - T_{i+1} \qquad (T_{n+1} = 0) \tag{12.5}$$

Calculation of the roots should start with the smallest time constant; i.e., with the highest subscript i. The accuracy of the root calculation formulas is better than 5 percent when the ratio of the time constants is higher than 10 to 20. For a smaller time constant ratio, the value of the roots can be corrected as follows:

$$T'_i = T_i \times \frac{T_{i-1} + T_i}{T_{i-1}} \tag{12.6}$$

However, it is an iteration step; after the correction of T_i, the value T_{i-1} need not be corrected.

Procedure 2: Complex Roots

It is well known that a complex conjugated pole-pair can be characterized by the time constant and the damping factor, ζ. If $D_k < 0$, the approximate values of the roots are:

-

-

$$T_{k+2}$$

$$T_k = \sqrt{\frac{a_{k+1}}{a_{k-1}}}$$

$$T_{k-1} = \frac{a_{k-1}}{a_{k-2}} - T_{k+2}$$

$$2\zeta_k = \frac{1}{T_k} \times \left(\frac{a_k}{a_{k-1}} - \frac{T_k^2}{T_{k-1}} - T_{k+2} \right)$$

$$T_{k-2}$$

$$\tag{12.7}$$

-

-

The accuracy of the root calculation formulas is better than 10 percent when the ratio of the time constants is higher than 10 to 20 and the damping factor is in the range of 0.2 to 1. In the case of smaller time constant ratio and smaller damping factor, the value of the roots can be corrected as follows:

$$T'_{k+2} = T_{k+2} \times \left(1 + 2\zeta_k \times \frac{T_{k+2}}{T_k}\right)$$

$$2\zeta'_k = 2z_k \times \left(1 + 2\xi_k \times \frac{T'_{k+2}}{T_k}\right)$$

$$T'_{k-1} = T_{k-1} - 2\zeta'_k T_k \tag{12.8}$$

For

$$\frac{T_{k-2}}{T'_{k-1}} < 10...20$$

the value of time constant T'_{k-1} can be further corrected using Eq. (12.6). After applying the correction formulas, the accuracy of the calculated roots will be better than 5 to 10 percent, even for root arrangements that seldom appear in engineering practice.

The root calculation formulas described above can be used for more complex roots if there is at least one real root between them.

For two complex roots without a real root between them (i.e., for $D_{k+1} < 0$, $D_{k-1} < 0$, and $D_k > 0$), the roots can be calculated as:

•

•

$$T_{k+2}$$

$$T_{k+1} = \sqrt{\frac{a_{k+2}}{a_k}}$$

$$T_{k-1} = \sqrt{\frac{a_k}{a_{k-2}}}$$

$$2\zeta_{k-1} = \frac{1}{T_{k-1}} \times \left(\frac{a_{k-1}}{a_{k-2}} - \frac{a_{k+1}}{a_k} - \frac{T_{k-1}^2}{T_{k-2}} \right)$$

$$2\zeta_{k+1} = \frac{1}{T_{k+1}} \times \left(\frac{a_{k+1}}{a_k} - T_{k+2} - 2\zeta_{k-1} \times T_{k-1} \times \frac{T_{k+1}^2}{T_{k-1}^2} \right)$$

$$T_{k-3}$$

•

• (12.9)

$T_{k+2} = 0$ and $T_{k-2} = \infty$ should be substituted in Eq. (12.9) when either value is missing. The accuracy of the value of the time constants is better than 5 percent, and that of the damping factors is better than 10 percent. When the complex root time constants or the damping factors differ from each other better than one order, the roots can be corrected as follows:

$$2\zeta_{k-1} = \frac{1}{T_{k-1}} \times \left(\frac{a_{k-1}}{a_{k+1}} - K \times \frac{a_{k+1}}{a_k} - \frac{T_{k-1}^2}{T_{k-2}} \right)$$

$$2\zeta_{k+1} = \frac{1}{T_{k+1}} \times \left(K \times \frac{a_{k+1}}{a_k} - 2\zeta_{k-1} \times \frac{T_{k+1}^2}{T_{k-1}} - T_{k+2} - T_{k-3} - \frac{T_{k+1}^2}{T_{k-2}} \right)$$

$$T'_{k+2} = T_{k+2} \times \left(1 + 2\zeta_{k+1} \times \frac{T_{k+2}}{T_{k+1}} \right)$$

$$2\zeta'_{k+1} = 2\zeta_{k+1} \times \left(1 + 2\zeta_{k+1} \times \frac{T_{k+2}}{T_{k+1}} \right)$$

$$2\zeta'_{k-1} = 2\zeta_{k-1} \times \left(1 + 2\zeta_{k-1} \times \frac{T_{k-2}}{T_{k-1}} \right)$$

$$T'_{k-2} = (T_{k-2} - 2\zeta_{k-1} \times T_{k-1}) \times \frac{T_{k-3} - T_{k-2}}{T_{k-3}}$$

(12.10)

where

$$K = \frac{T_{k-1}^2}{T_{k-1}^2 + T_{k+1}^2}$$

The above correction formulas give the values of the complex roots with better than five percent accuracy.

Procedure 3: Multiple Roots

Two successive negative determinants indicate a triple root; three successive negative determinants mean a fourfold root, and so on. For $D_n < 0$ and $D_{n-1} < 0$, the time constant of the triple root is as follows:

$$T_k = \sqrt[3]{\frac{a_{k+1}}{a_{k-2}}} \tag{12.11}$$

As appropriate, the time constant of the fourfold root is the fourth root of the ratio a_{n+2} and a_{n-2}.

We will use two examples here for calculating the roots of Laplace transform in power series form.

Example 1

Let be the factors in Eq. (12.2) be as follows:

$$a_1 = 1.125$$

$$a_2 = 0.171125$$

$$a_3 = 4.43375 \times 10^{-2}$$

$$a_4 = 1.2175 \times 10^{-3}$$

$$a_5 = 5 \times 10^{-6} \tag{12.12}$$

The type of the roots can be determined by Eq. (12.4):

$$D_1 = 1.2^2 - 4 \times 0.17 > 0$$

$$D_2 = 1.7^2 \times 10^{-2} - 4 \times 1.1 \times 4.4 \times 10^{-2} < 0$$

$$D_3 = 4.4 \times 10^{-4} - 4 \times 0.17 \times 1.2 \times 10^{-3} > 0$$

$$D_4 = 1.2^2 \times 10^{-6} - 4.4 \times 4 \times 10^{-2} \times 5 \times 10^{-6} > 0 \tag{12.13}$$

As Eq. (12.13) tells us, the Laplace transform has three real roots and a complex conjugated one (D_2 is negative, but D_3 is positive). Thus, the roots must be calculated using Eq. (12.7) as follows:

$$T_5 = \frac{5 \times 10^{-6}}{1.21 \times 10^{-3}} = 4.11 \times 10^{-3}$$

$$T_4 = \frac{1.21 \times 0^{-3}}{4.43 \times 10^{-2}} - 4.11 \times 10^{-3} = 2.33 \times 10^{-2}$$

$$T_2 = \sqrt{\frac{4.43 \times 10^{-2}}{1.125}} = 0.198$$

$$T_1 = 1.125 - 0.0233 = 1.1$$

$$2\zeta_2 = \frac{1}{0.198} \times (\frac{0.171}{1.125} - \frac{0.198^2}{1.1} - 2.33 \times 10^{-2}) = 0.47 \tag{12.14}$$

For more precise calculations, the roots may be corrected; hence, the time constants T_1 and T_2, as well as T_3 and T_4, are in close proximity. The corrected values are as follows:

$$T'_5 = 4.48 \times 10^{-3} \qquad \text{[Eq. (12.6)]}$$

$$T'_4 = 2.46 \times 10^{-2} \qquad [T'_{k+2} \text{ in Eq. (12.8)}]$$

$$2\zeta'_2 = 0.497 \qquad [2\zeta'_k \text{ in Eq. (12.8)}]$$

$$T'_1 = 1 \qquad [T'_{k-1} \text{ in Eq. (12.8)}] \qquad (12.15)$$

For illustrating the accuracy of the approximate root calculation formulas, the accurate values of the roots are:

$$T_1 = 1$$

$$T_2 = 0.2$$

$$2\zeta_2 = 0.5$$

$$T_4 = 2.5 \times 10^{-2}$$

$$T_5 = 5 \times 10^{-3} \qquad (12.16)$$

Comparing the calculated and the accurate values of the roots, it can be seen that the accuracy of the root calculation method is better than 10 percent, even for a time constant ratio of 5. By correcting the roots, the accuracy has become within a few percent.

Example 2

Calculate the roots of the Laplace transform, in the form of Eq. (12.2), when the factors are:

$$a_1 = 1.044 \qquad\qquad a_4 = 1.413 \times 10^{-5}$$

$$a_2 = 4.898 \times 10^{-2} \qquad a_5 = 5.009 \times 10^{-7} \qquad a_7 = 9.188 \times 10^{-14}$$

$$a_3 = 5.094 \times 10^{-3} \qquad a_6 = 6.873 \times 10^{-10} \qquad\qquad (12.17)$$

For determining the types of the roots:

$D_1 > 0$ (real root)

$D_2 < 0$ $\quad$ $D_3 > 0$ (complex root)

$D_4 < 0$ $\quad$ $D_5 > 0$ (complex root)

$D_6 > 0$ (two real roots—see D_5) $\quad$ (12.18)

The real roots must be calculated using Eq. (12.5), and the two complex conjugated roots using Eq. (12.9), since any real root is between them:

$T_7 = 1.337 \times 10^{-4}$ $\qquad$ [Eq. (12.5)]

$T_6 = 1.238 \times 10^{-3}$ $\qquad$ [T_{k+1} in Eq. (12.9)]

$T_4 = 0.992 \times 10^{-2} \approx 10^{-2}$ $\qquad$ [T_{k-1} in Eq. (12.9)]

$T_2 = 6.985 \times 10^{-2}$ $\qquad$ [T_{k+1} in Eq. (12.9)]

$T_1 = 1.044$ $\qquad$ [T_{k-2} in Eq. (12.9)]

$2\zeta_2 = 0.565$ $\qquad$ [$2\zeta_{k-1}$ in Eq. (12.9)]

$2\zeta_4 = 0.044$ $\qquad$ [$(2\zeta_{k+1})$ in Eq. (12.9)] $\quad$ (12.19)

The ratio of time constants T_2 and T_4 is smaller than 10 (i.e., they are within a decade), so the value may be corrected using (12.6) and (12.10):

$T_7' = 1.48 \times 10^{-4}$ $\qquad$ [Eq. (12.6)]

$2\zeta_2' = 0.588$ $\qquad$ [Eq. (12.10)]

$T_1' = 1.0044 \approx 1$ $\qquad$ [Eq. (12.6)] $\quad$ (12.20)

We can use the accurate values of the roots to illustrate the accuracy of the approximate root calculation formulas as follows:

$$T_7 = 1.5 \times 10^{-4} \qquad T_2 = 7 \times 10^{-2} \qquad 2\zeta_4 = 0.05$$

$$T_6 = 1.25 \times 10^{-3} \qquad T_1 = 1 \qquad 2\zeta_2 = 0.6$$

$$T_4 = 10^{-2}$$

$$\tag{12.21}$$

The uncorrected values are good for engineering calculations. As a result of the correction, the calculated roots agree with the accurate values to within a few percentage points.

12.2.2 Energy Calculation Formulas for Impulse-Like Signals

To analyze an electronic device's susceptibility to impulse-like disturbances, the first step is to determine the approximate amplitude of the noise that reaches sensitive circuit components (see Section 12.3.1). If the disturbance amplitude is much higher than the acceptable value, the energy content of the disturbance should be also determined. The acceptable value for semiconductor components can be found in the data sheet as the maximum repetitive peak voltage (or current) that the component can withstand. Lacking that information, by default we should select a value slightly above the dc (or peak ac) power supply voltage.

The energy content of an aperiodic signal can be calculated by means of the definitive expression of Eq. (4.36) but, in general, this method of energy content calculation is rather difficult. Working in the frequency domain, explicit formulas can be found for calculating the energy content. However, these are usually rather complex, and thus their employment is difficult, even in the case of just a few energy sources in the transmission network.

Instead of calculating the correct energy content value, approximation formulas may be used for engineering design [313]. These formulas contain only data which can be taken from the broken-line envelope of the amplitude density function (spectrum).

In electrical engineering, the time function f(t) can mean voltage or current, so the value calculated by Eq. (4.36) has the characteristics of V^2s and A^2s, respectively; i.e., it can be explained as the energy on a 1 Ω resistor, sometimes referred to as "energy integral." In the following text, it is notated as "E."

If f(t) describes the voltage occurring on a resistor with resistance R, the energy transformed into heat on the resistor, E_U, is as follows:

$$E_U = \frac{E}{R} \tag{12.22}$$

Measuring the voltage in V and the resistance in Ω, the energy will be measured in units of watt seconds (Ws).

Time function f(t) may also represent impulse-like current. It has been observed that, in this case, the approximation formulas give the I^2t rating or, in other words, the $\int i^2 dt$ rating, which well characterizes the behavior of semiconductors in a transient overload condition. If the transient current is flowing through a resistor with resistance R, the energy transferred into heat on the resistor is:

$$E_i = E \times R \tag{12.23}$$

Measuring the current in A and the resistance in Ω, the energy measurement unit again will be watt seconds (Ws).

For periodic repetition of the impulse-like signal, the resistance transformed into heat on the resistance is measured in Ω:

$$P = E_u \times f \quad \text{or} \quad P = E_I \times f \tag{12.24}$$

where

$f =$ the repetition rate in hertz

By means of the approximation relationships, we can determine the energy content of transient signals that have spectra of the types seen in Fig. 12.2. For applying the approximation energy calculation formulas, we must read certain values from the spectrum. These are the value of the horizontal spectrum section, some cutoff angular frequency, and (for spectra with peaks) the damping factors. We will use four procedures based on Fig. 12.2.

Procedure 1: Spectrum with Only Falling Sections (Fig. 12.2a)

In case of a spectrum consisting of falling sections in addition to the horizontal sections, the approximate energy content is:

$$E_a = \frac{A_r^2 \times \omega_r}{2} \tag{12.25}$$

The above formula provides a proper approach only if the ratio of ω_r/ω_1 is less than 10. For shorter horizontal spectrum sections, the relative error of Eq. (12.25) depends on the ratio number:

$$n = \omega_1 / \omega_r \tag{12.26}$$

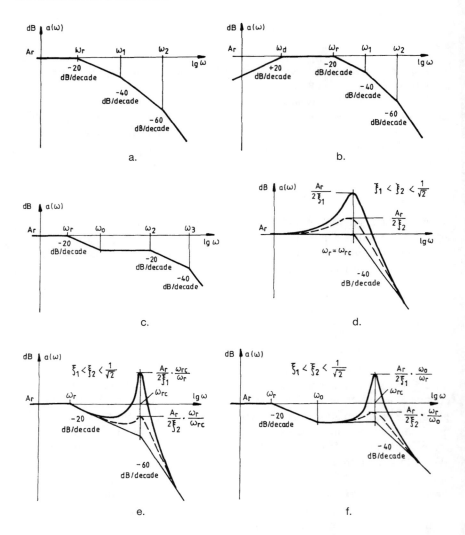

FIGURE 12.2 Broken-line envelopes of amplitude density functions to which formulas for energy content approximation can be applied

Using the above ratio number, the relative error function of the energy calculation approximation formula is as follows:

$$e_r = \frac{E_a}{E} = 1 + \frac{1}{n} \tag{12.27}$$

The relative error is often calculated and given in decibels where the reference quantity is the accurate energy content. The relative error function also shows that falling spectrum sections that follow the first falling section have very little influence on the energy content value. Therefore, for energy content approximation, they can be neglected.

Procedure 2: Spectrum with a Rising Initial Section (Fig. 12.2b)

Even in a case where the spectrum contains a rising section in the angular frequency range of $\omega < \omega_r$, the approximate energy content can be calculated by means of Eq. (12.25). The inaccuracy due to the rising initial section is given by the relative error function of Eq. (12.27), except that the ratio number n should be redefined as:

$$n = \frac{\omega_r}{\omega_d} \tag{12.28}$$

If falling sections greater than -20 dB/decade slope also exist within the angular frequency range of $\omega > \omega_r$, the maximum error caused by using the approximate Eq. (12.25) is the product of the values calculated by Eq. (12.27) and the ratio numbers as given by Eqs. (12.26) and (12.28):

$$e_r = \left(1 + \frac{\omega_d}{\omega_r}\right) \times \left(1 + \frac{\omega_r}{\omega_1}\right) \tag{12.29}$$

For calculating the approximation error in dB units, the error factors should be added.

Procedure 3: Spectrum Containing Two Horizontal Sections (Fig. 12.2c)

To calculate the approximate energy content in this case, the angular frequencies ω_r, ω_o, and ω_2 should be read from the spectrum, in addition to value A_r. In the knowledge of these values, the approximate energy content is:

$$E_a = \frac{A_r^2 \times \omega_r}{2} \times \left(1 + \frac{\omega_r \times \omega_2}{\omega_o^2}\right) \tag{12.30}$$

Comparing Eqs. (12.29) and (12.25), it can be seen that in the angular frequency range of $\omega > \omega_r$, the horizontal spectrum section will noticeably increase the approximate energy content only if:

$$\frac{\omega_r \times \omega_2}{\omega_o^2} > 1 \tag{12.31}$$

The maximum error of formula (12.30) can be calculated on the basis of Eq. (12.29):

$$e_r = \left(1 + \frac{\omega_d}{\omega_r}\right) \times \left(1 + \frac{\omega_r}{\omega_1}\right) \times \left(1 + \frac{\omega_2}{\omega_3}\right) \tag{12.32}$$

When there is no initial rising spectrum section, $\omega_d = 0$ should be substituted in the above expression; i.e., the first factor of the product will not increase the relative error.

Procedure 4: Spectrum Containing a Peak (Figs. 12.2d, 12.2e, and 12.2f)

The peak in the amplitude density function indicates that the aperiodic signal will have a periodic component as well. The periodic component can be described by means of the angular frequency at the peak, while the aperiodic component can be characterized by the damping factor, ζ. An oscillating system will exist in the case of a damping factor smaller than

$$\zeta = \frac{1}{\sqrt{2}}$$

The relationship between peaks in the spectrum and the damping factor ζ is shown in Figs. 12.2d, 12.2e, and 12.2f.

In the case of a spectrum of the type seen in Fig. 12.2d, the approximate energy content can be calculated by the formula:

$$E_a = \frac{A_r^2 \times \omega_r}{4\zeta} \tag{12.33}$$

If the spectrum does not contain further falling sections, the approximate and the exact energy content are equal.

The above approximation formula gives a higher value than the accurate energy for spectra that continue with a section of –20 dB/decade slope after the first –40 dB/decade slope. Indicating the cutoff angular frequency between the –40 dB/decade and –20 dB/decade slope with ω_{or}, the relative approximation error is:

$$e_{r_-} = \frac{\omega_{or}^2}{\omega_{or}^2 + \omega_r^2} \tag{12.34}$$

The maximum approximation error will occur for spectra with an initial rising section. In this case, the relative error of Eq. (12.33) becomes:

$$e_{r_+} = 1 + \frac{2\zeta \times \omega_d}{\omega_r} + \frac{\omega_d^2}{\omega_r^2} \qquad (12.35)$$

In engineering practice, the spectrum section with a –40 dB/decade slope will be followed by sections with –60 dB/decade or higher slopes. Referring to the cut-off frequency as ω_{1r}, the relative approximation error of the energy calculation formula of Eq. (12.33) is as follows:

$$e_r = 1 + \frac{\omega_r^2}{2\zeta \times \omega_{1r} \times \omega_r + \omega_{1r}^2} \qquad (12.36)$$

In the angular frequency range of $\omega > \omega_r$, neglecting subsequent spectrum sections with greater than –40 dB/decade slope will cause an error of less than 1 dB.

In the case of a spectrum of the type seen in Fig. 12.2e, the approximate energy content is:

$$E_a = \frac{A_r^2 \times \omega_r}{2} \times \left(1 + \frac{\omega_r}{\omega_{rc}} \times \frac{1}{2\zeta}\right) \qquad (12.37)$$

This formula gives nearly the exact energy content, with the deviation relative to the exact value being less than 1 dB. The peaks will increase the approximate energy content only in the case of:

$$\frac{\omega_r}{\omega_{rc}} \times \frac{1}{2\zeta} > 1 \qquad (12.38)$$

In this energy content calculation, falling sections of –80 dB/decade slope or greater can be neglected. The maximum inaccuracy in the approximation can be expected for spectrums with initial rising section for which the relative error is:

$$e_r = \left(1 + \frac{\omega_d}{\omega_r}\right) \times \left(1 + \frac{2\zeta \times \omega_r}{\omega_{rc}} + \frac{\omega_r^2}{\omega_{rc}^2}\right) \qquad (12.39)$$

In the case of a spectrum of the type shown in Fig. 12.2f, the angular frequencies ω_r, ω_o, and ω_{rc}, should be read from the spectrum, in addition to the values of A_r and ζ. We then can calculate the approximate energy content using the following formula:

$$E_a = \frac{A_r^2 \times \omega_r}{2} \times \left(1 + \frac{\omega_r \times \omega_{rc}}{\omega_o^2} \times \frac{1}{2\zeta}\right) \tag{12.40}$$

In this energy content approximation, the falling spectrum sections of −60 dB/decade slope or greater should not be taken into consideration. The peaks will increase the energy content only if:

$$\frac{\omega_r \times \omega_{rc}}{\omega_o^2} \times \frac{1}{2\zeta} > 1 \tag{12.41}$$

The accuracy of the approximation formula Eq. (12.40) is in the range of −2 to +4 percent (−0.4 to +0.4 dB), depending on the form of the spectrum, except in the case of a spectrum with a rising initial section. In this case, the approximation error can be calculated using Eqs. (12.27) and (12.28).

To illustrate the application of these energy calculation formulas, let us look at an example. The task is to check the disturbance susceptibility of an electronic system. Let us suppose that a transient disturbance (for instance, a switching transient) reaches the equipment input through a long pair of wires. Let us also assume that, although the circuit connected to this input does not contain sensitive parts, a capacitive coupling exists through the wiring toward some sensitive electronics. The problem is shown in Fig. 12.3a. The input is protected against HF noise by the filter capacitor, C_f. The question is whether, in the case of the assumed disturbance signal $u_g(t)$ as shown in the figure, a failure in the electronic system can be predicted.

To analyze the EMS performances of electronic unit, we must plot the spectrum of the transient signal appearing on resistor R_1. The electrical equivalent circuit of the signal transmission network can be seen in Fig. 12.3b. The circuit elements R_n and L_n represent the impedances of the wire pair, and resistor R_e represents the input resistance of the electronic unit.

When plotting the frequency function of the circuit, we can neglect the direct load resistance across the wire pair, R_i, (presumably > 25 Ω) as with resistance R_n, it provides a rather small attenuation, and it does not damp the resonant circuit $L_n C_f$. Thus:

$$Tr(s) = \frac{U_{in}(s)}{U_{out}(s)}$$

$$= \frac{s \times R_e C_c}{1 + s\,(R_e C_c + R_n C_f + R_n C_c) + s^2\,(L_n C_f + R_n R_e C_f C_c + L_n C_c) + s^3 R_e L_n C_f C_c} \tag{12.42}$$

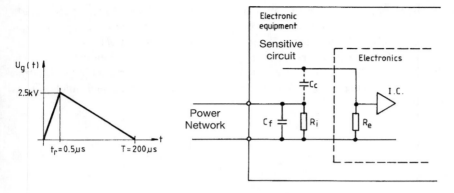

a. Schematic representation of the problem

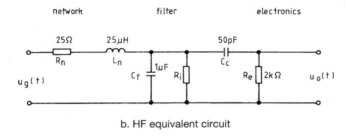

b. HF equivalent circuit

FIGURE 12.3 An electronic unit's EMS performance in the case of an impulse-like disturbance

In our example, let us assume that the resistance R_e, forming the internal resistance of the electronic unit, is 2 kΩ. After the appropriate substitutions, the Laplace transform yields:

$$Tr(s) = \frac{s \times 10^{-6}}{1 + s \times 1.5 \times 10^{-6} + s^2 \times 2.5 \times 10^{-11} + s^3 \times 2.5 \times 10^{-17}} \quad (12.43)$$

By means of the root calculation method described in Section 12.2.1, it can be shown that the roots of Eq. (12.43) are:

$$T_3 = 10^{-6} \quad s^{-1}$$

$$T_2 = 5 \times 10^{-6} \quad s^{-1}$$

$$2\zeta_2 = 0.1 \quad (12.44)$$

The Bode diagram of the Laplace transform has the shape of Fig. 12.2b; it starts with rising initial section, and after the horizontal one there are two sections with -20 and -40 dB/decade slope. The peak in the spectrum is at $1/T_2$.

It should be noted that because capacitors C_f and C_c differ greatly in value, the frequency function of the transmission network can be determined by the plotting rules of the Bode diagram. With fair approximation, the frequency function is as follows [313]:

$$Tr(s) = \frac{1}{1 + s \times R_n C_f + s^2 \times L_n C_f} \times \frac{s \times R_e C_c}{1 + s \times R_e C_c} \quad (12.45)$$

The spectrum of the excitation signal, $u_g(t)$, can be determined by graphic spectrum analysis as described in Section 5.3.1 (see the example). The value of the horizontal spectrum section is:

$$A_g = \int u_g(t)\, dt = 0.25 \text{ Vs} \quad (12.46)$$

The fractional angular frequencies are:

$$\omega_{g1} = \frac{4}{T} = 2 \times 10^4 \ \text{s}^{-1}$$

$$\omega_{g2} = \frac{1}{t_r} = 2 \times 10^6 \ \text{s}^{-1} \quad (12.47)$$

The spectrum of the transient disturbance signal appearing on the output of the transmission network is shown by the broken line in Fig. 12.4. This spectrum (thick line) was obtained by adding up the spectrum of the excitation signal $a_g(\omega)$ and the frequency function $Tr(\omega)$. The spectrum of the disturbance coupled into the electronic unit is similar to Fig. 12.2d.

Since the peak value of the noise signal is about 100 V (see Section 12.3.1), for EMS analysis the energy content also should be determined. For calculating the disturbing energy, the value of A_r has to be converted to the referenced unit; i.e., to Vs:

$$A_r = -6 - 40 \ \text{dB(Vs)} = 0.5 \times 10^{-2} \text{ Vs} \quad (12.48)$$

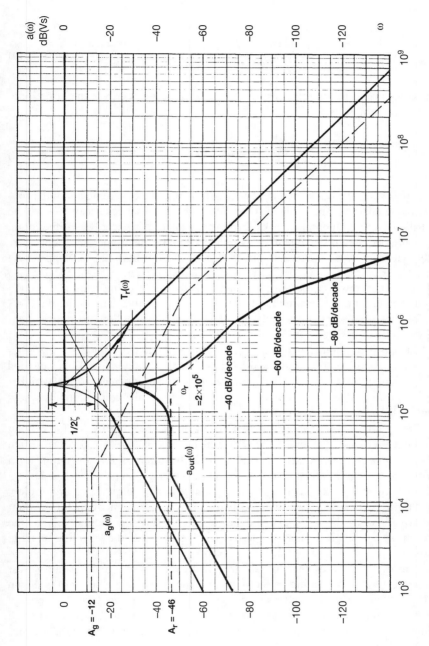

FIGURE 12.4 Spectrum of the disturbance signal coupled into the electronic unit of Figure 12.3

Given that $\omega_r = 2 \times 10^5 \text{ s}^{-1}$, we have all the required data to calculate the energy content using Eq. (12.33):

$$E_a = \frac{(0.5 \times 10^{-2})^2 \times 2 \times 10^5}{2 \times 0.1} = 25 \text{ V}^2\text{s} \tag{12.49}$$

Thus, according to the relationship of Eq. (12.22), the energy transformed into heat on resistor R_e is:

$$E_U = \frac{25 \text{ V}^2\text{s}}{2000 \text{ } \Omega} = 1.25 \times 10^{-2} \text{ Ws} \quad \text{i.e., } 12.5 \text{ mJ} \tag{12.50}$$

This means that the transient disturbance getting into the electronic unit will probably cause a failure (see Table 12.1), although the disturbance energy, E_U, will be distributed among the impedances forming the total resistance, R_e. Furthermore, the energy content of the disturbance coupled into the electronic unit is considerably increased by the oscillating member, as seen from Eq. (12.33). Suppressing the peaks by forming a high-frequency damping would decrease the value of E_U by one order; however, because we neglected the multiplying factor $1/(2\zeta)$, the disturbing energy remains higher than the acceptable value. Therefore, the EMS performance can be predicted only by a detailed analysis of how the transient energy will be distributed within the electronic unit.

The approximation error of Eq. (12.33) can be calculated using Eq. (12.35) because the actual spectrum (see Fig. 12.4) has a rising initial section. The calculated energy content of the disturbing signal is a maximum of 2 percent (0.2 dB) higher than the exact value; i.e., the inaccuracy of the energy calculation formula is now negligible.

The approximation energy calculation formulas are quite useful for solving transient problems that fall somewhat outside the usual EMC field but often emerge in engineering practice. EMC experts are often asked to examine such problems. It is well known that, at the time of switching on an ac or dc supply, a transient current peak can arise and can damage power semiconductors or blow out the fuses. Because the behavior of fuses and power semiconductors at transient overload can be well characterized by the I^2t rating (or, alternatively, by the $\int i^2 dt$ rating), an engineer may frequently need to calculate the I^2t rating of surge currents.

Consider the task of checking the transient overload of a rectifier in a power supply, caused by switching on the ac mains. The layout of the problem is shown in Fig. 12.5a. The constant current consumption can be neglected when the energy of the current peak is examined. The highest current transient can be expected when switching on at maximum mains voltage. Let the peak value of the secondary transformer voltage be 60 V.

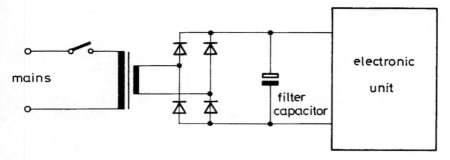

a. Schematic Layout of the Problem

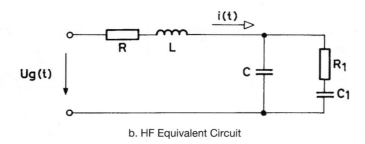

b. HF Equivalent Circuit

FIGURE 12.5   Examination of transient overload in semiconductors and fuses

The first step is to set up the electrical equivalent circuit and devise the Laplace transform of the current flowing through the rectifier. The equivalent circuit is shown in Fig. 12.5b, where we will let the values of the elements be as follows:

$$u_g(t) = 60 \times 1(t) \qquad R = 2\ \Omega \qquad R_1 = 5\ \Omega$$

$$C = 3000\ \mu F \qquad L = 1\ mH \qquad C_1 = 500\ \mu F \qquad (12.51)$$

The elements R and L represent the impedance of the transformer. The capacitance C_1 and series resistance R_1 are parts of the electronic unit. The Laplace transform of the current i(t) is:

$$I(s) = \frac{U_g \times (C + C_1) \times \left(1 + R_1 \times \dfrac{C_1 C}{C_1 + C}\right)}{1 + s\,(RC + R_1 C_1 + R C_1) + s^2\,(LC + LC_1 + RR_1 CC_1) + s^3 R_1 LCC_1} \qquad (12.52)$$

After appropriate substitutions, the above Laplace transform yields:

$$I(s) = \frac{0.21\,(1 + s \times 2.14 \times 10^{-3})}{1 + s \times 9.5 \times 10^{-3} + s^2 \times 1.1 \times 10^{-5} + s^3 \times 7.5 \times 10^{-9}} \tag{12.53}$$

Using the root calculation method described in Section 12.2.1, it can be shown that the spectrum of the above Laplace transform is like that of Fig. 12.2f, where:

$$A_r = 0.21 \text{ As} \qquad \omega_r = 105 \text{ s}^{-1} \qquad \omega_o = 465 \text{ s}^{-1}$$

$$\omega_{rc} = 1.125 \text{ s}^{-1} \qquad 2\zeta = 1.23 \tag{12.54}$$

The transient overload of the rectifier can be calculated from Eq. (12.40). The energy calculation formula gives the value of the $\int i^2 dt$ rating:

$$E_a = \int i^2 dt = \frac{0.21^2 \times 105}{2} \times (1 + \frac{465^2}{105 \times 1125} \times \frac{1}{1.23}) = 5.7 \text{ A}^2 \text{s} \tag{12.55}$$

This $\int i^2 dt$ rating is not dangerous for rectifiers of the lowest power ratings.

The above example was worth examination because the equivalent circuit in Fig. 12.5b and the Laplace transform of Eq. (12.52) are also suitable for studying other similar problems.

Most power electronic equipment supplied from a dc network (e.g., from a battery) contains a buffer capacitor on the input. But it is not permissible to allow the inrush current to blow the fuse. To calculate the transient overload on the fuse, the same equivalent circuit could be set up as shown in Fig. 12.5b. Let the value of the elements be:

$$u_g(t) = 50 \times 1(t) \qquad R = 1\ \Omega \qquad R_1 = 2\ \Omega$$

$$C = 25000\ \mu F \qquad L = 250\ \mu H \qquad C_1 = 2000\ \mu F \tag{12.56}$$

The elements R and L now represent the impedance of the cabling between the equipment and the battery. In this case, the Laplace transform of the inrush current, i(t), is:

$$I(s) = \frac{1.35\,(1 + s \times 4 \times 10^{-3})}{1 + s \times 6.7 \times 10^{-3} + s^2 \times 10^{-4} + s^3 \times 24 \times 10^{-8}} \tag{12.57}$$

The spectrum of this Laplace transform is the type of Fig. 12.2d, in spite of the differentiating effect. The values needed to calculate the energy of the inrush current are:

$$A_r = 1.35 \text{ As} \qquad \omega_r = 100 \text{ s}^{-1} \qquad 2\zeta = 0.65 \tag{12.58}$$

According to Eq. (12.33), the transient overload of the fuse is:

$$E_a = \int i^2 dt = \frac{1.35^2 \times 100}{2 \times 0.65} A^2 s = 140 \text{ A}^2 s \tag{12.59}$$

Thus we must choose a fuse with an $I^2 t$ rating greater than 140 $A^2 s$.

At present, the inaccuracy of the energy calculation formula is about 6 percent (0.5 dB) as calculated by Eq. (12.35), because of the differentiating member in the numerator. However, it is small enough that it does not affect any conclusions we may draw from this approximate energy calculation method.

12.2.3 Energy Density Function Plotting and Application

In some electrical apparatus (e.g., in telecommunication equipment), the transient disturbance is coupled onto the sensitive parts through a bandpass filter. Attempts have been made to analyze the EMS performance of such apparatus on the basis of energy concept [60, 61, 218, 219]. However, the approximation energy calculation formulas given in Section 12.2.2 are for calculating the energy content of impulse-like signals. Therefore, they are difficult to apply to a transmission path with a bandpass filter. For ease of calculation, the concept of cumulative energy distribution has been introduced [324].

An otherwise circuitous and time-consuming EMS analysis of an electronic unit with a bandpass filter can be made easier by plotting the approximate curve of the energy density function. The knowledge of the approximate energy density function offers much information that is essential in solving EMC-related operational disturbance problems.

For the energy calculation formulas, let the energy density function be defined as:

$$e(\omega) = \frac{1}{\pi} \times \int a^2(\omega) \, d\omega \tag{12.60}$$

The energy density function as defined above is basically analogous to the cumulative energy distribution.

The energy content of a transient signal can be calculated by means of the energy density function as follows:

$$E = e(\infty) - e(0) = e(\infty) \tag{12.61}$$

The energy density function is a very useful tool for calculating the energy content of signal components in a given frequency range:

$$E_{band} = e(\omega_{upper}) - e(\omega_{lower}) \tag{12.62}$$

According to whether the amplitude density function (spectrum) refers to current or voltage (Vs or As), the unit of the energy density function is V^2s or A^2s.

The energy density function is practical for plotting in a log-log coordinate system. In this case, the energy density function, just like the amplitude density function, can be approximated by straight lines. The plotting rules of the envelope are worked out for the spectra shown in Fig. 12.2 [314]. For the sake of simplicity, the following symbols will be used:

AADC	approximate amplitude density curve (broken-line envelope of the amplitude density function)
ADC	amplitude density function
AEDC	approximate energy density curve (broken-line of the energy density function)
AFR	angular frequency range
EDF	energy density function
SAC	section of the AADC
SEC	section of the AEDC

The value of the EDF (or AEDC) can be given in dB only if the reference base is defined. Per the designations of Fig. 12.2, let the reference energy be as follows:

$$E_r = \frac{A_r^2 \times \omega_r}{\pi} \tag{12.63}$$

After calculating the reference energy, we must first plot the SEC belonging to the SAC A_r. This SEC is a straight line of +20 dB/decade slope, crossing the 0 dB axis at the angular frequency of ω_r.

The further plotting procedure depends on the nature of the AADC. Four procedures are detailed in the following paragraphs.

Procedure 1: AADC with Falling Section (Fig. 12.2a)

In the case of AADCs consisting only of falling SACs in the AFR of $\omega > \omega_r$, the AEDC goes on with a straight line of +10 dB/decade slope. This SEC starts from the point $e_r(\omega_r)$ (i.e., from point ω_r of the 0 dB axis) and is valid to the horizontal SEC $e_a(\omega)$.

The value of the SEC $e_a(\omega)$ that gives the approximate energy content of the signal can be calculated in accordance with Eq. (12.25) as follows:

$$E_a = e_a(\omega) = E_r \times \frac{\pi}{2} \qquad (12.64)$$

The AEDC is plotted in Fig. 12.6.

The SACs with slopes greater than -20 dB/decade have no effect on the shape of the AEDC, as was mentioned in Section 12.2.2.

Procedure 2: AADC with a Rising Initial Section (Fig. 12.2b)

In spite of the initial rising SAC, the SECs in the AFR $\omega > \omega_r$ should be plotted as described in Procedure 1. In the AFR $\omega < \omega_r$, the SEC $e_r(\omega)$ extends only to the angular frequency of:

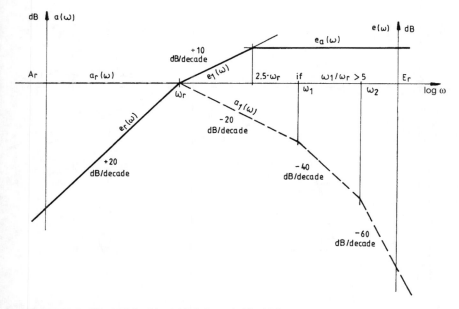

FIGURE 12.6   The AEDC of the AADC shown in Fig. 12.2a

$$\omega_{t1} = \sqrt{3} \times \omega_d \qquad (12.65)$$

The AEDC in the AFR $\omega < \omega_{t1}$ is a straight line with +60 dB/decade slope, going through the point $e_r(\omega_{t1})$. This SEC, $e_d(\omega)$, is plotted in Fig. 12.7.

For a more accurate approximation, a transitional SEC may be drawn near the angular frequency ω_{t1}. This SEC (labeled $e_{dt}(\omega)$ in Fig. 12.7) is a straight line of +40 dB/decade slope intersecting SEC $e_r(\omega)$ at angular frequency of:

$$\omega_{rt} = 4\omega_d \qquad (12.66)$$

If SEC $e_{dt}(\omega)$ is plotted and the horizontal SAC is shorter than one decade, the SECs $e_r(\omega)$ and $e_1(\omega)$ may be eliminated. When $\omega_{rt} > \omega_r$, SEC $e_{dt}(\omega)$ extends only to angular frequency w_r. The AEDC goes on from the point $e_{dt}(\omega_r)$ with a straight line of +20 dB/decade slope and with SEC $e_a(\omega)$.

Procedure 3: AADC Containing Two Horizontal Sections (Fig. 12.2c)

Depending on the type of AADC, the initial SECs have to be plotted as described in Procedure 1 or 2.

SEC $e_a(\omega)$ extends only to SEC $e_o(\omega)$, which is a straight line of +20 dB/decade slope starting at the point ω_m of the 0 dB axis. The AF ω_m can be calculated as follows:

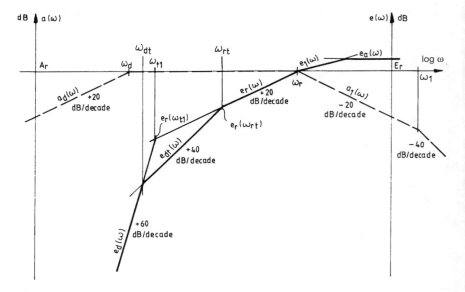

FIGURE 12.7 The AEDC of the AADC shown in Fig. 12.2b

$$\omega_m = \frac{\omega_o^2}{\omega_r} \qquad (12.67)$$

SEC $e_o(\omega)$ is valid to AF ω_2. If AFR $\omega < \omega_2$, the AEDC has to be plotted as if the 0 dB axis would run through the point $e_o(\omega_2)$. SEC $e_2(\omega)$ starts with a +10 dB/decade slope from the point $e_o(\omega_2)$ and continues to the horizontal SEC, giving the energy content in accordance with Eq. (12.30):

$$E_t = e_t(\omega) = E_r \times \frac{\pi}{2} \times \left(1 + \frac{\omega_r \times \omega_2}{\omega_o^2}\right) \qquad (12.68)$$

The AEDC is plotted in Fig. 12.8.

The approximation error near the angular frequency ω_m can be decreased or even eliminated by plotting a transitional SEC between the SECs $e_a(\omega)$ and $e_o(\omega)$. This transitional SEC, labeled $e_{ot}(\omega)$ in Fig. 12.8, starts at the point $\omega_m/10$ of the 0 dB axis with +10 dB/decade slope. If SEC $e_{ot}(\omega)$ is plotted, and the ratio of ω_m to ω_2 is small, SEC $e_o(\omega)$ may vanish so that SEC $e_{ot}(\omega)$ runs to the horizontal SEC, $e_t(\omega)$.

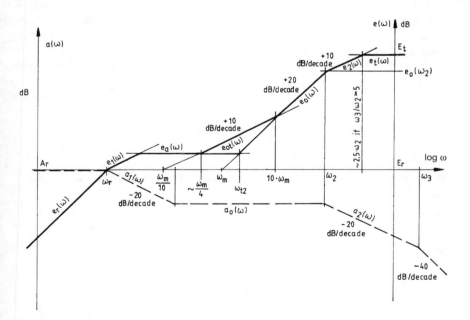

FIGURE 12.8 The AEDC of the AADC shown in Fig. 12.2c

Procedure 4: AADC Containing a Peak (Figs. 12.2d, 12.2e. and 12.2f)

In the case of an ADF containing a single pole (i.e., when the AADC is like that shown in Fig. 12.2d), SEC $e_r(\omega)$ is valid to the lower cutoff angular frequency of:

$$\omega_1 = \omega_r \times \frac{1 - \sqrt{2} \times \zeta}{2} \tag{12.69}$$

The approximate energy content [i.e., the horizontal SEC, $e_a(\omega)$] can be calculated in accordance with Eq. (12.33) as follows:

$$E_a = e_a(\omega) = E_r \times \frac{\pi}{2} \times \frac{1}{2\zeta} \tag{12.70}$$

SEC $e_a(\omega)$ starts at the upper cutoff angular frequency, given by the expression:

$$\omega_u = \omega_r \times (1 + \zeta) \tag{12.71}$$

In the AFR $\omega_1 < \omega < \omega_u$, the AEDC is a straight line connecting points $e_r(\omega_1)$ and $e_a(\omega_u)$. The plot of AEDC is shown in Fig. 12.9a.

If $\zeta < 0.2$, the slope of the SEC in AFR $\omega_1 < \omega < \omega_u$ can be greater than +40 dB/decade. To increase the accuracy of the AEDC, two SECs have to be plotted in this AFR. The one starts at point $e_r(\omega_1)$ with a +40 dB/decade slope and extends to the other one, which is a straight line connecting points $e_r(\omega_r / \sqrt{2}$ and $e_a(\omega_u)$. Such an AEDC is shown in Fig. 12.9b.

Figure 12.10 shows the plot of the AEDC belonging to an AADC as shown in Fig. 12.2e. The SEC $e_a(\omega)$ runs to the lower cutoff angular frequency of:

$$\omega_{rl} = \omega_r \times \left[1 - \sqrt{2} \times \zeta \times \left(1 + \frac{\omega_r}{\zeta \times \omega_{rc}} \right) \right] \tag{12.72}$$

The approximate energy content of the impulse-like signal can be calculated according to Eq. (12.37) as follows:

$$E_t = e_t(\omega) = E_r \times \frac{\pi}{2} \times \left(1 + \frac{\omega_r}{2\zeta \times \omega_{rc}} \right) \tag{12.73}$$

The SEC $e_t(\omega)$ is valid from the angular frequency:

$$\omega_{ru} = \omega_{rc}(1 + \zeta) \tag{12.74}$$

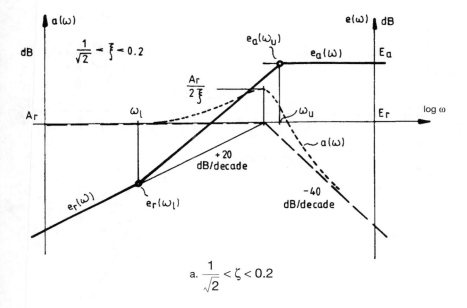

$$a. \ \frac{1}{\sqrt{2}} < \zeta < 0.2$$

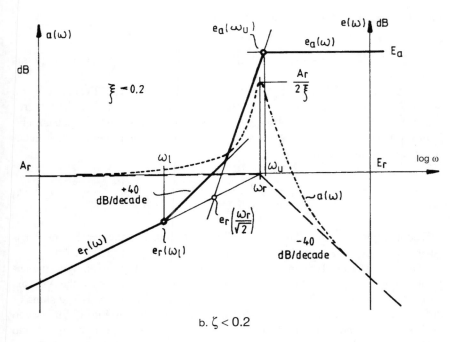

$$b. \ \zeta < 0.2$$

FIGURE 12.9 The AEDC of the AADC shown in Fig. 12.2d

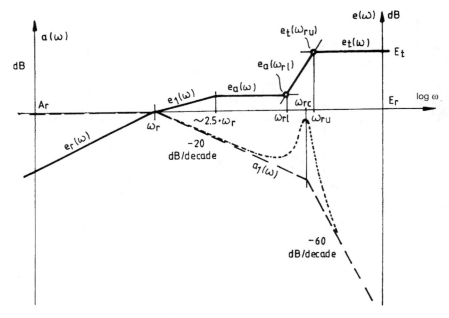

FIGURE 12.10 The AEDC of the AACD shown in Fig. 12.2e

In the AFR $\omega_{rl} < \omega < \omega_{ru}$, the AEDC is a straight line connecting the points $e_a(\omega_{rl})$ and $e_t(\omega_{ru})$.

In a case of AADCs like Fig. 12.2f, the plotting method for SECs up to $e_o(\omega)$ or $e_{ot}(\omega)$ is the same as it would be if there were no peak; i.e., as described in Procedures 1 and 2. To plot the subsequent SECs, we first must calculate the angular frequency ω_{rl} using Eq. (12.72), as well as the value of the horizontal SEC in accordance with Eq. (12.40):

$$E_t = e_t(\omega) = E_r \times \frac{\pi}{2} \times \left(1 + \frac{\omega_r \times \omega_2}{\omega_o^2} \times \frac{1}{2\zeta}\right) \qquad (12.75)$$

The next plotting procedure to be followed depends on whether the cutoff angular frequency $\omega_{rl}/2$ is on the SEC $e_o(\omega)$ or $e_{ot}(\omega)$.

If $\omega_{rl}/2 > 10 \times \omega_m$ [i.e., $\omega_{rl}/2$ is located on SEC $e_o(\omega)$], it runs only to the angular frequency $\omega_{ru}/2$, and the SEC $e_t(\omega)$ starts at ω_{rl}. The AEDC in the AFR $\omega_{rl} < \omega < \omega_{ru}$ is a straight line connecting points $e_o(\omega_{rl}/2)$ and $e_t(\omega_{ru})$. The slope of this SEC may exceed the value of +40 dB/decade, so similar to Fig. 12.9b, we can draw a straight line with +40 dB/decade slope starting at the point $e_o(\omega_{rl}/2)$, and another straight line connecting the points $e_o(\omega_{rl})$ and $et(\omega_{ru})$. Such an AEDC is shown in Fig. 12.11a.

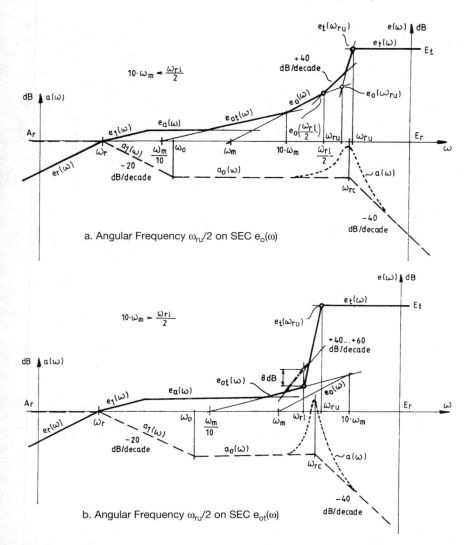

a. Angular Frequency $\omega_{ru}/2$ on SEC $e_o(\omega)$

b. Angular Frequency $\omega_{ru}/2$ on SEC $e_{ot}(\omega)$

FIGURE 12.11   The AEDC of the AADC shown in Fig. 12.2f

If $\omega_{rl}/2 < 10 \times \omega_m$ [i.e., $\omega_{rl}/2$ is located on SEC $e_{ot}(\omega)$], SEC $e_{ot}(\omega)$ runs to the angular frequency ω_{rl}, and the AEDC is formed in the AFR $\omega_{rl} < \omega < \omega_{ru}$ by the straight line connecting the points $e_{ot}(\omega_{rl})$ and $e_t(\omega_{ru})$. In the case of small damping factors when the slope of this SEC exceeds +80 dB/decade, an additional SEC may be plotted in this AFR, running with a +40 to +60 dB/decade slope through the point that is +8 dB higher at the cutoff angular frequency. This additional SEC is shown in Fig. 12.11b.

We will now consider an illustrative example of how to use the AEDC for EMS analysis [314]. Let us suppose that an impulse-like disturbance reaches a long wire pair connected to the input of an electronic unit. The disturbance is coupled into the electronics via the stray wiring capacitance. The coupled disturbance passes through a bandpass filter and into the sensitive parts as shown in Fig. 12.12a. The question is whether, in the case of the assumed disturbance signal $u_g(t)$, we should anticipate a failure in the electronic unit. The energy of the disturbance reaching the sensitive parts must be less than 10^{-7} Ws to avoid failure.

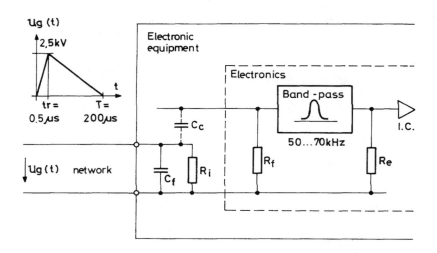

a. Schematic layout of the problem

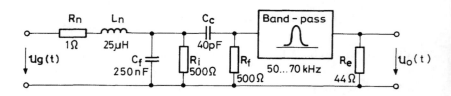

b. High-frequency equivalent circuit

FIGURE 12.12 Examination of the EMS of an electronic unit with a bandpass filter, in the case of exposure to an impulse-like current

The electrical equivalent circuit is shown in Fig. 12.12b. Elements R_n and L_n represent the impedance of the input wiring, and C_f is a filter capacitor protecting the input against HF noise. R_i is the power supply input resistance, whereas R_f is that of the front end electronic unit. R_e designates the total resistance of the electronics after the filter, and C_c is the stray capacitance of the wiring.

To analyze the EMS performance, we first must determine the spectrum of the disturbance getting through the coupling capacitance, C_c. The AADC of the transient signal $u_o(t)$ on the resistor R_e (marked $a_{out}(\omega)$ in Fig. 12.13) is the sum of the AADC of input signal $a_g(\omega)$ and the frequency function of transmission path $T_r(\omega)$ (see the example in Section 12.2.2). For calculating $T_r(\omega)$, only R_f has been taken into consideration because $R_f < R_e$. The spectra are shown in Fig. 12.13.

Since the peak value of the coupled transient disturbance is about 50 V, we also need to know the energy content of the noise signal passing through the bandpass filter and into the sensitive parts. This can be calculated best by means of the energy density function. The AADC $a_{out}(\omega)$ has a rising initial SAC and a peak, so the initial SECs up to $e_r(\omega)$ must be plotted as described in Procedure 2. The subsequent SECs are calculated according to Procedure 4.

To set up the coordinate system for plotting the AEDC, first calculate the reference energy using Eq. (12.63). Here, the value A_r, expressed in dB, needs to be converted:

$$A_r = -80 \text{ dB (Vs)} = 10^{-4} \text{ Vs} \tag{12.76}$$

Knowing that $\omega_r = 4 \times 10^5 \text{ s}^{-1}$, the reference energy is:

$$E_r = \frac{(10^{-4})^2 \times 4 \times 10^5}{\pi} = V^2s = 1.27 \times 10^{-3} \text{ V}^2\text{s} \tag{12.77}$$

For plotting the AEDC of the output signal $u_o(t)$, the value following values must be determined:

$\omega_{t1} = 3.46 \times 10^4 \text{ s}^{-1}$		Eq. (12.65)
$\omega_{rt} = 8 \times 10^4 \text{ s}^{-1}$		Eq. (12.66)
$\omega_l = 1.86 \times 10^5 \text{ s}^{-1}$		Eq. (12.69)
$\dfrac{\omega_r}{\sqrt{2}} = 2.86 \times 10^5 \text{ s}^{-1}$		$\zeta < 0.2$
$\omega_u = 4.2 \times 10^5 \text{ s}^{-1}$		Eq. (12.71)
$E_a = 15.7 \times E_r \quad +24 \text{ dB}$		Eq. (12.70)

$$(12.78)$$

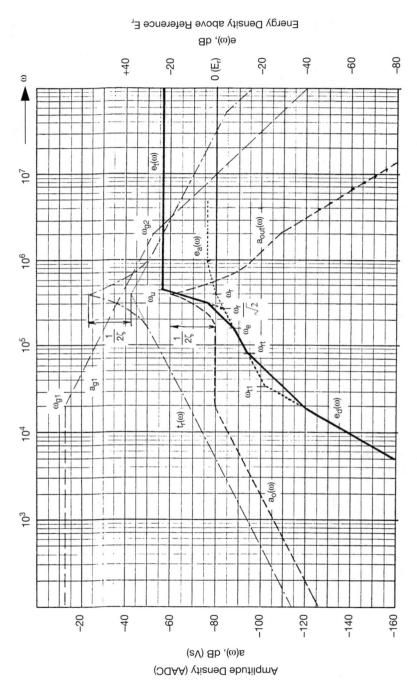

FIGURE 12.13 The AADC and the AEDC of the disturbance on resistor R_e in Fig. 12.12

The plot begins at the SEC $e_r(\omega)$ and extends as a straight line with +60 dB/decade slope through $e_r(\omega_{t1})$ in the AFR of $\omega < \omega_{t1}$. For a more accurate approximation, the transitional SEC $e_{dt}(\omega)$ may be drawn with a +40 dB/decade through the point $e_r(\omega_{rt})$. Then, mark points ω_l and $\omega_r/\sqrt{2}$ on the SEC $e_r(\omega)$ and plot the two sections in the AFR $\omega_l < \omega < \omega_u$. Finally, plot the horizontal SEC in the AFR $\omega_u > \omega$.

The AEDC is drawn with a thick line in Fig. 12.13. For calculating the energy content of the disturbance that passes through the bandpass filter to resistor R_e, we read the value of the AEDC at the filter's cutoff frequencies. These angular frequencies are $\omega_2 = 2\pi \times 70 \times 10^3 = 4.4 \times 10^5$ s^{-1} and $\omega_1 = 2\pi \times 50 \times 10^3 = 3.14 \times 10^5$ s^{-1}. Thus, taken from the approximate curve, which gives the energy in dB above the referency energy content E_r:

$$e(\omega_2) = +24 \text{ dB above } E_r = 15.7 \times 1.27 \times 10^{-3} = 2 \times 10^{-2} \quad \text{V}^2\text{s}$$

$$e(\omega_1) = +6 \text{ dB above } E_r = 2 \times 1.27 \times 10^{-3} = 2.5 \times 10^{-3} \quad \text{V}^2\text{s} \tag{12.79}$$

The approximate energy of the signal components in the frequency range of the passband can be calculated using Eq. (12.62):

$$E_e = \frac{e(\omega_2) - e(\omega_1)}{R_e} = \frac{(2 - 0.25) \times 10^{-2}}{4 \times 10^3} \text{ Ws} = 4.375 \times 10^{-6} \text{ Ws} \tag{12.80}$$

Although this disturbance energy will be distributed among the impedances forming the total resistance, R_e, a failure can be expected because E_e is much greater than 10^{-7} Ws.

We should note that conclusions may be drawn from the AEDC about the relationship between the circuit elements and the energy of the disturbance passing through the bandpass filter. This provides some initial data on how to improve EMS performance.

First, let us examine which circuit elements have the greatest effect on the coupled disturbance energy. As seen from Eq. (12.63), from which we derive the referency energy, the horizontal spectrum section must be decreased as much as possible. At present, the horizontal SAC is determined by the rising initial SAC of $T_r(\omega)$; i.e., the best way to reduce the disturbance energy is to decrease the product of $R_e C_c$. The other way is to eliminate the peak in the frequency function by creating some HF damping. Without a peak in AADC, the SEC $e_a(\omega)$ will be much lower than with the peak. To illustrate the effect of HF damping, the AEDC belonging to an AADC without a peak is also plotted in Fig. 12.13.

Taking the dotted-line AEDC into consideration, the AEDC values at the cutoff angular frequencies are $e'(\omega_2) = +0.5$ dB and $e'(\omega_1) = -2$ dB. Therefore, the energy of the coupled disturbance without a peak would be as follows:

$$E_e = \frac{(1.05 - 0.79) \times 1.27 \times 0^{-3}}{4 \times 10^3} = 10^{-7} \quad \text{Ws} \qquad (12.81)$$

This indicates that the EMS problem surely could be solved by developing some HF damping.

With simple calculation, we can also find how the disturbance energy on the resistor R_e would change if the bandpass filter were tuned to the 60 to 80 kHz range. To find an answer, only the values of the AEDC must be read at new angular frequencies. The angular frequency ω'_2 is on the horizontal SEC [i.e., $e_a(\omega'_2)$ = E_a = +23 dB (2×10^{-2} V^2s)], and at $\omega'_1 = 3.75 \times 10^5$ s^{-1} the value of AEDC is +13 dB (6.10^{-3} V^2s). Therefore, the disturbing energy on resistor R_e in the case of the tuned bandpass filter is approximately:

$$E' = \frac{e(\omega_2') - e(\omega_1')}{R_e} = \frac{(2 - 0.6) \times 10^{-2}}{4 \times 10^3} = 3.5 \times 10^{-6} \quad \text{Ws} \qquad (12.82)$$

This disturbance energy is still higher than the acceptable value of 10^{-7} Ws. Therefore, by tuning the band-pass filter to 60 to 80 kHz, the EMS performance would not be improved enough. Studying the AEDC shows that by lowering the cutoff frequency of the bandpass filter to 70 kHz (to angular frequency 4.2×10^5), the disturbance energy on resistor R_e would be reduced practically to zero.

Finally, let us adopt the task of designing a noise filter for the electronic unit's input that will solve the EMS problem without any HF damping or bandpass filter tuning. The disturbance energy on resistor R_e is certain to be less than 10^{-7} Ws if the AEDC does not reach the value 2.5×10^{-3} V^2s [i.e., compared to the reference energy, the value of $2E_r$ = +6 dB (R_e = 4 kΩ). As we can see from the figure, the angular frequency is about 3.1×10^5 s^{-1}; i.e., a frequency of about 50 kHz belongs to this energy value. This means that by incorporating an input noise filter with a cutoff frequency of less than 50 kHz, we can prevent HF noise from coupling via the wiring stray capacitances, and the EMS problem will be solved.

12.3 IMPULSE CHARACTERISTICS AND NOISE FILTER DESIGN

Impulse-like disturbances are usually qualified on the basis of amplitude, and noise filtering usually aims only at reducing peak amplitude values. This approach has become largely unsupportable with the knowledge that EMS performance depends not only on peak value but also on the energy content, frequency content, and time duration of the disturbance. In many cases, we must consider these impulse characteristics to properly characterize the susceptibility of electronics to transient noises. Methods have been devised to analyze EMS performance on the basis of energy concept alone, as discussed in Section 12.2.

For most practical applications, we adequately can predict breakdown levels from energy content only. But for predicting operational malfunctions that fall short of breakdown, it is much better to examine the impulse characteristics introduced in Section 4.7.1. The main obstacle to employing a wider variety of impulse considerations in EMS analysis and noise filter design is that these parameters can be difficult to measure and time consuming to calculate. However, some simple approximation formulas have been worked out for engineering practice, so it is now feasible to include a more comprehensive range of impulse characteristics in filter design.

12.3.1 Calculation of Impulse Characteristics

In engineering practice, it is usually too complicated to calculate impulse characteristics using Eqs. (4.35) through (4.38). But with a knowledge of the energy density function as well as approximation formulas for energy calculation, simple-to-use expressions can be worked out for this purpose, as illustrated in the following two examples.

Example 1: Peak Value, Impulse Strength, and Maximum Slope
The impulse strength, peak value, and maximum slope of an impulse-like signal can be taken directly from the spectrum: the impulse strength is proportional to the horizontal spectrum section, the peak value is proportional to the first cutoff frequency of the horizontal spectrum section, and the maximum slope is proportional to the cutoff frequency between the –20 and –40 dB/decade slopes [7]. The relationship between data taken from the spectrum and the above-mentioned impulse characteristics of the time function is summarized in Table 12.2 for spectra as shown in Fig. 12.2. The given relationships are approximate, but the inaccuracy of the impulse characteristic calculation formulas are less than 5 percent. For more accurate calculation, the approximate peak value and maximum slope can be corrected as follows:

For spectra as in Figs. 12.2a and 12.2b:
The peak value and the maximum slope should be corrected, much as for the energy calculation formulas, only if the spectrum section with –20 dB/decade slope is relative short. The corrected peak value of the time function is:

$$PV_c = PV \times \left(\frac{\omega_1}{\omega_1 + \omega_r} \right)^2 \tag{12.83}$$

TABLE 12.2 Peak Value, Strength, and Maximum Slope of Impulse-Like Disturbances

	Peak Value PV	Impulse Strength I	Max. Slope $|df/dt|_{max}$
Fig. 12.4a	$A_r \times \omega_r$	A_r	$A_r \times \omega_r \times \omega_1$
Fig. 12.4b	$A_r \times \omega_r$	(1) 0	$A_r \times \omega_r \times \omega_1$
Fig. 12.4c	$A_r \times \dfrac{\omega_r}{\omega_o} \times \omega_2$	(2) $A_r \times \left(1 + \dfrac{\omega_r}{\omega_o}\right)$	$A_r \times \dfrac{\omega_r}{\omega_o} \times \omega_2 \times \omega_3$
Fig. 12.4d	$A_r \times \omega_r \times (1 - \zeta \times \sqrt{1 - \zeta^2})$	—	$A_r \times \omega_r^2 \times \sqrt{1 - \zeta^2}$
Fig. 12.4e	$A_r \times \omega_r \times (1 - \zeta \times \sqrt{1 - \zeta^2})$	(3) A_r	$A_r \times \omega_r \times \omega_{rc} \times \sqrt{1 - \zeta^2}$
Fig. 12.4f	$A_r \times \dfrac{\omega_r}{\omega_o} \times (1 - \zeta \times \sqrt{1 - \zeta^2})$	(4) $A_r \times \left(1 + \dfrac{\omega_r}{\omega_o}\right)$	$A_r \times \dfrac{\omega_r}{\omega_o} \times \omega_r^2 \times \sqrt{1 - \zeta^2}$

Notes:
1. Interpreting the impulse strength as $\int |f(t)|\, dt$, the value of absolute impulse strength, when $n = \omega_d/\omega_r$, is as follows:

$$|I| = 2A_r \times \left(\frac{n}{n + 1}\right)^2$$

2. For a rising initial spectrum section, $I = 0$; otherwise, see note 1 with $n = \omega_o/\omega_r$.
3. If $\omega_r < 2\zeta \times \omega_{rc}$, otherwise not interpretable. For a rising initial spectrum section, $I = 0$; otherwise, see note 1.
4. If $\omega_r \times \omega_r < 2\zeta \times \omega_o^2$; otherwise not interpretable. For a rising initial spectrum section, $I = 0$; otherwise see note 1 with $n = \omega_{rc}/\omega_r$.

The correction function in the above expression is the same as the reciprocal of Eq. (12.27), serving to correct the approximate energy content, except that the cut-off angular frequencies of the spectrum section with -20 dB/decade slope have been substituted.

The approximation accuracy of maximum slope depends on the length of the spectrum section with -40 dB/decade slope. The corrected value of the highest slope of the time function is:

$$\left(\frac{df}{dt}\right)'_{max} = \left(\frac{df}{dt}\right)_{max} \times \left(\frac{\omega_2}{\omega_2 + \omega_1}\right)^2 \tag{12.84}$$

We should note that the maximum slope correction function is analogous to that of peak value, except that in Eq. (12.27) the cutoff angular frequencies of the spectrum section with –40 dB/decade slope have been taken into consideration.

For spectra as in Fig. 12.2c:

The peak value and maximum slope of the time function may be corrected as for spectra like Figs. 12.2a and 12.2b, only the falling spectrum sections following the second horizontal section should be taken into consideration. The corrected peak value of the time function is:

$$PV_c = PV \times \left(\frac{\omega_3}{\omega_3 + \omega_2} \right)^2 \tag{12.85}$$

And the corrected maximum slope is:

$$\left(\frac{df}{dt} \right)'_{max} = \left(\frac{df}{dt} \right)_{max} \times \left(\frac{\omega_4}{\omega_4 + \omega_3} \right)^2 \tag{12.86}$$

For spectra as in Fig. 12.2d:

The formulas used to calculate the impulse characteristics are accurate enough for engineering requirements so that no correction is needed.

For spectra as in Fig. 12.2e:

The formula for calculating the maximum slope is accurate enough so that only the peak value should be corrected if the spectrum section with –40 dB/decade slope is relatively short. The corrected peak value of the time function is:

$$PV_c = PV \times \left(\frac{\omega_{rc}}{\omega_{rc} + \omega_r} \right)^2 \tag{12.87}$$

For spectra as in Fig. 12.2f:

As with spectra shown in Fig. 12.2e, only the peak value should be corrected when the second horizontal spectrum section is relatively short. The corrected peak value of the time function is:

$$PV_c = PV \times \left(\frac{\omega_{rc}}{\omega_{rc} + \omega_o} \right)^2 \tag{12.88}$$

It should be noted that impulse strength, except for spectra like Fig. 12.2, is not a suitable characteristic for analyzing impulse-like disturbances. For spectra with a rising initial section, as well as for a time function with an oscillating component, the impulse strength is zero or not equal to the impulse strength of the aperiodic component only. The absolute impulse strength, defined as in Table 12.2 notes, seems to be a better impulse characteristic than impulse strength as defined by Eq. (4.35). The absolute impulse strength is more closely related to the disturbing effect of a transient signal, but an easy-to-use formula to calculate its approximate value might be worked out only for spectra with a rising initial section. Since the real time function can be approximated by an equivalent triangle signal (see Section 12.3.3), for engineering calculations, the absolute impulse strength may be approximated as the sum of the impulse strength of the positive and negative parts of the equivalent time function.

Example 2: Energy Content, RMS Frequency Duration, and RMS Time Duration

In addition to the question of energy content (see Section 12.2), we have introduced rms frequency duration and rms time duration for characterizing impulse-like signals. These characteristics are defined by Eqs. (4.37) and (4.38). For engineering practice, simple formulas have been devised. These are summarized in Table 12.3. The energy calculation formulas corresponding to the expressions given in Section 12.2.2 are included only for the sake of uniformity.

Section 12.2.2 quantified the approximation error of the energy content formulas. The formulas for calculating the approximate value of rms frequency duration and rms time duration are accurate enough for general engineering purposes.

12.3.2 Impulse Characteristics Involved in Noise Filter

In most cases, the purpose of a noise filter is to protect sensitive parts from the effects of RF signals and impulse-like transients. To maximize the electronic device's EMS performance, rather than focusing on the attenuation requirement, it is best to approach filter design by considering signal characteristics that correspond to impulses that are most likely to be encountered in actual operation.

As shown in Section 12.3.1, a close connection exists between impulse characteristics and some major spectrum parameters. These major parameters are (1) the value of the horizontal spectrum section, (2) some cutoff frequency, and (3) for oscillating systems, the damping factor. In consideration of these relationships, we may impose certain requirements on the spectrum of a transient disturbance that reaches sensitive parts. In the knowledge of disturbance spectrum requirements and the assumed disturbance, we may determine the required frequency function of the transmission line, including the noise filter. And having determined the required frequency function of the noise filter, we can design the filter elements and configuration using methods that are commonly known in electrical and telecommunication technology.

TABLE 12.3   Energy Content, RMS Frequency Duration, And RMS Time Duration of Impulse-Like Disturbances

	Energy E	RMS Frequency Duration Ω	RMS Time Duration T
Fig. 12.2a	$\dfrac{A_r^2 \omega_r}{2}$	$\sqrt{\omega_r \omega_1}$	$\dfrac{1}{\sqrt{2}\,\omega_r}$
Fig. 12.2b	$\dfrac{A_r^2 \omega_r}{2}$	$\sqrt{\omega_r \omega_1}$	$\dfrac{1}{\sqrt{2}\,\omega_r}$
Fig. 12.2c	$\dfrac{A_r^2 \omega_r}{2}\left[1 + \dfrac{\omega_2}{\omega_m}\right]$	$\sqrt{\omega_r \omega_1} \times \sqrt{\dfrac{\omega_2}{\omega_2 + \omega_m}}$	$\dfrac{1}{\sqrt{2}\,\omega_r} \times \sqrt{\dfrac{\omega_m}{\omega_2 + \omega_m}}$
Fig. 12.2d	$\dfrac{A_r^2 \omega_r}{4\zeta}$	ω_r	$\dfrac{1}{\sqrt{2}\,\omega_r} \times \dfrac{1}{\zeta}$
Fig. 12.2e	$\dfrac{A_r^2 \omega_r}{2}\left[1 + \dfrac{\omega_r}{2\zeta\omega_{rc}}\right]$	$\omega_{rc}\sqrt{\dfrac{\omega_r}{\omega_r + 2\zeta\omega_{rc}}}$	$\dfrac{1}{\sqrt{2}\,\omega_r} \times \sqrt{\dfrac{2\zeta\omega_{rc}}{\omega_r + 2\zeta\omega_{rc}}}$
Fig. 12.2f	$\dfrac{A_r^2 \omega_r}{2}\left[1 + \dfrac{\omega_{rc}}{2\zeta\omega_m}\right]$	$\omega_{rc}\sqrt{\dfrac{\omega_{rc}}{\omega_{rc} + 2\zeta\omega_m}}$	$\dfrac{1}{\sqrt{2}\,\omega_r} \times \sqrt{\dfrac{2\zeta\omega_m}{\omega_{rc} + 2\zeta\omega_m}}$

where $\omega_m = \dfrac{\omega_o^2}{\omega_r}$

In EMS performance analysis, the task is often to study the relationship between a particular circuit element and characteristics of the disturbance. This can be the case when the exact value of a circuit element is not known, only the order of magnitude may be estimated, and the task is to check whether a malfunction can be expected in a case of some assumed transient disturbance. A similar problem arises when a very simple noise filter configuration is used, given that some element (usually serial impedance) is considered as an element of the noise filter (see Section 7.5 on capacitor filters). In this case, the noise filter element should be designed so that the value of the chosen impulse characteristic does not exceed a given limit for assumed transient disturbance. These problems can be solved by means of the root calculation method described in Section 12.2.1 as well as the impulse characteristic calculation formulas given in Section 12.3.1. Using these two calculation methods, explicit expressions can be set up based on the value of a circuit element and the most important impulse characteristic limit. Although both methods are approximations, the resulting expressions are usually accurate enough for engineering practice and often provide very useful information for optimizing filter performance.

Because of the wide variety of EMI problems encountered, a universal method to link impulse characteristics to noise filter design cannot be devised. But to demonstrate the proper design steps, let us solve three typical noise filtering problems. These were chosen because they clearly illustrate the use of the root calculation method and impulse characteristic calculation formulas for EMS analysis.

Example 1: Designing Noise Filters for Limited Disturbance Energy

First we will return to the problem analyzed in Section 12.2.2 and illustrated in Fig. 12.3. Let the task be to determine whether the disturbance energy passing into the electronics can be limited to 10^{-7} Ws by increasing the serial resistance of R_n. For operational reasons, this resistance may not be higher than 1,000 Ω. (By revisiting this previously discussed problem, we can illustrate the use of energy calculation formulas without the need to plot the spectrum of the disturbing signal.)

In Eq. (12.42), which gives the Laplace transform of the frequency function of the transmission network, the variable resistance R_n can be represented by a variable, K, as follows:

$$R_n = K \times 1\Omega \qquad\qquad K = 1 \ldots 1000 \tag{12.89}$$

Thus, the Laplace transform of the frequency function with allowable neglections yields:

$$Tr(s) = \frac{s \times 10^{-6}}{1 + s(1+K) \times 10^{-6} + s^2(K+25) \times 10^{-12} + s^3 2.5 \times 10^{-18}} \tag{12.90}$$

To calculate the energy content of disturbance reaching the resistor R_e, the nature of the spectrum, and some of its major parameters, must be known. To obtain these data, we first calculate the roots of the Laplace transform. The nature of the roots can be derived from the determinant of Eq. (12.4):

$$D_1 = (K+1)^2 \times 10^{-12} - (4K+100) \times 10^{-12} \qquad D_1 > 0 \text{ if } K > 10$$

$$D_2 = (K+25)^2 \times 10^{-24} - 4 \times (K+1) \times 10^{-23} \qquad D_2 > 0 \tag{12.91}$$

For $1 < K < 10$, the frequency function has one complex pole pair and one real pole, and for $K > 10$, it has three real poles. This means that the nature of the spectrum of the disturbing signal depends on the variable K, since the spectrum of the excitation signal (shown in Fig. 12.3a) is independent of the variable.

Since the coupled disturbance energy was much higher than the acceptable value, let us begin the calculations in the variable range of $K > 10$. In this variable range, the Laplace transform of the frequency function has only real poles; therefore, they can be calculated by Eq. (12.5). With some degree of approximation:

$$T_3 = \frac{2.5}{K + 25} \times 10^{-6}$$

$$T_2 = \frac{K + 25}{K} \times 10^{-6}$$

$$T_1 = \frac{K^2 - 25}{K} \times 10^{-6} \tag{12.92}$$

Keep in mind that the spectrum of the excitation signal consists of a horizontal section of $A_g = 0.25$ Vs to the cutoff angular frequency $\omega_{g1} = 2 \times 10^4$ s^{-1}, then continues with a -20 dB/decade slope, and then from the cutoff angular frequency $\omega_{g2} = 2 \times 10^6$ s^{-1} runs with a -40 dB/decade slope.

The frequency function has a rising initial spectrum section (see the s-term in the numerator), so the value of the horizontal spectrum section will be determined by the lowest cutoff angular frequency. As seen in Eq. (12.92), for $K > 10$, the cutoff angular frequency $\omega_d = 1/T_1$ will be the lowest among the time constants of the frequency function, and it increases with the variable K. First, calculate the variable value at which the lowest cutoff angular frequency of the excitation signal and the frequency function will be equal. Solving $\omega_{g1} = \omega_1$ gives $K = 50$; i.e., for $R_n > 50$ Ω. The value of cutoff angular frequency ω_d will be not constant, and so the horizontal spectrum section will be dependent on variable K. Between the two ranges, first let us examine this case. For this resistance value, the first cutoff angular frequency is as follows:

$$\omega_d = \omega_1 = \frac{10^{-6}}{K} \tag{12.93}$$

The value of the horizontal spectrum section of the disturbing signal can be calculated with the approximation that, at the lowest cutoff angular frequency, the value of the denominator is equal to 1. Thus:

$$A_r = A_g \times Y(\omega_d) = 0.25 \times \frac{10^{-6}}{K} \times 10^{-6} = \frac{0.25}{K} \quad \text{Vs} \quad (12.94)$$

In the examined variable range, the Laplace transform has only a real root. Thus, the spectrum of the output signal will be as in Fig. 12.2a. For using the energy calculation formula, we still must determine the cutoff angular frequency ω_r. The next lowest cutoff angular frequency, following ω_d, is ω_{g1}. Substituting Eq. (12.94) and $\omega_r = \omega_{g1}$ in Eq. (12.25), the disturbing energy becomes:

$$E_e = \left(\frac{2.5}{K}\right)^2 \quad V^2 s \quad (12.95)$$

This expression also shows the relationship between the resistance R_n and the disturbing energy, since the energy appearing across R_n can be calculated by multiplying with a constant R_n. For the task of reducing this energy below the limit of 10^{-7} Ws, the acceptable energy content can be calculated based on Eq. (12.22):

$$E_{max} = E_e \times R_e = 10^{-7} \times 2000 = 2 \times 10^4 \quad V^2 s \quad (12.96)$$

The required value of resistor R_n can be calculated by solving the equation $E_{max} = E_e$. The result is:

$$K \geq 1.75 \times 10^3 \quad (12.97)$$

The energy content of the transient disturbance appearing across resistor R_e will be less than 10^{-7} Ws if serial resistance R_n is higher than 1.75 kΩ. Before stating that the problem cannot be solved by increasing the serial resistance, the peak value of the disturbance coupled into the electronic, should be also checked. The peak value can be calculated using a formula taken from Table 12.2:

$$\hat{U}_e = \frac{0.25}{K} \times 2 \times 10^4 = \frac{5000}{K} \quad V \quad (12.98)$$

Assuming that the sensitive parts in the electronic unit are capable of withstanding a voltage peak of 20 V, the minimum required value of the serial resistance is 250 Ω. In the case of a serial resistance lower than 250 Ω, no EMS problem can be expected, but when the serial resistance is between 250 and 1,750 Ω, EMS problems can occur because of the high energy. Because the serial resistance can be a maximum of 1,000 Ω, the problem should be analyzed further by decreasing the energy content of the disturbance. One possible solution is to limit the peak value of the disturbance with some type of transient protector.

Example 2: Designing Noise Filters for Limited RMS Frequency Duration

In this example, the HF equivalent of the transmission network between the transient disturbance source and the sensitive parts in the electronic unit will be as shown in Fig. 12.14. Assume that the electronic unit receives analog signal. The task is to compute the capacitance, C, which serves to suppress HF noise, so that the operational signal of approximately 100 Hz can pass through the transmission network without attenuation. The time function of the predicted transient disturbance is shown in the figure. After designing the capacitance value, we also need to check whether some breakdown may be expected in the electronic device if the disturbance reaches a small signal transistor.

The above requirement also may be formulated such that the rms frequency band of the output signal should be, for example, 2 kHz. In this case, the analog operational signal will be not attenuated, but the HF noise components will be highly suppressed.

The Laplace transform of the frequency function of the transmission network is:

$$Y(s) = \frac{1}{1 + s\,(RC + R_1C + R_1C_1) + s^2\,(RR_1CC_1 + L_1C + L_1C_1) + s^3\,(RL_1CC_1)} \tag{12.99}$$

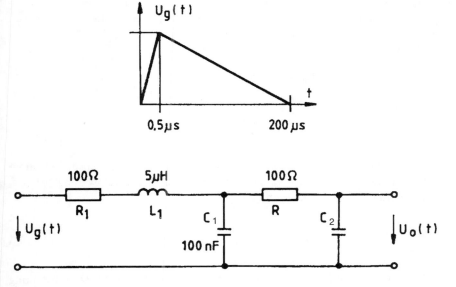

FIGURE 12.14 Noise filter design for limited rms frequency duration

Let us consider the unknown capacitance much as calculated in Eq. (12.89); that is, $C = K \times 10^{-9}$ F. Substituting this term and the element values given in the figure, the coefficients of the denominator are:

$$a_1 = (1 + 0.03 \times K) \times 10^{-5}$$

$$a_2 = (2 \times K + 0.5) \times 10^{-12}$$

$$a_3 = K \times 10^{-19} \tag{12.100}$$

For determining the nature of the roots, the determinants from Eq. (12.4) are as follows:

$$D_1 = 10^{-12} (100 - 2 \times K + 0.09 \times K^2) > 0$$

$$D_2 = 10^{-24} (4 \times K^2 - 4 \times K - 0.12 \times K^2) < 0 \tag{12.101}$$

The Laplace transform has three real poles. Since the spectrum of the excitation signal has horizontal and falling sections, the spectrum of the output signal is as shown in Fig. 12.2a. For such spectra, only the value of the lowest two cutoff angular frequencies has to be known (see Table 12.3).

The cutoff angular frequencies of the excitation signal spectrum are $\omega_{g1} = 2 \times 10^4$ s^{-1} and $\omega_{g2} = 2 \times 10^6$ s^{-1} (see the examples in Section 12.2.2 and above). It is enough to calculate the two lowest cutoff frequencies:

$$\omega_1 = \frac{10^5}{1 + 0.03 \times K}$$

$$\omega_2 = \frac{1 + 0.03 \times K}{2 \times K + 0.5} \times 10^7 \tag{12.102}$$

Among ω_1, ω_2, ω_{g1}, and ω_{g2}, the two lowest are ω_1 and ω_{g1} since, with an increase in the variable K, the value of ω_2 will not decrease below about 10^5 s^{-1}. Thus the rms frequency duration is:

$$\Omega = \sqrt{\omega_1 \times \omega_{g1}} = 10^5 \sqrt{\frac{2}{1 + 0.03 \times K}} \tag{12.103}$$

The requirement is to reduce the rms frequency duration to 2 kHz. Substituting the condition $\Omega = 10^4$ s^{-1} into the above expression, the solution of the equation is K = 6667; i.e., the capacitance of filter capacitor C should be less than 6.6 μF to suppress HF noise but avoid attenuation of the operational signal.

Let us further suppose that the output of the transmission network is connected through a serial resistor of 10 kΩ to the basis of a small signal transistor as shown in Fig. 12.15. First, the peak value of the disturbance at the output of the transmission network has to be determined. $A_r = 0.25$ Vs and

$$\omega_r = \omega_1 \, (K = 6667) = 500 \quad s^{-1} \qquad (12.104)$$

Therefore, the peak value of the impulse-like disturbance is:

$$\hat{U}_{out} = 0.25 \times 500 \ V = 125 \ V \qquad (12.105)$$

The peak value of the transient disturbance is high enough to require a study of the energy content also. The transient energy developed in the internal base-emitter diode can be calculated with fair approximation as the product of the impulse strength of the transient current and the base-emitter voltage. The spectrum of the disturbing current is equal to the spectrum of the disturbing voltage divided by the serial resistance. The impulse strength is given by the horizontal section of the spectrum (see Sections 5.3.1 and 12.3.1), thus:

$$I_i = \frac{A_g}{R_s} = \frac{0.25}{10^4} \frac{Vs}{\Omega} = 2.5 \times 10^{-5} \quad As \qquad (12.106)$$

Considering the transistor base-emitter voltage of 0.8 V, the disturbing energy is about 2×10^{-5} Ws, which is acceptable for most small signal transistors (see Table 12.1).

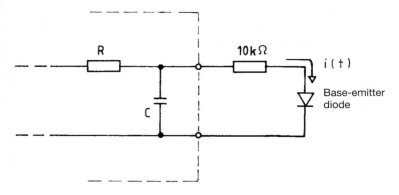

FIGURE 12.15 Equivalent circuit to calculate the disturbance energy for the problem shown in Fig. 12.14

This example also shows that, as demonstrated in Example 1, it is not sufficient to determine only the peak value or only the energy content for EMS analysis.

Example 3: Designing Noise Filters for Limited RMS Time Duration

Assume that the HF equivalent circuit between the assumed transient disturbance and the electronic unit is as shown in Fig. 12.16. The time function of the predicted disturbance is also shown in the figure. The filter capacitor, C, is for suppressing HF noise components. The capacitance of C should be designed so that the output signal cannot be longer than 1 ms in duration.

The above may also be formulated so that the rms time duration of the disturbance on the output of the transmission network is shorter than 1 ms.

The Laplace transform of the frequency function of the transmission network is as follows:

$$
Y(s) = \cfrac{\cfrac{R}{R_1 + R_2 + R}}{1 + s\left[\cfrac{R_1 R}{R_1 + R_2 + R} \times C + \cfrac{R_1 C_1}{R_1 + R_2 + R} + \cfrac{L_1}{R_1 + R_2 + R}\right]}
$$

$$
+ s^2\left[\cfrac{RL_1}{R_1 + R_2 + R} \times (C + C_1) + \cfrac{R_2 C_2}{R_1 + R_2 + R} \times (RR_1 C + L_1)\right]
$$

$$
+ s^3 \times \cfrac{R_2}{R_1 + R_2 + R} \times RC \times L_1 C_1
$$

$$(12.107)$$

In the above Laplace transform for the capacitor to be designed we substitute $C = K \times 10^{-7}$ F ($C = K \times 100$ nF) and the circuit element values given in the figure, and the coefficients of the denominator are:

$$a_1 = (2 + 0.2 \times K) \times 10^{-4}$$

$$a_2 = 1.2 \times K \times 10^{-9}$$

$$a_3 = 3 \times K \times 10^{-14}$$

$$(12.108)$$

To determine the nature of the poles, the determinants have to be calculated using Eq. (12.4). The determinants yield:

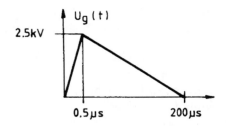

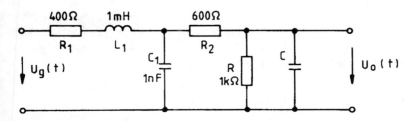

FIGURE 12.16 Noise filter design for limited rms time duration

$$D_1 = 4 \times (1 - 0.3 \times K + 0.01 \times K^2) \times 10^{-8} > 0$$

$$D_2 = (-0.9 \times K^2 - 24 \times K) \times 10^{-18} < 0 \qquad (12.109)$$

The frequency function of the transmission network has one real and one complex pole pair, independent of the variable K; thus, the roots can be calculated by Eq. (12.7):

$$T_{rc} = \sqrt{\frac{1.5 \times K}{1 + 0.1 \times K}} \qquad\qquad 2\zeta = \frac{0.6}{\sqrt{0.15 + \dfrac{1.5}{K}}}$$

$$T_1 = 2 \times (1 + 0.1 \times K) \times 10^{-4} \qquad (12.110)$$

For K > 1, the cutoff angular frequencies are in order from the lowest to the highest as $1/T_1$, ω_{g1}, the peak in frequency function at $1/T_{rc}$ and ω_{g2}. Since the peak are very low for K > 1, the spectrum of the output signal is more akin to Fig. 12.2a than Fig. 12.2e, the rms time duration is (see Table 12.3):

$$T_{out} = \frac{1}{\sqrt{2} \times \omega_r} = \sqrt{2} \times (1 + 0.1 \times K) \times 10^{-4} \qquad (12.111)$$

The acceptable rms time duration is set to 1 ms. Thus, the above equation yields that K = 60; i.e., the capacitance of the filter capacitor might not be higher than 6 μF.

Finally, let us determine the peak value of the output signal in the case of the assumed disturbance. To calculate this, the horizontal spectrum section and the first cutoff angular frequency have to be known, as seen from Table 12.2. Since $A_r = 0.125$ Vs, and for C = 6 μF the first cutoff angular frequency is $\omega_r = \omega_1 = 700 \text{ s}^{-1}$, the peak value is about 100 V. This is quite high, so the energy content also should be examined. Consider the electronic unit as a resistor of 10 kΩ connected in parallel to capacitor C. This parallel resistor has only a negligible effect on the frequency function; thus the disturbance energy on this resistor, calculated by Eqs. (12.25) and (12.22), is about 5×10^{-4} Ws. This energy is rather high, and the sensitive circuit element could break down. As a possible solution, we can:

1. use a parallel resistor on the output of the circuit shown in Fig. 12.16, smaller than the impedance of the electronic circuit
2. reevaluate the situation using the same method but a greater inductance
3. reevaluate the situation using the same method but a 10 μF capacitor rather than a 1 nF capacitor

12.3.3 Approximate Plotting of the Time Function of Impulse-Like Transient Disturbances

For a detailed analysis of electronic system malfunctions, we often must know something about the time function of the transient disturbance. Although the exact time function can be determined easily by computer analysis, such exactitude usually is not essential to study EMS performance. In engineering practice, a knowledge of the approximate time function normally is sufficient for predicting whether a malfunction is likely to appear.

To generate an approximate plot of the time function of an impulse-like signal, only the spectrum of the transient signal has to be known. The plotting methodology was set up on the assumption that the impulse characteristics of the exact and the approximation time function should correspond.

First, calculate the peak value, the maximum slope, and the rms time duration using the formulas of Tables 12.2 and 12.3. Note also that the approximation is only valid if the time function is impulsive in nature. This is the case not only when the excitation signal is a transient but also when the highest time constant of the transmission network is much greater than the time duration of the excitation signal. The plotting rules for various spectra are as follows:

For Spectra as Shown in Figs. 12.2a and 12.2b

In such cases, the approximation time function has a triangular shape as shown in Fig. 12.17. The rising section has a slope of one-half the value of the calculated maximum slope. This straight line extends to the calculated peak value. The base

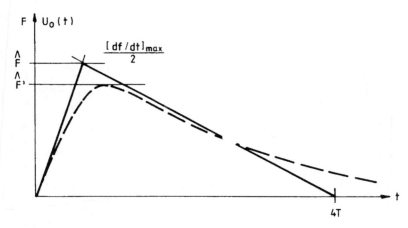

FIGURE 12.17   Plotting the approximate time function of spectra as in Figs. 12.2a and 12.2b

point of the approximation triangle is at four times the rms time duration. For a more exact approximation, the peak value may be corrected by cutting off the peak of the triangle at the corrected peak value. It should be understood that the exact time function of spectra with a rising initial section changes the sign (i.e., goes to negative—see the impulse strength of 0 in Table 12.2). However, this negative amplitude, compared to the positive one, is low enough for EMS analysis.

For Spectra as in Fig. 12.2c

In this case, the exact time function can be plotted as the envelope of two tri-angles, placed on each other. The one with lower but longer peak value should be plotted as if were any second horizontal spectrum section ($\omega_1 = \infty$). The peak value of this longer triangle is U_0, and the rms time duration is T_0. The second trian-gle, with high peak value and short duration, is to be plotted as if there were any horizontal first spectrum section (i.e., the impulse characteristics should be calcu-lated with the substitutions of $A_r = a_0(\omega)$ and $\omega_r = \omega_2$. The peak value of this tri-angle is U and the time duration is T. The plotting of the approximation time function is shown in Fig. 12.18.

For Spectra as in Fig. 12.2d

Here, the approximation time function must be plotted by neglecting the initial rising section (if there is one) as well as the other spectrum section following the one of –40 dB/decade slope. The time function of a spectrum with peak and –40 dB/decade slope (oscillating two-member system) is known in electrical tech-nology: the peak value can be calculated by Table 12.2, the frequency of the os-cillation can be determined from ω_r, and the envelope of the damped oscillation is related to ζ.

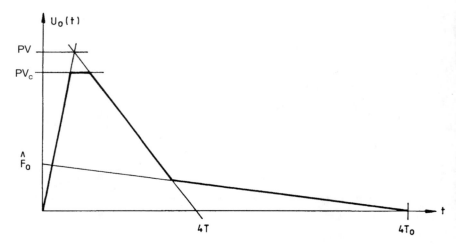

FIGURE 12.18   Plotting the approximate time function of spectra as in Fig.12.2c

For spectra as in Fig. 12.2e

This plot is shown in Fig. 12.19. The approximation time function consists of two parts. The first is a triangle plotted by ignoring the peak (i.e., as described for spectra as in Fig. 12.2a). The second part is an oscillating one. To plot it, first calculate the peak value on the basis of Table 12.2, then mark it on the vertical axis. The envelope of the damped oscillation should be plotted if its initial amplitude would be the difference between the actual peak value (F_0) and that of the triangle (F). The oscillation frequency is determined by ω_{rc}, and the damping by ζ.

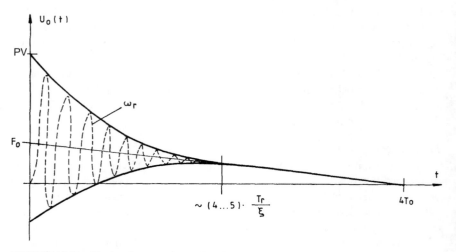

FIGURE 12.19   Plotting the approximate time function of spectra as in Fig. 12.2e

For Spectra as in Fig. 12.2f

In this case, general-use plotting rules could not be devised. However, for cutoff frequency ratios higher than one order of magnitude, the first horizontal spectrum section should not be considered, and the approximation time function may be plotted given that the spectrum would consist of only the horizontal section of the value $A_r = a_0(\omega)$ and the peak at ω_{rc}. The explanation for this is that the exact time function, just as for spectra shown in Figs. 12.2c and 12.2f, may be approximated as the envelope of two signal parts, but the triangular part, plotted by the first horizontal spectrum section, is small enough ignore for EMS analysis.

To illustrate the use of the plotting rules, let us determine the approximation time function of the output signals of the transmission network discussed in Section 12.3.2, Example 3, and Fig. 12.16. The spectrum of the output signal is like that of Fig. 12.2a; thus, for plotting the approximation triangle, the peak value has to be calculated. Since $A_r = 0.125$ Vs and $\omega_r = 700$ s^{-1}, the peak value on the basis of Table 12.2 is as follows:

$$U_m = 0.125 \times 700 \text{ V} = 90 \text{ V} \tag{12.112}$$

The other cutoff angular frequency, again neglecting the very low peak, can be calculated using Eq. (12.110). For $K = 60$, the second cutoff angular frequency is $\omega_2 = 2,700$ s^{-1}. Thus, the maximum slope on the basis of Table 12.2 is:

$$\left[\frac{du}{dt}\right]_{max} = 90 \times 2700 \text{ V/s} = 230 \text{ V/ms} \tag{12.113}$$

The rms time duration is 1 ms. The approximate time function for the excitation signal shown in Fig. 12.16 is plotted in Fig. 12.20. The plotted approximation time function is a good predictor of whether the circuit protection will be activated.

The above-described approximation rules can be used not only for EMS performance analysis but also for noise filter design, since these rules create a close connection between the most important parameters of the spectrum and the time function. Using the root calculation formulas, relationships can be set up between certain circuit element and time function characteristics that must be limited for solving EMC problems. The design steps are similar to those discussed in Section 12.3.2.

12.4 SURGE PROTECTION DEVICES

Electrical equipments are usually protected from high-amplitude, short-duration transients by surge suppressors. These devices have a nonlinear voltage-current characteristic. The impedance of the surge suppression element is very high at normal voltages but becomes very low at high voltages. Surge suppressors are

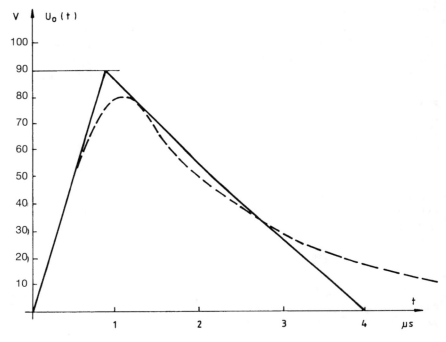

FIGURE 12.20 Approximate disturbance time function on the output of the transmission network shown in Fig. 12.16

connected between the terminal of the circuit to be protected and earth, using the shortest possible wire and a large cross-section to provide a low dc and HF impedance.

The basic requirement for a surge protection device is to keep high-amplitude transient voltages and currents away from the equipment to be protected. This is fulfilled if the voltage-current characteristic of the protector bends abruptly at the desired voltage, and there is but a short time delay after the protector reacts to a transient with a fast rise time. The other important characteristic of a surge suppressor is its ability to absorb high transient energy levels.

For absorbing transient surges generated by switching operations or lightning, gas-filled absorbers are widely used. The gas-filled units are special spark gaps in a chamber filled with a noble gas. As the voltage on the electrodes increases to a particular value, a spark-over occurs. The details of the breakdown characteristics depend on the electrode material, type of gas, gas pressure, electrode spacing, and the rate of the applied voltage. The arc voltage is about 10 to 40 V and is determined by the discharge current. When the voltage drops to a particular level, the arcing will stop and there will be some delay before the voltage again builds up to a level sufficient to generate another arc. This characteristic should be considered for ac circuits as well, as the arc may stop automatically at every current turn.

Gas absorbers are fabricated for a wide range of nominal voltages. Three-element gas absorber units are also available. In this type of device, a third electrode is placed between the other two. The central one has a small hole and is kept at ground potential. When one side of the chamber is broken down, the charged carriers are driven through the hold into the other chamber to hasten its breakdown. This type of gas absorber is particularly useful in protecting balanced two-wire circuits.

The impulsive spark-over voltage is considerably higher than that of the static spark-over. The time delay and spark-over characteristics of a typical 350 V gas absorber are shown in Fig. 12.21. The figure clearly shows that the spark-over voltage also changes in a relatively wide range for static voltage, and the range will grow wider as the rise time of the surge voltage increases. This range can be explained by unavoidable differences in manufacturing processes. Figure 12.21 also shows that the time delay for a small rise time is great, and it is shorter only for higher rise times.

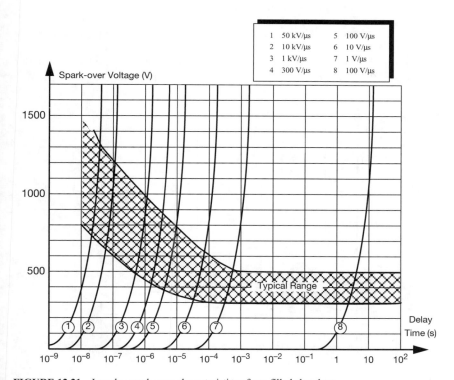

FIGURE 12.21 Impulse spark-over characteristics of gas-filled absorbers

Relatively small gas-filled protectors can absorb high currents and display very small (typically, a few picofarads) capacitance between terminals. As a result, they are appropriate for high-frequency applications. In dc equipment, a special solution may be required to prevent the clamped voltage from dropping below the operating voltage, to limit the follow-on current, and to ensure that the arc will be extinguished in power frequency circuits.

for applications where a more precise surge protection level is needed, devices with nonlinear characteristics are used. These elements are called *varistors*. The voltage-current characteristic of a varistor is similar to that of the well known zener (Z) diode and can be approximated by the following equation:

$$I = C \times U^{\alpha} \tag{12.114}$$

where

C = an experimental constant

I = current across the varistor

U = voltage on the varistor

α = a constant characterizing the nonlinearity of the element

Figure 12.22 shows the typical exponent of some standard varistors. For better comparison, the characteristics are referred to 200 V and 1 mA.

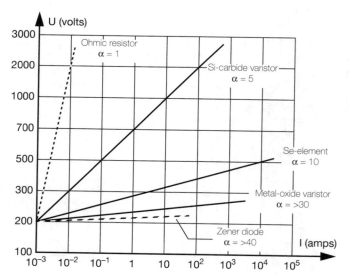

FIGURE 12.22 Typical nonlinearity of varistors

The varistor's response for a high rise time surge voltage is very short—typically about 10 ns—and is not dependent on the rise time of the transient. The response characteristics are primarily influenced by the varistor capacitance and the lead inductance. In contrast to the gas absorber, varistors give more precise surge protection, have a greater life expectancy, and allow no follow-on current after the transient surge. Varistors have relatively high capacitance—two to three orders higher than gas absorbers. When using varistors to protect power lines, this capacitance is between phase and earth. In some applications the leakage current should also be considered (see section 7.5.1).

For decreasing the leakage current, small gaps are connected in series with varistors. The gaps decrease the leakage current for 127 and 220 V power frequency voltages up to some 10 µA, but the time delay is increased. Using ZnO varistors in these surge protection devices, the exponent a in Eq. (12.114) is in the range of 40 to 80; i.e., the nonlinearity of these protection elements is very good.

For protection of sensitive electronic units, a faster surge protection element, the so-called *suppression diode,* has been developed. It has a typical time delay of some 10 picoseconds, but their heat capacity is lower than that of varistors.

Surge protection that combines very low time delays, accurate operation, and relatively high transient energy capacity can be provided only by hybrid circuitry. These devices are composed of a variety of surge transient absorbers and additional circuitry, usually in the form of lowpass filters or serial impedances designed for the particular application.

A solution for hybrid protection against a surge voltage is shown in Figure 12.23. The gas tube is used for clearing the bulk of the high-energy transient (especially the common-mode components, with lightning-induced surge). The arc

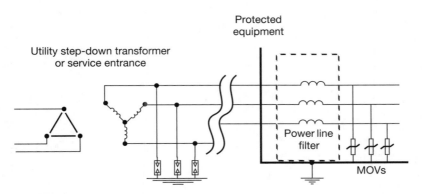

FIGURE 12.23 Hybrid surge protection for power entry (for three-phase input)

extinguishes itself at the ac voltage zero-crossing. For a low-energy pulse with short duration (such as the residue of the gas tube action), varistors are installed after the equipment or powerline filter, whose series inductors limit the varistor sink current.

Bibliography

BOOKS

[1]   *A Handbook Series on Electromagnetic Interference and Compatibility.* Vol. 1–6. ICT, Gainesville, Virginia, 1971.

[2]   *Electrical Filters.* ICT, Gainesville, Virginia, 1970.

[3]   *Electromagnetic Shielding Materials and Performance.* ICT, Gainesville, Virginia, 1975.

[4]   Ficchi, R. F. *Practical Design for Electromagnetic Compatibility.* Hayden Book Co. Inc., New York, 1975.

[5]   Kaufman, M.; Seidman, A. H. *Handbook of Electronic Calculations.* McGraw-Hill Book Co., New York, 1979.

[6]   Ott, H. W. *Noise Reduction Techniques in Electronic Systems.* John Wiley and Sons, New York, 1976.

[7]   Stoll, D. *Elektromagnetische Vertraglichkeit.* Elitera, Berlin, 1976.

[8]   Tornau, F. *VEM Handbuch: Elektrische Stoerbeeinflussung in Automatisierungs und Datenverarbeitungsanlagen.* VEB Verlag Technik, Berlin, 1972.

[9]   Gal, T. *Az elektromagneses zavarvedelem nehany problemaja.* (Dissertation in Hungarian language: Some problems of EMC), BME Villamosmernoki Kar, Automatizalasi Tanszek (Technical University of Budapest, Faculty of Electrical Engineering), 1977.

[10] Siemens Sonderheft *Funk-Entoerung.* 1975, Bestell-Nr. B 26/1323.

ARTICLES/PAPERS

[11] (1983) "Arbeitsblatt Nr. 166: Gesetze und Normen zum Themen EDV." *Elektronik,* No. 10. 1983, pp. 101–106.

[12]   (1970) "Bibliography on Surge Voltages in AC Power Circuits Rated 600 Volts and Less." *IEEE Transactions on Power Apparatus and Systems,* Vol. PAS-89, No. 6. Jul/Aug. 1970, pp. 1056–1061.

[13]   (1974) "Circuit Consideration in Using Wire and Cable." *Electronic Packaging and Production,* Oct. 1974, pp. w20–w26.

[14]   (1980) "EMI Filters Guard Electronic Circuits." *JEE.* Vol. 17, No. 168. Dec. 1980, pp. 64–66

[15]   (1980) "Filter Typen zur Stoerunterdruckung." *Electronik Entwicklung,* No. 11. 1980, pp. 78–89.

[16]   (1976) "Focus on Network Analyzers." *Electonic Design,* Vol. 24, No. 16. Aug. 1976, pp. 50–55.

[17]   (1979) "Noise Elemination Becomes Essential for Electrinics." *JEE,* Vol. 16, No. 149. May 1979, pp. 66–69.

[18]   (1972) "Ringkern-Funk-Enstoerdrosseln fuer Thyristorgerate." *Siemens-Bauteile-Informationen,* Vol. 10, No. 2. 1972, pp. 25–28.

[19]   (1983) "VDE-Vorschriften und DIN-Normen zur Themen EDV." *Elektronik,* No. 10. 1983, p. 84.

[20]   (1983) "VG-Normen fuer MC." *Elektronik, No.* 10. 1983, p. 84.

[21]   Acuna, V. E. (1975) "Transfer Impedance Measurement as a Test for Eleectromagnetic Compatibility." *1975 IEEE International EMC Sysmposium Record.* New York, pp. 5AIc1–c5.

[22]   Adamovics, M.; Sroka, J. (1976) "The Elaboration of Ceramic Capacitors for Radio Interference Suppression." *3rd International Wroclaw Symposium on EMC,* 1976, pp. 94–100.

[23]   Adolphs, D. (1972) "Eigenschaften und Stoersicherheit von Hysterese-Gattern." *NTZ,* No. 10. 1972, pp. 457–462.

[24]   Akima, H. (1967) "Charge Time of a Linear Detector." *IEEE Transactions on EMC,* Vol. EMC-9, No. 2. Sept. 1967, pp. 58–65.

[25]   Annanpalo, J. (1986) "Common-Mode Interference Rejection in Electrically Short Twisted Pairs." *EMC Technology,* Vol. 5, No. 5. Sept/Oct. 1986, pp. 29–35.

[26]   Appleby, T. H. (1973) "An Investigation of Integrated Circuit Destruction by Noise Pulses." *Radio and Electronic Engineer.* Vol. 43, No. 4. Apr. 1973, pp. 279–287.

[27]   Aprille, T. J.; Trick, T. N. (1972) "Steady-State Analyze of Nonlinear Circuits with Periodic Inputs." *Proc. IEE,* Vol. 62. Jan. 1972, pp. 108–114.

[28]   Ari, N. (1985) "Computerprogramsystem zur Auslegung der Schirmung gegen elektromagnetische Stoerfelder. *N. ETZ-B,* Vol. 106, No. 9. 1985, pp. 440–443.

[29]   Audone, B. (1973) "Graphical Harmonic Analyses." *IEEE Transactions on EMC,* Vol. EMC-15, No. 2. May 1973, pp. 72–74.

[30]   Audone, B.; Bolla, L. (1976) "Characterisation of Transient Interference Signals." *3rd International Wroclaw Symposium on EMC,* 1976, pp. 34–39.

[31]   Audone, B.; Bolla, L. (1978) "Insertion Loss of Mismatched EMI Suppressors." *IEEE Transactions on EMC,* Vol. EMC-20, No. 3. Aug. 1978, pp. 384–389.

[32]   Audone, B.; Franzini-Tibaldeo, G. (1979) "Broad-Band and Narrow-Band Measurements." *IEEE Transactions on EMC,* Vol. EMC- 21, No. 1. Feb. 1979, pp. 9–12.

[33]   Autesserre, L. (1976) "La compatibilite electromagnetique et la normalisation." *RGE,* No. 3. Marc. 1976, pp. 191–195.

[34] Babcock, L,F. (1967) "Shielding Circuits from EMP." *IEEE Transactions on EMC,* Vol. EMC-9, No. 2. Sept. 1967, pp. 45–48.

[35] Babcock, L. F.; Sagasta, P. J. (1966) "Simplified Prediction of Conducted and Radiated Interference Levels for Pulses and Step Functions." *IEEE Transactions on EMC,* Vol. EMC-8, No. 2. Jun. 1966, pp. 97–101.

[36] Bach, W. (1986) "Gestoerter Betrieb." Teil I: Elektrotechnik, Vol. 68, No. 15. 29. Sept. 1986, pp. 35–38. *Teil II: Elektrotechnik,* Vol. 68, No. 17. 20. Oct. 1986, pp. 22–28.

[37] Baert, D. H. J. (1974) "Solution for Component Values of Radio Frequency Interference Filters for Phase-Controlled Switching Systems." *Radio and Electronic Engineer.* Vol. 44, No. 12. Dec. 1974, pp. 652–658.

[38] Baert, D. H. J. (1975) "The RFI Suppression of Phase Controlled Systems with Nonlinear Filters." *IEEE Transactions on EMC,* Vol. EMC-17, No. 3. Aug. 1975, pp. 132–139.

[39] Baumann, M. (1984) "Pruefgeneratoren zur Simulation von breitbandigen Stoergroessen." *Bulletin ASE,* Vol. 75, No. 7. 1984, pp. 374–379.

[40] Bennison, E.; Ghazi, A. J.; Ferland, P. (1973) "Lighting Surges in Open Wire, Coaxial and Paired Cables." *IEEE Transactions on Communications,* Vol. COM-21, No. 10. Oct. 1973, pp. 1136–1143.

[41] Bernstein, S. C. (1969) "Insertion Loss Measurements in a 5-ohm System." 1969 *IEEE International EMC Symposium Record.* New York: IEEE, pp. 111–115.

[42] Bertram, R. (1989) "Aktives Filter schuetz gegen Stoerimpulse." *Elektronik,* No. 14. 7. Jul. 1989, pp. 140–141.

[43] Besson, R. (1988) "La compatibilite electro-magnetique: Concepts et composants." *TLE,* No. 531. 1988, pp. 20–25.

[44] Bienek K. H. P. (1975) "Elektrische Stoerbeeinflussung und ihre Beseitigung in elektrischen Geraten und Anlagen." *Funk Technik,* No. 12. 1975, pp. 360–362.

[45] Bloom, S. D.; Massey, R. P.; Zweig, N. L. (1977) "Design and Testing for Improved Radiated Susceptibility of DC–DC Converters." 1977 IEEE International EMC Symposium Record. New York: IEEE, pp. 433–437.

[46] Bodle, D. W.; Gresh, P. A. (1961) "Lighting Surges in Paired Telephone Cable Facilities." *The Bell System Technical Journal,* March. 1961, pp. 547–576.

[47] Boean, V. (1973) "Designing Logic Cuircuits for High Noise Immunity." *IEEE Spectrum,* Vol. 10. 1973, pp. 53–59.

[48] Boll, R.; Bretthauer, K. (1970) "Magnetische Bauelemente mit Bandringkern aus weichmagnetischen Legierungen." *Siemens Zeitschrift,* Vol. 44, No. 3. 1970, pp. 142–149.

[49] Bond, N. (1971) "RFI: One the Undesirables." *Electronic Engineering,* Aug. 1971, pp. 32–45.

[50] Boyd, R. E.; Malack, J. A.; Rosenberger, I. E. (1977) "Technical Consideration for Establishing Narrow-Band EMI Requirements for Dataprocessing Equipment and Office Machines." *IEEE Transactions on EMC,* Vol. EMC-19, No. 1. Feb. 1977, pp. 22–29.

[51] Bridges, J. E. (1975) "Determination of Filter Performance for Any Arbitrary Source and Load Impedance Based on Experimental Measurements." *1975 IEEE International EMC Symposium Record.* New York: IEEE, pp. 5AIIc1–c5.

[52] Bridges, E. (1962) "A Low Frequency Filter Inherently Free from Spurious Respons-
      es." VIII. *Tri-Service EMC Conference Record,* 1962, pp. 564–589.
[53] Brooks, J. (1965) "A Low Frequency Current Probe System for Making Conducted
      Noise Power Measurements." *IEEE Transactions on EMC,* Vol. EMC-7, No. 2. Jun.
      1965, p. 207.
[54] Broyde, F. (1988) "Evolution des normes et des essais de susceptibilite." *TLE,* No.
      539. 1988, pp. 41–43.
[55] Budzilowich, P. N. (1969) "Electrical Noise: Its Nature, Causes, Solution." *Control
      Engineering,* May 1969, pp. 74–78.
[56] Buerskens, H. (1970) "Funktions- und Stoersicherheit elektromagnetischer
      Steuerungs- und Regelungseinrichtungen." *VDE Fachberichte* 1970, pp. 112–115.
[57] Bull, J. H. (1972) "Electrical Interference on Mains Supplies." *Electrical Review,*
      No. 1. Dec. 1972, pp. 745–748.
[58] Burruano, S. J. (1965) "Curing Interference in Relay Systems—Look to the Source,
      then Suppress." *Electronic Design,* 29. Nov. 1965. (Special Issue on Relay Applica-
      tion), pp. 37–43.
[59] Canavero, F. G.; Pignari, S.; Daniele. V. (1990) "A Parametric Study of EMI Cou-
      pling with Electronic Subsystems." *10th International Wroclaw Symposium on
      EMC,* 1990, pp. 286–290.
[60] Caprio, S. J. (1980) "Estimating the Effect of Filtering on Pulse Power." *IEEE
      Transactions on EMC,* Vol. EMC-22, No. 4. Nov. 1980, pp. 266–268.
[61] Caprio, S. J. (1980) "Comparison of the Response of a Heterodyne Receiver to Vid-
      eo-Pulse and Impulse-Type Signals." *IEEE Transactions on EMC,* Vol. EMC-22,
      No. 1. Feb. 1980, pp. 68–72.
[62] Carsten, B. (1986) "High Frequency Conductor Losses in Switch-Mode Magnetics."
      PCIM. Nov. 1986, pp. 34–46.
[63] Casey, K. F.; Vance, E. F. (1978) "EMP Coupling Through Cable Shields." *IEEE
      Transactions on EMC,* Vol. EMC-20, No. 1. Feb. 1978, pp. 100–106.
[64] Chambers, J. (1986) "Surface Mounting Feed-Through Suppression Filter Perform-
      ance." *EMC Technology,* Vol. 5, No. 3. May/Jun. 1986, pp. 21–30.
[65] Chan, M. H. L.; Donaldson, R. W. (1989) "Amplitude, Widht, and Interarrival Dis-
      tribution for Noise Impulses on Intrabuilding Power Line Communication Net-
      works." *IEEE Transactions on EMC,* Vol. EMC-31, No. 3. Aug. 1989, pp. 320–323.
[66] Chen, C. L. (1975) "Transient Protection Devices." *1975 IEEE International EMC
      Symposium Record.* New York: IEEE, pp. 3AIa1–a3.
[67] Chesworth, E. T. (1986) "Electromagnetic Interference Control in Structures and
      Shieldinds." *EMC Technology,* Vol. 5, No. 1. Jan/Feb. 1986, pp. 39–48, 54.
[68] Chowdhuri, P. (1973) "Transient-Voltage Characteristic of Silicon Power Rectifi-
      ers." *IEEE Transactions on Industrial Applications* Vol. IA-9. k, No. 5. Sept/Oct.
      1973, pp. 582–592.
[69] Clark, O. M. (1975) "Devices and Methods for EMP Transient Suppression." *1975
      IEEE International EMC Symposium Record.* New York: IEEE, pp. 3AIIb1–b6.
[70] Clark, D. B.; Benning, R. D.; Kersten, P. R.; Chaffee, D. L. (1968) "Power Filter In-
      sertion Loss in Operational-Type Circuits." *IEEE Transactions on EMC,* Vol. EMC-
      10, No. 2. Jun. 1968. (Special Filter Issue), pp. 243–255.

[71] Colon, F. R.; Trick, T. N. (1973) "Fast Periodic Steady-State Analysis for Large-Signal Electronic Circuits." *IEEE Journal on Solid-State Circuits,* Vol. SC-8. Aug. 1973, pp. 260–269.

[72] Cook, J. H. (1979) "Quasi-Peak-to-RMS Voltage Conversion." *IEEE Transactions on EMC,* Vol. EMC-21, No. 1. Feb. 1979, pp. 9–12.

[73] Cooperstein, B. (1981) "Swept-Frequency Method for EMI Filter Insertion Loss Measurements." *1981 IEEE International EMC Symposium Record.* New York: IEEE, pp. 11–15.

[74] Cordes, H. (1985) "Netzstoersimulation und Entstoertechnik zur Sicherstellung der EMV—Nicht nur ein notwendiges Uebel." *ETZ-B,* Vol. 106, No. 9. 1985, pp. 432–433.

[75] Cory, W. E. (1977) "The Importance of Impedance in Conduction Measurements." *IEEE Transactions on EMC,* Vol. 19, No. 3. Aug. 1977, pp. 153–154.

[76] Cowdell, R. B. (1976) "DC to Daylight Filter." *1976 IEEE International EMC Symposium Record.* New York: IEEE, pp. 136–141.

[77] Cowdell, R. B. (1968) "Charts Simplify Pulse Analysis." *Electronics,* 2. Sept. 1968, pp. 62–69.

[78] Cowdell, R. B. (1969) "Filter Effect Power Line Voltages." *1969 IEEE International EMC Symposium Record.* New York: IEEE, pp. 250–256.

[79] Cowdell, R. B. (1972) "Nomograms Simplfy Calculations of Magnetic Shielding Effectiveness." *Electronic Design News,* 1. Sept. 1982, p. 44.

[80] Cowdell, R. B. (1979) "Unscrambling the Mysteries About Twisted Wire." *1979 IEEE International EMC Symposium Record.* New York: IEEE, pp. 180–187.

[81] Danzeisen, K.; Stecher, M. (1985) "Stoermessungen nach CISPR." *Markt und Technik,* No. 42. 18. Oct. 1985, pp. 86–91, 93.

[82] Dandens, P. L.; Watson, W.; Dillard, J. K. (1958) "Transient Recovery Voltages on Power Systems." *IEEE Transactions on Power Apparatus and Systems,* Vol. PAS-77. Aug. 1958, pp. 581–592.

[83] Deb, G. K.; Mukherjee, M.; Kumar, P. S. (1988) "Study of RF Coupling Behaivour of Digital Logic Devices." *9th International Wroclaw Symposium on EMC,* 1988, pp. 679–683.

[84] DeLoux, G. (1976) "La normalisation internacionale et les perturbations reciproques des equipments electriques et des reseaus." *RGE,* No. 4. Apr. 1976.

[85] Demetry, J. P.; Pasquina, L. N. (1969) "Acheiving Filter Effectiveness and Insuring Reliability." *1969 IEEE International EMC Symposium Record.* New York: IEEE, pp. 290–297.

[86] Denny, H. W.; Warren, W. (1968) "Lossy Transmission Line Filter." *IEEE Transactions on EMC,* Vol. EMC-10, No. 4. Dec. 1968, pp. 363–370.

[87] diMarzio, A. (1968) "Graphical Solution to Harmonic Analysis." *IEEE Transactions on Aerospace and Electronic Systems,* Vol. AES-4, No. 5. Sept. 1968, pp. 693–707.

[88] Donovan, J. C. (1971) "A Source of Transient Error Voltage in Switched Analog Signal Transmission." *IEEE Transactions on Industrial Electronics and Control Instrumentation,* Vol. IECI-18, No. 3. Aug. 1971, pp. 109–115.

[89] Donzel, M.; Massat, M. (1976) "Asymmetrical Impedance of Supplies." *3rd International Wroclaw Symposium on EMC,* 1976, pp. 7–10.

[90] DuBow, J. (1974) "Passive Component Makers." *Electronics,* May 1974, pp. 89–104.

[91]   Duncan, R. S.; Stone, H. A. (1956) A Survey of the Application of Ferrites to Inductor Design." *Proc. IRE,* Jan. 1956, pp. 3–13.

[92]   Dvorak, T. (1974) "Measurement of Electromagnetic Field Immunity." *IEEE Transactions on EMC,* Vol. EMC-16, No. 3. 1974, pp. 154–160.

[93]   Dvorak, T. (1989) "EMV-Normen im Wandel der Technologie." *Technische Rundschau,* No. 9. 1989, pp. 30–33.

[94]   Edelman, L. (1969) "Prediction of Power Supply Generated Interference." *IEEE Transactions on EMC,* Vol. EMC-11, No. 3. Aug. 1969, pp. 117–122.

[95]   Edson, W. A. (1955) "The Single Layer Solenoid RF Transformer." *Proc. IRE,* Aug. 1955, pp. 932–936.

[96]   Eisbruck, S. H.; Giordano, F. A. (1968) "A Survey of Power-Line Filter Measurement Techniques." *IEEE Transactions on EMC,* Vol. EMC-10, No. 2. Jun. 1968. (Special Filter Issue), pp. 238–242.

[97]   Ethenoz, P. (1981) "Begrenzung der Spannungssteilheit mit RC-Dampfungsglied." *Industrie-Elektrik + Elektronik,* No. 7. Marc. 1981, pp. 36–40.

[98]   Feist, K. H. (1985) "Einflussgroessen bei der Vermeidung unzulassiger elektromagnetischer Beeinflussung." *ETZ-B,* Vol. 106, No. 9. 1985, pp. 434–439.

[99]   Feser, K. (1985) "Mikroelektronik und elektromagnetische Vertraglichkeit." *ETZ-B,* Vol. 106, No. 9. 1985, pp. 422–423.

[100] Feser, K. (1982) "Entstehung transienter Stoerspannungen und -stroeme." *Schweizer Maschinenmarkt,* No. 33. 1982, pp. 48–53.

[101] Feser, K.; Lutz, M. (1980) "Pruefung elektronischer Systeme und Gerate auf elektromagnetische Vertraglichkeit." *Industrie-Elektrik + Elektronik,* No. 25, 1980, pp. 371–374.

[102] Ficchi, R. F. (1965) "Spectra Analysis—an RFI Prediction Tool." *IEEE International Conference Record,* Vol. 13, No. 2, 1965, pp. 1–13.

[103] Finger, T. A. (1970) "NST Factor Aids Design of Stable AC Power System." *Instrumentation Technology,* March 1970, pp. 53–56.

[104] Fisher, F. A.; Martzloff, F. D. (1976) "Transient Control Levels, a Proposal for Insulation Coordination in Low-Voltage Systems." *IEEE Transactions on Power Apparatus and Systems,* Vol. PAS-95, No. 1. Jan/Feb, 1976, pp. 120–129.

[105] Forti, N.; Liberatore, A.; Millanta, L. (1990) "Transient Disturbance Types and Cases in the Power Distribution Network of Civil Buildings." *10th International Wroclaw Symposium on EMC,* 1990, pp. 207–211.

[106] Fritze, F. (1976) "Vierpol-Alumunium-Elektrolytkonden-satoren." *Bauteile Report,* Vol. 14, No. 3, 1976, pp. 79–82.

[107] Gauger, E. (1988) "Measuring Methods for EMI-Testing According to CISPR 22 Recommendation." *9th International Wroclaw Symposium on EMC,* 1988, pp. 491–495.

[108] Gellissen, H. D. (1986) "Signalleitung fuer kleine Spannungsimpulse in Hochspannungsanlagen." *etzBd,* Vol. 107, No. 18/19, 1986, pp. 870–875.

[109] Georgopoulos, C. J. (1986) "EMI Control in the Installation and Interconnection of Digital Equipment." *EMC Technology,* Vol. 5, No. 2. March/Apr., 1986, pp. 55–63.

[110] Geselowitz, D. B. (1961) "Response of Ideal Radio Noise Meter to Continuous Sine Wave, Recurrent Impulses and Random Noise." *IEEE Transactions on EMC,* Vol. EMC-3, No. 1, 1961, pp. 2–11.

[111] Girndt, A. (1978) "Erfassung von netzseitigen Ueber-spannungen in den Einsatzbereich der Halbleiterbauelemente." *Der Elektro-Praktiker,* Vol. 32, No. 7, 1978, pp. 224–227.

[112] Goedbloed, J. J. (1987) "Transients in Low-Voltage Supply Networks." *IEEE Transactions on EMC,* Vol. EMC-29, No. 2, 1987, pp. 104–115.

[113] Goessl, G. (1970) "Stromkompensierte SIFERRIT-Drosseln zur Funkentstoerung." *Siemens Bauteile Informationen,* No. 8, 1970. (Sonderheft fuer Funkentstoerung), pp. 6–8.

[114] Goldberg, G. (1990) "Status of the Standarisation Work on EMC in IEC, CENELEC and Other Organisations." *10th International Wroclaw Symposium on EMC,* 1990, pp. 919–923.

[115] Goldberg, G. (1990) "Standards on Transient Disturbances in Low Voltage Power Networks." *10th International Wroclaw Symposium on EMC, 1990,* pp. 932–936.

[116] Gonshor, D. V. (1978) "Filter Pin Insertion Loss Under Nonideal Conditions." *1978 IEEE International EMC Symposium Record.* New York: IEEE, pp. 323–327.

[117] Gonshor, D. V.; Osburn, J. (1982) "Improved Graphical Interference Analysis." *1978 IEEE International EMC Symposium Record.* New York: IEEE, pp. 74–80.

[118] Greenstein, L. J.; Tobin, H. G. (1963) "Analysis of Cable-Coupled Interference." *IEEE Transactions on EMC,* Vol. EMC-5, No. 1. March 1963, pp. 43–55.

[119] Greenwood, A.; Lee, T. H. (1963) "Generalized Damping Curves and Their Use in Solving Power-Switching Transients." *IEEE Transactions on Power Apparatus and Systems,* Vol. PAS-83, No. 1. Aug, 1963, pp. 527–535.

[120] Griscom, I. B.; Sauter, D. M.; Ellis, H. M. (1958) "Transient Recovery Voltages on Power System." *IEEE Transactions on Power Apparatus and Systems,* Vol. PAS-77. Aug. 1958, pp. 592–606.

[121] Grossman, W. K.; Fischer, J. T. (1967) "Evaluating Filters in Situ Under Heavy Load Currents and Normal Working Impedances." *1967 IEEE International EMC Symposium Record.* New York:IEEE, pp. 352–364

[122] Gutmann, R. J. (1980) "Application of RF Circuit Design Principles." *IEEE Transactions on Industrial Electronic and Control Instrumentation,* Vol. IECI-27. k. Aug. 1980, pp. 156–164.

[123] Gutsche, S. (1971) "A Spectrum Prediction Technique for AM Pulses of Arbitrary Shape." *IEEE Transactions on EMC,* Vol. EMC-13. May 1971, pp. 64–69.

[124] Ha, I. W.; Yarbrough R. (1976) "A Lossy Element for EMC Filters." *IEEE Transactions on EMC,* Vol. EMC-18, No. 4. Nov. 1976, pp. 141–148.

[125] Haber, F. (1967) "Response of Quasi-Peak Detector to Periodic Impulse with Random Amplitudes." *IEEE Transactions on EMC,* Vol. EMC-9, No. 1. March 1967, pp. 1–6.

[126] Habiger, E. (1984) "Elektromagnetische Vertraglichkeit in der Automatisierungstechnik." *Messen+Steuern+Regeln,* Vol. 27, No. 6. 1984, pp. 242–246.

[127] Hagn, G. H. (1977) "Definitions of Electromagnetic Noise and Interference." *1977 IEEE International EMC Symposium Record.* New York: IEEE, pp. 122–127.

[128] Hall, J. K.; Palmer. D. S. (1976) "Electrical Noise Generated by Thyristor Control." *Proc. IEE,* Vol. 123, No. 8. Aug. 1976, pp. 781–786.

[129] Harada, K.; Ninomiya, T. (1978) "Noise Generation of a Switching Regulator." *IEEE Transactions on Aerospace and Electronic Systems,* Vol. AES-14, No. 1. Jan. 1978, pp. 178–184.

[130] Harms, R. (1986) "An EMI Generator to Improve Equipment Test Results." *EMC Technology,* Vol. 4, No. 4. Jul/Aug. 1986, pp. 31–37.

[131] Harper, G. (1971) "Progress in Broad Band Choke Design." *Comp. Conf. Proc.,* 1971, pp. 59–66.

[132] Hata, C. (1989) "EMI Filters Feature More Sophistication, Better Noise Reduction." *JEE,* Aug. 1989, pp. 45–47.

[133] Hata, T. (1989) "Matsushita EMI Filter Incorporates Ferrite Bead Core for Greater Noise Reduction." *JEE,* Jun. 1989, pp. 36–39.

[134] Heirman, P. N. (1983) "Proposed Immunity Measurement Techniques Portion of ANSI C. 63. 4." *1983 IEEE International EMC Symposium Record.* New York: IEEE, pp. 566–569.

[135] Hinz, G. (1972) "Funk-Entstoerdrossel fuer Thyristor-steuerungen." *Siemens Zeitschrift,* No. 4. 1972, pp. 312–314.

[136] Hoeft, L. O. (1986) "Using Transfer Impedance to Solve EMP Design and Analysis Problems." *EMC Technology,* Vol. 5, No. 6. Nov/Dec. 1986, pp. 27–34.

[137] Hoegberg, R.; Loetberg, E.; Scuka, V. (1986) "Blitzschutz von elektronischen Einrichtungen." *etzBd,* Vol. 107, No. 1. 1986, pp. 24–27.

[138] Hoffart, H. M. (1968) "Electromagnetic Interference Reduction Filters." *IEEE Transactions on EMC,* Vol. EMC-10, No. 2. Jun. 1968. (Special Filter Issue), pp. 225–232.

[139] Hohmann, D. B. (1977) "How to Pigeonhole EMC Problems." *2nd International Symposium on EMC,* Montreux 1977, pp. 305–310.

[140] Hohmann, W. P. (1978) "Stoerprobleme bei Schaltnetzteilen." *Electronik Industrie,* No. 5. 1978, pp. 14–17.

[141] Hohmann, W. P. (1984) "Entstehung einer neuen Netzentstoer-filter-Generation." *Elektronik Industrie,* No. 1. 1984, pp. 60–67.

[142] Holmstrom, F. R. (1986) "The Model of Conductive Interference in Rapid Transit Signalling Systems." *IEEE Transactions on Industrial Applications,* Vol. IA-22, No. 4. Jul/Aug. 1986, pp. 756–762.

[143] Holownia, J. (1976) "Conducted Interference Measurement." *3rd International Wroclaw Symposium on EMC,* 1976, pp. 55–62.

[144] Hofstra, J. S.; Dinallo, M. A.; Hoeft, L. O. (1982) "Measured Transfer Impedance of Braid and Convoluted Shield." *1982 International EMC Symposium Record.* New York: IEEE, pp. 482–488.

[145] Horn, T. (1984) "Mikroprocessoren und Stoersicherheit." *Elektronik,* No. 5. 1982, pp. 91–94.

[146] Hornsby, J. (1978) "Predicting Insertion Loss of Common-Mode Power Line Filter." *IEEE Transactions on EMC,* Vol. EMC-20, No. 3. Aug. 1978, pp. 320–335.

[147] Howell, E. K. (1979) "How Switches Produce Electrical Noise." *IEEE Tranasactions on EMC,* Vol. EMC-21, No. 3. Aug. 1979, pp. 162–170.

[148] Ida, E. S. (1962) "Reducing Electrical Interference." *Control Engineering,* Feb. 1962, pp. 107–111.

[149] Ideguchi, T.; Hattori, M.; Koga, H. (1983) "Advanced Induction Voltage Reduction Method for Telecommunication Cable by Arrester and Ferromagnetic Loading." *1983 IEEE International EMC Symposium Record.* New York: IEEE, pp. 291–295.

[150] Immesberger, B. (1982) "Netzstoerungen und Stoer-sicherheit." *Elektronik,* No. 23. 1984, pp. 118–122.

[151] Jackson, G. A. (1990) "An Overview of CISPR Work on EMC." *10th International Wroclaw Symposium on EMC,* 1990, pp. 924–926.

[152] Jambor, L. D.; Schukantz, J.; Haber, E. (1966) Parallel Wire Susceptibility Testing for Signal Lines." *IEEE Transactions on EMC,* Vol. EMC-8, No. 2. Jun. 1962, pp. 111–117.

[153] Jecko, B.; Dafif, O.; Reineix, A. (1988) "Mechanisme de l'interaction d'une impulsion electro-magnetique (I. E. M.) avec des structures filaires." *L'Onde Electrique,* Vol. 68, No. 1. 1988, pp. 98–110.

[154] Jerabek, A. (1875) "Gerate fuer Stoerspannungsunter-suchungen." *Brown Bovery Mitteilungen,* No. 3. 1975, pp. 105–109.

[155] Jobe, D. J.; Jespersen, C. P. (1969) "Selection and Test of Power Line Filters for Use in Equioment Designed to Meet Government EMC Specifications." *1969 IEEE International EMC Symposium Record.* New York: IEEE, pp. 283–289.

[156] Johnston, J. E. (1983) "The Role of Integrated Cicuits Decoupling in Electromagnetic Compatibility." *EMC Technology,* Vol. 2, No. 4. Oct./Dec. 1983, pp. 18–24.

[157] Jonnada, R. K. R. (1965) "The Effect of Filter Capacitors on Interference Characteristics of Three-Phase Power Supply Networks." *IEEE Transactions on EMC,* Vol. EMC-7, No. 3. Sept. 1965, pp. 319–326.

[158] Jursik, J. (1963) "Rejecting Common Mode Noise." *Control Engineering,* Aug. 1963, pp. 61–66.

[159] Kaiser, W. (1964) "Entstoerung elektrischer Steuerungen bei industriellem Einsatz." *Siemens Zeitschrift,* No. 38. 1964, pp. 914–917.

[160] Kaiserswerth, H. K. (1966) "Funk-Entstoerung." *Siemens-Bauteile-Informationen.* 4. sz. 1966, pp. 117–119.

[161] Kawabe, N. (1989) "High-Density Parts Mounting and Anti-EMI Measures Among Interface Connectors." *JEE,* Feb. 1989, pp. 40–43.

[162] Key, T. S. (1979) "Diagnosing Power Quality Related Computer Problems." *IEEE Transactions on Industrial Applications,* Vol. IA-15, No. 4. 1979, pp. 381–393.

[163] Kirk, W. J.; Carter, L. S.; Wadell, M. L. (1976) "Eliminate Static Damage to Circuit." *Electronic Design,* 7. March 1976.

[164] Koberger, K. (1986) "EMI-" Kobold Killer." *Elektronikschau,* No. 12. 1986, pp. 16–30.

[165] Kohling, A. (1985) "EMV-Planung fuer Krankenhausneubauten." *ETZ-B,* Vol. 106, No. 9. 1985, pp. 428–430.

[166] Kohling, A.; Steinmeyer, G. (1985) "Planung der elektromagnetischen Vertraglichkeit von Systemen." *ETZ-B,* Vol. 106, No. 9. 1985, pp. 424–426.

[167] Konno, T.; Sato, H. (1989) "TDK Introduces EMI Filter Series." *JEE,* Jun. 1989, pp. 32–34.

[168] Kowalkowski, K. (1985) "Standard Filters Fight EMI Contamination Caused by Switched-Mode Power Supplies." *EMC Technology,* Vol. 4, No. 3. Jul/Sept. 1985, pp. 59–62, 69.

[169] Kuebel, V. (1970) "Der Ableitstrom in der Funk-Entstoertechnik." *Siemens Bauteile Informationen,* No. 8. 1970. (Sonderheft Funk-Entstoerung), pp. 32–34.

[170] Kuebel, V. (1975) "Eigenschaften und Anwendung von Funk-Entstoerfiltern mit stromkompensierten Drosseln." *Bauteile Report,* Vol. 13, No. 4. 1975, pp. 108–111.

[171] Kuebler, W.; Cameroon, S. (1979) "The Definition of Frequency Dependent Rejection." *IEEE Transactions on EMC,* VOL. EMC-21, No. 4. Nov. 1979, pp. 349–350.

[172] Kujalowicz, J. (1988) "Correlation between Terminal Interference of an Artifical Mains Network and Disturbance Voltage in Mains." *9th International Wroclaw Symposium on EMC,* 1988, pp. 561–563.

[173] Kukula, L.; Zielinski, J. (1988) "Evaluation of the Atmospheric Interference Effects on Control Systems." *9th International Wroclaw Symposium on EMC,* 1988, pp. 159–162.

[174] Kunkel, G. M. (1977) "Utilization of Transformer for Meeting EMI Power Line Requirements." *1977 IEEE International EMC Symposium Record.* New York: IEEE, pp. 58–59.

[175] Lasitter, H. A. (1969) "Power Line Impedance Determination Using the '3-voltmeter' Measurement Method." *1969 IEEE International EMC Symposium Record.* New York: IEEE, pp. 128–136.

[176] Laugwitz, E. (1977) "Electromagnetic Compatibility Considerations in the Conceptual Development and Productions Phase of a Major System." *2nd International Symposium on EMC,* Montreux, 1977, pp. 301–304.

[177] Lawatch, W.; Weisshaar, E. (1972) "Ein Si-Spannungs-begrenzer zur Beschaltung von Leistungs-thyristoren." *Brown Bovery Mitteilungen,* Vol. 59, No. 9. 1972, pp. 476–482.

[178] Legel, G. W. (1975) "Acoustic Noise Emanation from EMI Power Line Filter." *IEEE Transactions on EMC,* Vol. EMC-17, No. 3. Aug. 1975, p. 193.

[179] Lucas, T. N. (1985) "Linear System Reduction by Impulse Energy Approximation." *IEEE Transactions on Automatic Control,* Vol. AC-30, No. 8. Aug. 1985, pp. 784–786.

[180] Madle, P. J. (1975) "Cable and Connector Shielding Attenuation and Transfer Impedance Measurements Using Quadraxial and Quintaxial Test Method." *1975 IEEE International EMC Symposium Record.* New York: IEEE, pp. 4BIb1–b4.

[181] Maier, C.; Sterk, T. (1985) "What's All This Noise About Switchers?." *EMC Technology,* Vol. 4, No. 3. Jul/Sept. 1985, pp. 27–35.

[182] Major, M. (1988) "Electromagnetic Interference in Low-Voltage Power Supply Networks." *9th International Wroclaw Symposium on EMC,* 1988, pp. 737–741.

[183] Malack J. A. (1978) "Statistical Correlation between Conducted Voltages on the Power Line and Those Measured with a Line-Impedance Stabilisation Network." *IEEE Transactions on EMC,* Vol. EMC-20, No. 2. May 1978, pp. 346–349.

[184] Malack, J. A.; Engstrom, J. R. (1976) "RF Impedance of US and European Power Lines." *IEEE Transactions on EMC,* Vol. EMC-18, No. 1. Sept. 1976, pp. 36–38.

[185] Malack, J. A.; Nicholson, J. R. (1973) "Effect of Measurement Devices on Conducted Interference Levels." *IEEE Transactions on EMC,* Vol. EMC-15, No. 2. May 1973, pp. 61–65.

[186] Marcus, R. B. (1967) "The Significance of Negative Frequencies in Spectrum Analysis." *IEEE Transactions on EMC,* Vol. EMC-9, No. 3. Dec. 1967, pp. 123–126.

[187] Mardiguian, M. (1982) "Transfer Impedances of Balanced Shielded Cables." *EMC Technology,* Vol. 1, No. 3. Jul. 1982, pp. 54–58.

[188] Martin, B. (1989) "Role de la C. E. M. et essais de la compatibilite electromagnetique." *TLE,* No. 542. 1989, pp. 30–35.

[189] Martin, A. R. (1982) "An Introduction to Surface Transfer Impedance." *EMC Technology,* Vol. 1, No. 3. Jul. 1982, pp. 44–52.

[190] Martin, A. R. (1983) "A New Concept for EMI Protection of Cables and Harnesses." *EMC Technology,* Vol. 2, No. 2. Apr/Jun. 1983, pp. 60–65.

[191] Martin, A. R. (1986) "Analysis and Solution to the Field-Induced EMI Problem." *EMC Technology,* Vol. 5, No. 3. May/Jun. 1986, pp. 67–75.

[192] Martin, A. R.; Emert, S. E. (1980) "Shielding Effectiveness of Long Cables." *IEEE Transactions on EMC,* Vol. EMC-22, No. 4. 1980, pp. 269–275.

[193] Martin, A. R.; Mendenhal, M. (1984) "A Fast, Accurate and, Sensitive Method for Measuring Surface Transfer Impedance." *IEEE Transactions on EMC,* Vol. EMC-26, No. 2. May 1984, pp. 66–70.

[194] Martzloff, F. D.; Hahn, G. J. (1970) "Surge Voltage in Residental and Industrial Power Circuits." *IEEE Transactions on Power Apparatus and Systems,* Vol. PAS-89, No. 6. Jul/Aug. 1970, pp. 1049–1056.

[195] Martzloff, F. D. (1983) "The Propagation and Attenuation of Surge Voltage and Surge Currents in Low-Voltage AC Circuits." *IEEE Transactions on Power Apparatus and Systems,* Vol. PAS-102, No. 5. May 1983, pp. 1163–1170.

[196] Martzloff, F. D.; Gauper, H. A. (1986) "Surge and High-Frequency Propagation in Industrial Power Lines." *IEEE Transactions on Industrial Appalications,* Vol. IA-22, No. 4. Jul/Aug. 1986, pp. 634–640.

[197] Martzloff, F. D.; Gruzs, T. M. (1988) "Power Quality Site Surveys: Facts, Fiction, and Fallacies." *IEEE Transactions on Indsutrial Applications,* Vol. IA-24, No. 6. 1988, pp. 1005–1018.

[198] Master, C. A. (1969) "Electromagnetic Interference Control Considerations of Solenoid Operated Control Devices." *IEEE Transactions on EMC,* Vol. EMC-11, No. 2. May 1968, pp. 53–57.

[199] Matejic, M. D. (1978) "Evaluation of Relay Suppression Circuits for Reducing EMI." *IEEE Transactions on EMC,* Vol. EMC-20, No. 1. Feb. 1978, pp. 207–210.

[200] Max, J. J. (1983) "Distributed Low-Pass Filters for EMI Filtering." *1983 IEEE International EMC Symposium Record.* New York: IEEE, pp. 223–228.

[201] Max, J. J.; Curtins, H.; Shah, A. V. (1983) "Cascaded Distributed Low-Pass Filters for EMI Filtering." *1983 IEEE International EMC Symposium Record.* New York: IEEE, pp. 161–166.

[202] Mayer, F. (1975) "Distributed Filters as RFI Suppression Components." *1975 IEEE International EMC Symposium Record.* New York: IEEE, pp. 5AIIe1–e3.

[203] Mayer, F. (1966) "EMC Anti-Interference Wires, Cables and Filters." *IEEE Transactions on EMC,* Vol. EMC-8. Sept. 1966, pp. 153–160.

[204] Mayer, F. (1976) "RFI Suppression Components: State of the Art; New Developments." *IEEE Transactions on EMC,* Vol. EMC-18, No. 2. May 1976, pp. 59–70.

[205] Mayer, F. (1986) "Absorptive Low-Pass Cables: State of the Art and an Outlook to the Future." *IEEE Transactions on EMC,* Vol. EMC-28, No. 1. Feb. 1986, pp. 7–17.

[206] McDonald, G. M. (1966) "Instrumentation Problems Caused by Common-Mode Interference Conversion." *IEEE Transactiuons on EMC,* Vol. EMC-8, No. 1. Marc. 1966, pp. 17–24.

[207] Medhurst, R. G. (1947) "HF Resistance and Self Capacitance of Single Layer Solenoids." Part I: *Wireless Engineer,* Vol. 24, No. 281. Feb. 1947, pp. 35–43; Part II: *Wireless Engineer,* Vol. 24, No. 282. Feb. 1947, pp. 80–92.

[208] Meissen, W. (1986) "Transiente Netzueberspannungen." *etzBd,* Vol. 107, No. 2. 1986, pp. 50–55.

[209] Meppelink, J. (1983) "Elektromagnetische Vertraglichkeit elektrischen Einrichtungen." *Elektronik,* No. 10. 1983, pp. 78–83.

[210] Merewether, D. G.; Ezell, T. F.; (1976) "The Effect of Mutual Inductance and Mutual Capacitance on the Transient Response of Braided-Shield Coaxial Cables." *IEEE Transactions on EMC,* Vol. EMC-18, No. 1. 1976, pp. 15–20.

[211] Metcalfe, R. E.; von Allmen, R. H.; Caprio, S. J. (1965) "Investigation of Spectrum Signature Instrumentation." *IEEE Transactions on EMC,* Vol. EMC-7, No. 2. Jun. 1965, pp. 218–232.

[212] Millanta, L. M.; Forti, M. M.; Maci, S. S. (1988) "A Broad-Band Network for Power-Line Disturbance Voltage Measurements." *IEEE Transactions on EMC,* Vol. EMC-30, No. 3. Aug. 1988, pp. 351–357.

[213] Millanta, L. M.; Forti, M. (1989) "A Notch-Filter Network for Wide-Band Measurements of Transient Voltages on the Power Line." *IEEE Transactions on EMC,* Vol. EMC-31, No. 3. Aug. 1989, pp. 245–253.

[214] Milton, J.; Greenwood, E. (1968) "Improving the Specification for Power-Line Filters." *IEEE Transactions on EMC,* VOL. EMC-10, No. 2. Jun. 1968. (Special Filter Issue), pp. 264–268.

[215] Mitchell, D. M. (1978) "Damped EMI Filters for Switching Regulators." *IEEE Transactions on EMC,* Vol. EMC-20, No. 3. Aug. 1978, pp. 384–389.

[216] Mitchell, W. T. (1982) "FCC and VDE Impose Tight Conducted RFI/EMI Specs." *Electronic Design,* 23. Dec. 1982, pp. 141–147.

[217] Mitsuya, Y. (1989) "Wide Bandwith Choke Coil Reduces High Frequency Noise." *JEE,* Jun. 1989, pp. 28–30.

[218] Modestino, J. W.; Jung, K. Y.; Matis, K. R. (1983) "Modeling, Analysis and Simulation of Receiver Performance in Impulse Noise." *1983 IEEE International EMC Symposium Record.* New York: IEEE, pp. 1598–1605.

[219] Modestino, J. W.: Sankur, B. (1981) "Analysis and Modeling of Impulsive Noise." *Archiv fuer Elektrotechnik and Uebertragung,* Vol. 35, No. 12. 1981, pp. 481–488.

[220] Mohr, R. J. (1967) "Coupling Between Lines at High Frequecies." *IEEE Transactions on EMC,* Vol. EMC-9, No. 3. Dec. 1967, pp. 127–129.

[221] Mohr, R. J. (1967) "Coupling Between Open and Shielded Wire Liones over a Ground Plane." *IEEE Transactions on EMC,* Vol. EMC-9, No. 2. Sept. 1967, pp. 34–45.

[222] Montadon, E. (1982) "EMV, Erdungs- und Installations-praxis." *Technische Mitteilungen PTT,* No. 10. 1982, pp. 434–446.

[223] Morizet-Mahoudeaux, P.; Gaillard, P. (1988) "Utilisation d'un systeme expert en traitment du signal: aide au choix d'estimateurs de la densite spectrale de puissance." *L'Onde Electrique,* Vol. 58, No. 1, pp. 90–97.

[224] Mowatt, A. Q. (1976) "RFI Generation is a Factor When Selecting AC Switching Relays." *Electronics,* 2. Aug. 1976, pp. 50–55.

[225] Nakahara, M.; Ninomiya, T.; Harada, K. (1985) "Surge and Noise Generation in a Forward DC-to-DC Converter." *IEEE Transactions on Aerospace and Electronic Systems,* Vol. AES-21, No. 5. Sept. 1985, pp. 619–629.

[226] Nano, E. (1075) "Correction Factors for Quasi-Peak Measurements with Spectrum Analyzer." *1975 International EMC Conference Record,* Montreaux, pp. 156–161.

[227] Nicholson, J. R.; Malack, J. A. (1973) "RF Impedance of Power Lines and Line Impedance Stabilisation Networks Used in Conducted Interference Measurements." *IEEE Transactions on EMC,* Vol. EMC-15, No. 2. May 1975, pp. 84–86.

[228] Ninomiya, T.; Harada, K. (1980) "Common-Mode Noise Generation in a DC-to-DC Converter." *IEEE Transactions on Aerospace and Elecetronic Systems.* Vol. AES-16, No. 2. Marc. 1980, pp. 130–137.

[229] Ninomiya, T.; Harada, K.; Mamon, M. (1979) "Common-Mode Noise Generation of a DC-to-DC Converter." *1979 IEEE International EMC Symposium Record.* New York: IEEE, pp. 256–263.

[230] Nitta, S.; Takechi, E. (1988) "The Noise Immunity of LS Series NAND Gates." *9th International Wroclaw Symposium on EMC,* 1988, pp. 695–699.

[231] Nitta, S.; Takechi, E.; Shimayama, T. (1988) "The Noise Radiated from Showering Noise." *9th International Wroclaw Symposium on EMC,* 1988, pp. 613–617.

[232] Ogroske, E. (1984) "Die Stoerfreie Stromversorgung moderner Elektronik." *Elektronik,* No. 2. 1984, pp. 66–77.

[233] Olsen, R. G. (1984) "A Simple Model for Weakly Coupled Lossy Transmission Lines of Finite Length." *IEEE Transactions on EMC,* Vol. EMC-26, No. 2. May 1984, pp. 79–83.

[234] Oranc, H. S. (1975) "Effects of Impulsive Noise on Phase-Locked-Loop FM Demoddulator." *IEEE Transactions on EMC,* Vol. EMC-17, No. 2. May 1975, pp. 65–71.

[235] Ortloff, M. (1966) "Technische Grenzen der Bemessung von Filterketten zur Funkentstoerung von Starkstromanlagen." *Siemens Zeitschrift,* No. 3. March 1975, pp. 220–228.

[236] Ortloff, M. (1964) "Verfahren zum Messen der Hochfrequenzeigenschaften von Funk-Entstoerbauelementen." *Siemens Zeitschrift,* No. 12. 1964, pp. 907–914.

[237] Osburn, J. (1982) "Evaluation of Coupling Between Adjacent Circuits." *1982 IEEE International EMC Symposium Record.* New York: IEEE, pp. 318–322.

[238] Osburn, D. M. (1986) "Integration of Facilities Grounding Systems with User Electronic Systems." *EMC Technology,* Vol. 5, No. 1. Jan/Feb. 1986, pp. 29–32, 36, 54.

[239] Osburn, J. D. (1986) "Estimating System Level EMC Safety Margins from MIL-STD-461 Limits or Test Data." *EMC Technology,* Vol. 5, No. 4. Jul/Aug. 1986, pp. 21–23, 25, 28, 64.

[240] Ott, H. W. (1981) "Digital Circuit Grounding and Interconnection." *1981 IEEE International EMC Symposium Record.* New York: IEEE, pp. 292–297.

[241] Parker, W. H. (1976) "How to Specify an EMI Filter." *1976 IEEE International EMC Symposium Record.* New York: IEEE, pp. 131–135.

[242] Parker, C.; Tolen, B.; Parker, R. (1985) "Prayer Beads Solve Many of Your EMI Problems." *EMC Technology,* Vol. 4, No. 2. Apr/Jul. 1985, pp. 39–45, 70.

[243] Pasel, K. (1986) "Oberschwingungsbelastbarkeit der oeffentlichen Niederspannungsnetze." *etzBd,* Vol. 107, No. 2. 1986, pp. 56–59.

[244] Paul, C. R. (1978) "Prediction of Crosstalk in Ribbon Cables." *1978 IEEE International EMC Symposium Record.* New York: IEEE, pp. 36–43.

[245] Paul, C. R. (1982) "On the Superposition of Inductive and Capacitive Coupling in Crosstalk-Prediction Models." *IEEE Transactions on EMC,* Vol. EMC-24, No. 3. Aug. 1982, pp. 335–343.

[246] Paul, C. R.; Feather, A. E. (1976) "Computation of the Transmission Line Inductance and Capacitance Matrices from the Generalized Capacitance Matrix." *IEEE Transactions on EMC,* Vol. EMC-18, No. 4. Nov. 1976, pp. 175–182.

[246] Paul, C. R.; McKnight, Z. W. (1978) "Prediction of Crosstalk Involving Twisted Pairs of Wires." *1978 IEEE International EMC Symposium Record.* New York: IEEE, pp. 92–114.

[248] Paul, C. R.; McKnight, Z. W. (1979) "Prediction of Crosstalk Involving Twisted Pairs of Wires." *IEEE Transaction on EMC,* Vol. EMC-21, No. 2. May 1979, pp. 92–114.

[249] Paul, C. R.; Hardin, K. B. (1988) "Diagnosing and Reduction of Conducted Noise Emissions." *IEEE Transactions on EMC,* Vol. EMC-30, No. 4. 1988, pp. 553–560.

[250] Pellegrini, G. (1990) "Databank for EMC Standards." *10th International Wroclaw Symposium on EMC,* 1990, pp. 943–947.

[251] Pietrzik, G. (1980) "Testing for Electromagnetic Components." *Telcon. Rep.,* Vol. 3, No. 4. Dec. 1980, pp. 224–227

[252] Pluck, J. H. (1970) "A Simple Approach to R. F. Suppression and Shielding." *Proceedings I.R.E.E. Australia,* Aug. 1970, pp. 286–289.

[253] Probst, A. (1979) "Simulation elektrostatischer Entladungen." *ETZ-Bd,* Vol. 100, No. 10. 1979, pp. 494–497.

[254] Queen, R. H. (1985) "Common or Differential-Mode Noise? It Makes a Difference in Your Filter." *EMC Technology,* Vol. 4, No. 3. Jul/Sept. 1985, pp. 51–58, 85.

[255] Rapoza, J. G. (1980) "Unterdrueckung leitungsgebundener Stoerspannungen auf der Netzleitung von Schaltreglern." *Elektronik Entwicklung,* Vol. 15, No. 12. Dec. 1980, pp. 10–22.

[256] Rasek, W. (1979) "Planung der elektromagnetischen Vertraglichkeit fuer Baumassnahmen." *ETZ-Bd,* Vol. 100, No. 5. 1979, pp. 221–225.

[257] Rehder, H. (1979) "Stoerspannungen in Niederspannungs-netzen." *ETZ-B,* Vol. 100, No. 5. 1979, pp. 216–220.

[258] Rhoades, W. T. (1979) "Rectifiers, Noise and Regulatories." *1979 IEEE International EMC Symposium Record.* New York: IEEE, pp. 469–479.

[259] Rhoades, W. T. (1980) "Development of Power Main Transient Protection for Commercial Equipment." *1980 IEEE International EMC Symposium Record.* New York: IEEE, pp. 235–244.

[260] Richmond, T. (1988) "Minimizing Inductive Crosstalk in the Design of Switch-Mode Supplies." *PCI Proceedings,* Jun. 1988, pp. 313–322.

[261] Rice, L. R. (1972) "Choosing the Best Suppression Network for Your SCR-Converter." *Electronics,* 6. Nov. 1972, pp. 120.

[262] Ripka, K. W. (1972) "Entstehung und Beseitigung von Hochfrequenz-Stoersignalen in statischen Umrichtern." *ELIN-Zeitschrift,* Vol. 24, No. 3. 1972, pp. 92–97.

[263] Ristig, E. (1980) "Neue Bauformen stromkompensierter Drosseln." *Siemens Components,* Vol. 18, No. 6. Dec. 1980, pp. 294–297.

[264] Rock, F. E. (1986) "EMC Issues Relative to a 1 MHz Switcher." *PCIM,* Sept. 1986, pp. 22–26.

[265] Rodewald, A. (1989) "A Model for Fast Switching Transients in Power Systems." *IEEE Transactions on EMC.* Vol. EMC-31, No. 2. 1989, pp. 148–156.

[266] Roehr, B. (1986) "An Effective Transient and Noise Barrier for Switching Power Supplies." *PCIM,* Sept. 1986, pp. 28–32.

[267] Rosenberg, B. E.; Schulz, R. B. (1965) "A Parallel-Strip Line for Testing the RF Susceptibility." *IEEE Transactions on EMC,* Vol. EMC-7, No. 2. Jun. 1965, pp. 142–150.

[268] Russel, B. D. (1986) "Switching Transients in Power Substation Present Measurement Challenges." *EMC Technology,* Vol. 5, No. 1. Jan/Feb. 1986, pp. 21–24.

[269] Sailors, D. B. (1980) "Estimation of the Mean and Standaed Deviation from Quantities in Interference Modeling." *1980 IEEE International EMC Symposium Record.* New York: IEEE, pp. 391–396.

[270] Sakamoto, Y. (1989) "Three-Terminal Capacitors Come of Age." *JEE,* Aug. 1989, pp. 32–35.

[271] Sanetra, E. (1979) "EMV-Untersuchungen an einem Prozessrechner-Versuchsaufbau." *ETZ-B,* Vol. 100, No. 5. 1979, pp. 232–235.

[272] Sankur, B.; Modestino, J. W. (1982) "Performance of Receivers in Impulsive Noise." *Archiv fuer Elektrotechnik und Uebertragung,* Vol. 36, No. 3. 1982, pp. 216–220.

[273] Schackwitz, E. (1983) "Entstoermassnahmen elektrische Netze." *Elektronik,* No. 10. 1983, pp. 87–91.

[274] Schade, O. H. (1943) "Analysis of Rectifier Operation." *Proc. IRE,* Jul. 1943, pp. 341–361.

[275] Schaffernak, A. F. (1972) "Zur Stoersicherheit elektronischer Steuersysteme." *ETZ-B,* No. 13. 1972, pp. 315–321.

[276] Schaller, R. (1967) "Funk-Entstoerung von Si-Gleichrichterdioden." *Siemens Bauteile Informationen,* No. 3. 1967, pp. 103–105.

[277] Scharfman, W. E.; Vance, E. F.; Graf, K. A. (1978) "EMP Coupling to Power Lines." *IEEE Transactions on EMC,* Vol. EMC-20, No. 1. Feb. 1978, pp. 129–135.

[278] Schiffres, P. (1964) "A Dissipative Coaxial RFI Filter." *IEEE Transactions on EMC,* Vol. EMC-6, No. 1. Jan. 1964, pp. 55–61.

[279] Schindler, H.; Vau, G. (1979) "Die Planung der elektromagnetischen Vertraglichkeit fuer Baumassnahmen." *ETZ-Bd,* Vol. 100, No. 5. 1979, pp. 229–231.

[280] Schlicke, H. M. (1956) "Cascaded Feedthrough Capacitors." *Proc. IRE,* May 1956, pp. 686–691.

[281] Schlicke, H. M. (1964) "Theory of Simulated-Skin-Effect Filters as a Thin Film Approach." *IEEE Transactions on EMC,* Vol. EMC-6, No. 1. Jan. 1964, pp. 47–54.

[282] Schlicke, H. M. (1976) "Assuredly Effective Filters." *IEEE Transactions on EMC,* Vol. EMC-18, No. 4. Nov. 1976, pp. 141–148.

[283] Schlicke, H. M.; Struger, O. J. (1973) "Getting Noise Immunity in Industrial Controls." *IEEE Spectrum,* Jun. 1976, pp. 30–35

[284] Schlicke, H. M.; Weidmann, H. (1967) "Compatible EMI Filters." *IEEE Spectrum,* Oct. 1967, pp. 59–68.

[285] Schneider, F. (1986) "Netzspannungsausfalle und -einbrueche." Teil I: etzBd, Vol. 107, No. 2. 1986, pp. 60–65.; Teil II: *etzBd,* Vol. 107, No. 4. 1986, pp. 152–155.

[286] Schneider, L. (1982) "Power-Line EMI Filter Insertion Loss." *1982 IEEE International EMC Symposium Record.* New York: IEEE, pp. 348–354.

[287] Schulz, H. (1972) "Funk-Entstoerung mit stromkompensierten Drosseln." *Siemens Bauteile Informationen,* No. 10. 1972, pp. 34–36.

[288] Seiler, V.; Wimmer, J. (1977) "Schnittstellenfilter fuer Daten- und Signalleitungen." *Siemens Zeitschrift,* Vol. 51, No. 8. 1977, pp. 620–624.

[289] Shifman, J. C. (1965) "A Graphical Method for the Analysis and Synthesis of Electromagnetic Interference Filters." *IEEE Transactions on EMC,* Vol. EMC-7, No. 3. Sept. 1965, pp. 297–318.

[290] Showalter, B.; McBrayer, P. (1979) "RF Compatibility Environment to Component Part." *1979 IEEE International EMC Symposium Record.* New York: IEEE, pp. 91–95.

[291] Shower, R. M.; Kocher, C. P. (1979) "Modelling of Harmonics on Power Systems." *1979 IEEE International EMC Symposium Record.* New York: IEEE, pp. 414–426.

[292] Shower, R. M.; Schulz, R. B.; Lin, S. Y. (1981) "Fundamental Limits on EMC." *Proceedings of the IEEE,* 1981, pp. 183–195.

[293] Shower, R. M. (1990) "Achieving EMC." *10th International Wroclaw Symposium on EMC,* 1990, pp. 3–6.

[294] Slonim, M. (1979) "Harmonic Analysis of Periodic Discontinuous Functions." *Proc. IEE,* Vol. 67, No. 6. Jun. 1979, pp. 952–954.

[295] Smith, H. (1979) "Everyday Analysis Using IEMCAP Models with Extended Range." *1979 IEEE International EMC Symposium Record.* New York: IEEE, pp. 291–296.

[296] Smith, I. D.; Aslin, H. (1978) "Pulsed Power for EMP Simulation." *IEEE Transactions on EMC,* Vol. EMC-20, No. 1. Feb. 1978, pp. 53–59.

[297] Southwick, R. A.; Dolle, W. C. (1971) "Line Impedance Measuring Instrumentation Utilising Current Probe Coupling." *IEEE Transactions on EMC,* Vol. EMC-13, No. 4. Dec. 1971, pp. 31–36.

[298] Sowa, A. E. (1990) "Testing of Susceptibility and Vulnerability Waveforms." *10th International Wroclaw Symposium on EMC,* 1990, pp. 910–914.

[299] Spencer, P. (1985) "Switched-Mode Power Supply Filters." *EMC Technology,* Vol. 4, No. 3. Jul/Sept. 1985, pp. 27–35.

[300] Stadelhofen, A. (1976) "Comments on RF Impedance of U. S. and European Power Lines." *IEEE Transactions on EMC,* Vol. EMC-18, No. 3. Aug. 1976, p. 130.

[301] Standler, R. B. (1988) "Equations for Some Transient Overvoltage Test Waveforms." *IEEE Transactions on EMC,* Vol. EMC-30, No. 1. Feb. 1988, pp. 69–71.

[302] Stecher, M. (1975) "Funkstoermessungen in VHF-Bereich." *Elektronik Anzeiger,* Vol. 7, No. 11. 1975, pp. 259–261.

[303] Stirrat, W. A. (1960) "A General Technique for Interference Filtering." *IRE Transactions on Radio Frequency Interference,* Vol. RFI-1, No. 2. May 1960, pp. 12–17.

[304] Struzak, R. G. (1978) "CISPR Quasi-Peak Measuring Channels with Extended Range." *IEEE Transactions on EMC,* Vol. EMC-20, No. 3. Aug. 1978, pp. 361–367.

[305] Stumpers, F. L. H. M. (1973) "The 1973 CISPR Plenary Assembly at Moumonth College, New Yersey." *IEEE Transactions on EMC,* Vol. EMC-15, No. 4. Nov. 1973, pp. 197–199.

[306] Stumpers, F. L. H. M. (1975) "New Developments in CISPR—Results of the EMC Symposium in Montreaux." *IEEE Transactions on EMC,* Vol. EMC-17, No. 4. Nov. 1975, pp. 269–270.

[307] Suckening, S. (1980) "Upper Bound on Amplitude and Rate Change of a Time Domain Waveform from its Spectral Density." *IEEE Transactions on EMC,* Vol. EMC-22, No. 1. 1980, pp. 65–68.

[308] Suesse, H. (1984) "Funkentstoerung von Schaltnetzteilen." *Elektronik,* No. 23. 1984, pp. 101–106.

[309] Suesse, H. (1986) "Erhoehung der Stoerfestigkeit von uP-Systemen." Teil I: *Elektronik,* No. 7. 4. Appr. 1986, pp. 103–106; Teil II: *Elektronik,* No. 8. 18. Apr. 1986, pp. 115–120.

[310] Tangermann, E. P. (1982) "EMP Kontra Elektronik." *Funkschau,* No. 26. 1982, pp. 71–74.

[311] Taylor, J. R.; Sunda, J. A. (1971) "Suppressing Harmonics Generated by Thzristor Controlled Circuitry." *Electronic Engineering,* Aug. 1971, pp. 37–41.

[312] Tharp S.; Cox, P. (1979) "Impulse Strength Measurement: a New Technique for Analysing Power-Supply System Performance Under Actual Field Condition." *1979 IEEE International EMC Symposium Record.* New York: IEEE, pp. 74–80.

[313] Tihanyi, L. (1989) "Calculation of the Energy Content of Transient Signals." *EMC Technology,* Vol. 8, No. 2. March 1989, pp. 36–44.

[314] Tihanyi, L. (1990) "The Energy Density Function: A Useful Tool for Solving EMC Problems." *10th International Wroclaw Symposium on EMC,* 1990, pp. 449–454.

[315] Tokonami, M.; Mamada, N. (1989) "Multifunction Ceramics Improve EMI Characteristics of Bypass Capacitor." *JEE,* Apr. 1988. p. 50–53.

[316] Toppeto, A. A. (1980) "Evaluation and Comparison of Two Noise Separation Methods." *3rd International Wroclaw Symposium on EMC,* 1976, pp. 164–171.

[317] Turesin, V. M. (1967) "EMC Guide for Design Engineers." *IEEE Transactions on EMC,* Vol. EMC-9, No. 3. Dec. 1967, pp. 139–145.

[318] Turesin, V. M. (1969) "Theoretical Analysis and Design for Grounding to Accomplish EMI Control." *1969 IEEE International EMC Symposium Record.* New York: IEEE, pp. 495–499.

[319] Tyni, M. (1976) "The Transfer Impedance of Coaxial Cables with Braided Outer Conductor." *3rd International Wroclaw Symposium on EMC,* 1976, pp. 410–418.

[320] Uchimura, K.; Aida, T. (1990) "Generation Mechanism and Properties of Noise Induced by Realys Switching." *10th International Wroclaw Symposium,* 1990, pp. 520–524.

[321] Ulrich, R. (1973) "Funktionsstoerungen von Geraten und Zerstoerung von integrierten Schaltungen durch elektrostatischen aufgeladenen Personen." *NTZ Bd.* Vol. 26, No. 10, 1973, pp. 454–461.

[322] Vakil, S. M. (1978) "A Technique for Determination of Filter Insertion Loss as a Function of Arbitrary Generator and Load Impedance." *IEEE Transactions on EMC,* Vol. EMC-20, No. 2. May 1978, pp. 273–278.

[323] Vakil, S. M. (1983) 'MIL-STD-220A versus Classical Measurement of Filter Insertion Loss in a 50 W System." *IEEE Transactions on EMC,* Vol. 25, No. 4. Nov. 1983, pp. 382–388.

[324] VanBlaricum, M. L.; Hunt A. R. (1980) "A System Performance Criterion for Comparing Electromagnetic Environmental Effects." *1980 IEEE International EMC Symposium Record.* New York: IEEE, pp. 189–193.

[325] Vance, E. F.; Uman, M. A. (1988) "Differences Between Lighting and Nuclear Electromagnetic Pulse Interactions." *IEEE Transactions on EMC,* Vol. 30, No. 1. Feb. 1988, pp. 54–62.

[326] VanKeuren, R. (1975) "Effects of EMP-Induced Transients on Integrated Circuits." *1975 IEEE International EMC Symposium Record.* New York: IEEE, pp. 3AI-Ie1–e5.

[327] Vees, E. (1985) "State-of-the-Art Filtering in I/O Connectors." *EMC Technology,* Vol. 4, No. 2. Apr/Jul. 1985, pp. 61–65.

[328] Vines, R. M.; Trussel, H. J.; Shuey, K. C.; O'Neal, J. B. (1984) "Noise on Residental Power Distribution Circuits." *IEEE Transactions on EMC,* Vol. EMC-26, No. 4. Nov. 1984, pp. 161–168.

[329] Vines, R. M.; Trussel, H. J.; Shuey, K. C.; O'Neal, J. B. (1985) "Impedance of the Residental Power-Distribution Circuit." *IEEE Transactions on EMC,* Vol. EMC-27, No. 1. Feb. 1985, pp. 6–12.

[330] Violette, M. F.; Violette J. L. N. (1986) "EMI Control in the Design and Layout of Printed Circuit." *EMC Technology,* Vol. 5, No. 2. March/Apr. 1986, pp. 19–32.

[331] Wagar, H. N. (1969) "Prediction of Voltage Surge When Switching Contacts Interrupt Inductive Loads." *IEEE Transactions on Parts, Material and Packaging,* Vol. PMP-5, No. 4. Dec. 1969, pp. 149–155.

[332] Weber, F. (1976) "Stoerspannungsanalyse mit Netz- und Stoersimulatoren." *Elektronik,* Vol. 25, No. 8. 1976, pp. 39–42.

[333] Weber, J. (1979) "Sachaufgaben der EMV-Normung." ETZ-Bd. Vol. 100, No. 5. 1979, pp. 226–229.

[334] Weidmann, H.; McMartin, W. J. (1968) "Two Worst-Case Insertion Loss Test Methods for Passive Power-Line Interference Filters." *IEEE Transactions on EMC,* Vol. 10, No. 2. Jun. 1968. (Special Filter Issue), pp. 257–263.

[335] Weidman, C. D.; Krider, E. P. (1986) "The Amplitude Spectra of Lighting Radiation Fields in the Interval From 1 to 20 MHz." *Radio Science,* Vol. 21, No. 6. Nov/Dec. 1986, pp. 964–970.

[336] Wheeler, H. A. (1950) "Inductance Chart for Solenoid Coil." *Proc. IRE,* Dec. 1950, pp. 1398–1400.

[337] White, D. R. J. (1982) "EMI Control in the Design of Printed Circuit Boards." *EMC Technology,* Vol. 1, No. 1. Jan. 1982, pp. 74–83.

[338] Willard, F. G. (1970) "Transient Noise Suppression in Control Systems." *Control Engineering,* Sept. 1970, pp. 59–64.

[339] Williamson, T. (1982) "Digitalelektronik—Optimal Entstoert." Teil I: *Elektronik,* No. 19. 1982, pp. 57–60; Teil II: *Elektronik,* No. 20. 1982, pp. 57–60.

[340] Willman, J. F.; VanSteenberg, C. N. (1969) "Directional Current Probe Technique for Conducted EMI Measurements." *IEEE Transactions on EMC,* Vol. EMC-11, No. 2. May 1969, p. 98.

[341] Woody, J. A.; Paludi, C. A. (1980) "Modelling Techniques for Discrete Passive Components to Include Parasitic Effects in EMC Analysis Design." *1980 IEEE International EMC Symposium Record.* New York: IEEE, pp. 39–45.

INTERNATIONAL STANDARDS

[342] CISPR Publ. 1: "Specification for CISPR Radio Interference Measuring Apparatus for the Frequency Range 0,15 to 30 MHz."

[343] CISPR Publ. 9: "CISPR Limits of Radio Interference and Report of National Limits."

[344] CISPR Publ. 11: "Limits and Methods of Measurement of Radio Interference Characteristics of Industrial, Scientific and Medical /ISM/ Radio-Frequency Equipment."

[345] CISPR Publ. 12: "Limits and Methods of Measurement of Radio Interference Characteristics for Vehicles, Motor Boats, and Spark-Ignited Engine-Driven Devices."

[346] CISPR Publ. 13: "Limits and Methods of Measurement of Radio Interference Characteristics of Sound and Television Receivers."

[347] CISPR Publ. 14: "Limits and Methods of Measurement of Radio Interference Characteristics Houshold Electrical Applience, Portable Tools and Similar Electrical Apparatus."

[348] CISPR Publ. 15: "Limits and Methods of Measurement of Radio Interference Characteristics of Fluorescent Lamps and Luminaires."

[349] CISPR Publ. 16: "Specification for Radio Interference Measuring Apparatures and Measurement Methods."

[350] CISPR Publ. 21: "Interference of Mobile Radiocommunications in the Presence of Impulsive Noise."

[351] CISPR Publ. 22: "Limits and Methods of Measurement of Radio Interference Characteristics of Information Technology Equipment."

[352] IEC Publ. 255.3. (1971) "Single Input Energizing Quantity Measuring Relays with Non-Specified Time or with Independent Specified Time."

[353] IEC Publ. 255.4. (1976) "Single Input Energizing Quantity Measuring Relays with Dependent Specified Time."

[354] IEC Publ. 255.5 (1977) "Insulation Test for Electrical Relays."

[355] IEC Publ. 255.6 (1978) "Measuring Realys with More than One Input Energizing Quantity."

[356] IEC Publ. 255.10 (1979) "Application of the IEC Quantity Assessment System for Electronic Components to All-Or-Nothing Relays."

[357] IEC Publ. 801: "Electromagnetic Compatibility for Industrial Process Measurement and Control Equipment."

[358] IEC Publ. 1000-2-1-1990: "Description of the Electromagnetic Environment for Low-Frequency Conducted Disturbances and Signalling in Public Power Supply Systems."

[359] IEC Publ. 1000-2-2-1990: "Compatibility Levels for Low-Frequency Conducted Disturbances and Signalling in Public Power Supply Systems."

[360] IEC Publ. 1000-4-1: "Overview on Electromagnetic Compatibility Immunity Tests."

[361] ANSI/IEEE C37.90-1978: "Standard Relays and Relay Systems Associated with Electric Power Apparatus."

[362] ANSI/IEEE Std C62.41-1980 (formerly: IEEE Std 587-1980) "Guide on Surge Voltages in Low-voltage AC Power Circuits"

[363] ANSI/IEEE Std 4-1978: "Standard Techniques for High-Voltage Testing."

[364] European Standard EN 50.006 (1975) "Limitation of Disturbances in Electricity Supply Networks Caused by Domestic and Similar Appliances Equipped with Electronic Devices." (European Committee for Electrotechnical Standarisation, CENELEC, Bruxelles, Belgium).

[365] SEN 361 503 - 1.4.1977: "Stoerumgebungklassen und Pruefbestimmungen fuer Elektronikgerate in Regel- und Steuerausruestungen fuer Kraftwerksanlagen."

Index

About the Author

László Tihanyi is a graduate of the Budapest Technical University. He earned an M.Sc. degree in electrical engineering in 1969.

After graduation, Mr. Tihanyi began working at the Research Institute for Electrical Industry (VKI) in Budapest. In 1972, he received the title of "research expert" and in 1978 became the "senior research expert." His activities included development of high-power thyristorized and transistorized dc power supplies and, later, dc/ac converters and uninterruptible power supplies.

In 1985, Mr. Tihanyi accepted the position of Chief Research Consultant in the Institute for Power Electronics. Here, he focused primarily on solving EMI problems in electronic systems and development of a dimensioning method for power-line filters. From 1988 until 1990, he headed up the Department of Power Electronics at the Institute.

The author presently works as a consultant and free-lance technical writer.